Chemistry 1

FOR OCR

David Acaster
Lawrie Ryan

Brian Ratcliff

Helen Eccles

David Johnson

John Nicholson

John Raffan

Alan Winfield

CAMBRIDGE
UNIVERSITY PRESS

CAMBRIDGE UNIVERSITY PRESS
Cambridge, New York, Melbourne, Madrid, Cape Town, Singapore, São Paulo, Delhi

Cambridge University Press
The Edinburgh Building, Cambridge CB2 8RU, UK

www.cambridge.org
Information on this title: www.cambridge.org/9780521724562

© Cambridge University Press 2008

First published 2008

Printed in the United Kingdom at the University Press, Cambridge

A catalogue record for this publication is available from the British Library

ISBN 978-0-521-72456-2 paperback

ACKNOWLEDGEMENTS
Project management: Sue Kearsey
Picture research: Vanessa Miles
Front cover photograph: Computer artwork of fullerenes/Victor Habbick/Science Photo Library

Contents

Advice

Contents

Introduction

Cambridge OCR Advanced Sciences

The new *Cambridge OCR Advanced Sciences* course provides complete coverage of the revised OCR AS and A Level science specifications (Biology, Chemistry A and Physics A) for teaching from September 2008. There are two books for each subject – one covering AS and one covering A2. Some material has been drawn from the existing *Cambridge Advanced Sciences* books; however the majority is new.

The course has been developed in an innovative format, featuring Cambridge's new interactive PDFs on CD-ROM in the back of the books, and free access to a dedicated website. The CD-ROM provides additional material, including detailed objectives, hints on answering questions, and extension material. It also provides access to web-based e-learning activities (A Level and background revision from GCSE) to help students visualise abstract concepts, understand calculations and simulate scientific processes.

The books contain all the material required for teaching the specifications, and can be used either on their own or in conjunction with the interactive PDFs and the website.

In addition, *Teacher Resource CD-ROMs* with book PDFs plus extra material such as worksheets, practical activities and tests, are available for each book. These CD-ROMs also provide access to the new *Cambridge OCR Advanced Sciences* Planner website with a week-by-week adaptable teaching schedule.

Introduction to Chemistry 1 for OCR – the chemistry AS text

This book covers the entire OCR AS Chemistry A specification for first examination in 2009. Chapters 1 to 9 correspond to Unit F321, Atoms, Bonds and Groups. Chapters 10 to 18 correspond to Unit F322, Chains, Energy and Resources. Each chapter covers one of the numbered sections within the three Modules in Unit F321, and within the four Modules in Unit F322. The content of the chapters is generally arranged in the same sequence as in the specification.

The book is designed to be used by students who have studied Double Award Science at GCSE, or who have a separate qualification in chemistry. The language is kept simple, to improve accessibility for all students, while still maintaining scientific rigour throughout. Care is taken to introduce and use all the specialist terms that students need to gain a complete understanding of the chemical concepts introduced. In the text, key terms are highlighted in bold.

The depth and breadth of treatment of each topic is pitched at the appropriate level for OCR AS students. The accompanying CD-ROM also contains some extension material that goes a little beyond the requirements of the specification, which should interest and stretch more able students.

Some of the text and illustrations are based on material from the endorsed text *Chemistry 1*, which covered the earlier OCR specification, while some is completely new. All of it has been reviewed and revised, ensuring that the new specification is fully covered. In addition to the main content in each chapter, there are also How Science Works boxes, describing issues, applications or events, which put the chemical content introduced into a social context.

Self-assessment questions (SAQs) in each chapter provide opportunities to check understanding. They often address misunderstandings that commonly appear in examination answers, and will help students to avoid such errors. Past examination questions at the end of each chapter allow students to practise answering exam-style questions. The answers to these, along with exam-style mark schemes and hints on answering questions, are found on the accompanying CD-ROM.

Acknowledgements

We would like to thank the following for permission to reproduce images:

pp. 1, 17, 89, 90, 101*r*, 103, 136, 142*br*, 145, 159, 167*l*, 201 Andrew Lambert; pp. 2, 5 University of Cambridge Cavendish Laboratory, Madingley Road, Cambridge; p. 3 Manchester University/Science and Society Picture Library; pp. 11, 72 *gold*, *sulfur* GeoScience Features; pp. 12, 220*l* Science Photo Library; p. 23 © mediablitzimages (uk) Limited/Alamy; pp. 24, 25, 26, 31, 32 David Acaster; p. 30 © Shenval/Alamy; pp. 42, 57*c*, 61*l* Charles D. Winters/Science Photo Library; p. 46*t* Photo Library International; p. 46*cl* Mark Sykes/Science Photo Library; p. 46*cr* Lawrence Lawry/Science Photo Library; pp. 46*bl*, 51*r* Roberto de Gugliemo/Science Photo Library; pp. 46*br*, 181*tl* Dirk Wiersma/Science Photo Library; pp. 47*t*, 133, 168*r*, 176 NASA/Science Photo Library; p. 47*b* ICI; pp. 48, 144 © Stockbyte/Alamy; p. 50 Natural History Museum; pp. 51*l*, 91*r* Brian Ratcliff; pp. 57*t*, 64, 118, 119*l*, 120 Andrew Lambert Photography/Science Photo Library; p. 57*b* Tom McHugh/Science Photo Library; p. 58*t*, *b* Ancient Art and Architecture collection; p. 61*r* Tick Ahearn; pp. 62*l*, 72 *bromine*, 93*b* Peter Gould; p. 62*r* Britstock-IFA/TPL; p. 63 Steve Allen/Science Photo Library; p. 65 © Paul Almasy/CORBIS; p. 66*l* courtesy of Dr Jonathan Goodman, Department of Chemistry, Cambridge University, using the program Eadfrith (©J.M. Goodman, Cambridge University, 1994)/photo by Cambridge University Chemistry Department Photographic Unit; p. 66*r* Laguna Design/Science Photo Library; p. 72 *lead* Erich Schrempp/SPL; p. 72 *mercury* Vaughan Fleming/SPL; p. 72 *plutonium*, *uranium* US Dept of Energy/SPL; p. 72 *zinc* Astrid & Hanns-Frieder Michler/SPL; p. 73 courtesy of the Library & Information Centre of the Royal Society of Chemistry; p. 74*b* Gordon Woods, Malvern School; p. 75 CERN/Science Photo Library; p. 87 Ria Novosti/Science Photo Library;

p. 91*l* Martin Bond/Science Photo Library; p. 93*t* Leslie Garland Picture Library; p. 100 © Didier Robcis/Corbis; p. 101*l* Steve Davy/La Belle Aurore; p. 119*r* Wellcome Library, London; p. 129 © Macduff Everton/CORBIS; pp. 130, 150*r* Paul Rapson/Science Photo Library; p. 131*l* © f1 online/Alamy; p. 131*r* David Nunuk/Science Photo Library; p. 134 Simon Fraser/Science Photo Library; p. 135*l* Crown Copyright courtesy of CSL/Science Photo Library; p. 135*r* © Christina Koci Hernandez/San Francisco Chronicle/Corbis; p. 142*l* © Shehzad Noorani/Still Pictures; p. 142*tr* © Nick McGowan-Lowe/Alamy; p. 142*cr* Mauritius Die Bildagentur Gmbh; p. 147 © David Bailey/Alamy; p. 149*tl* © Mark Boulton/Alamy; p. 149*tr* © Marcelo Rudini/Alamy; p. 149*b* © Ashley Cooper/Alamy; p. 150*l* © Harald A. Jahn, Harald Jahn/CORBIS; p. 156 © Eric Nathan/Alamy; p. 157 © Directphoto.org/Alamy; pp. 167*r*, 186 Michael Brooke; p. 168*l* courtesy of Civil Aviation Authority, International Fire Training Centre, UK, and Robert Hartness; p. 172 © Ashley Cooper/Corbis; p. 175 TEK Image/Science Photo Library; p. 180 Peter Menzel/Science Photo Library; p. 181*tr* Michael Nemeth; p. 181*b* GustoImages/Science Photo Library; p. 198*t* Jeff Foott; p. 198*b* © Richard McDowell/Alamy; p. 199 Stockbyte; p. 203 Uwe Krejci; p. 214 GrowHow UK Limited; p. 215 Arthur Morris; p. 220*tr* Joe Pasieka/Science Photo Library; p. 220*br* Wesley Bocxe/Science Photo Library; p. 222 The Guardian; p. 225 The Observer; p. 229 NASA; p. 230 © Mike Abrahams/Alamy; p. 235 courtesy of Johnson Matthey Emission Control Technologies; p. 236 © Krista Kennell/ZUMA/Corbis; p. 237 © Guenter Rossenbach/zefa/Corbis; p. 238*l* © Roger Ressmeyer/CORBIS; p. 238*r* © Peter Brogden/Alamy; p. 239 AFP/Getty Images.

We would like to thank OCR for permission to reproduce questions from past examination questions.

Chapter 1

Atomic structure

Objectives

Chemistry is a science of change. Over the centuries people have heated rocks, distilled juices and probed solids, liquids and gases with electricity. From all this activity we have gained a great wealth of new materials – metals, medicines, plastics, dyes, ceramics, fertilisers, fuels and many others (Figure 1.1). But this creation of new materials is only part of the science and technology of chemistry. Chemists also want to understand the changes, to find patterns of behaviour and to discover the innermost nature of the materials.

Figure 1.1 All of these useful products, and many more, contain chemicals that have been created by applying chemistry to natural materials. Chemists must also find answers to problems caused when people misuse chemicals.

Our 'explanations' of the chemical behaviour of matter come from reasoning and model-building based on the evidence available from experiments. The work of scientists has shown us the following.

- All known materials, however complicated and varied they appear, can be broken down into the fundamental substances we call **elements**. These elements cannot be broken down further into simpler substances. So far, about 115 elements are known. Most exist in combinations with other elements in **compounds** but some, such as gold, nitrogen, oxygen and sulfur, are also found in

an uncombined state. Some elements would not exist on Earth without the artificial use of nuclear reactions. Chemists have given each element a symbol. This symbol is usually the first one or two letters of the name of the element; some are derived from their names in Latin. Some examples are:

Element	Symbol
carbon	C
lithium	Li
iron	Fe (from the Latin *ferrum*)
lead	Pb (from the Latin *plumbum*)

- Groups of elements show patterns of behaviour related to their atomic masses. A Russian chemist, Dmitri Ivanovich Mendeleev, summarised these patterns by arranging the elements into a 'Periodic Table'. Modern versions of the Periodic Table are widely used in chemistry. (A Periodic Table is shown in the Appendix and explained, much more fully, in Chapter 7.)
- All matter is composed of extremely small particles, called atoms. About 100 years ago, the accepted model for atoms included the assumptions that (i) atoms were tiny particles, which could not be divided further or destroyed, and (ii) all atoms of the same element were identical. This model was very helpful, but gave way to better models, as science and technology produced new evidence. This evidence has shown scientists that atoms have other particles inside them – they have an internal structure.

Scientists now believe that there are two basic types of particles – 'quarks' and 'leptons'. These are the building blocks from which everything is made, from microbes to galaxies. For many explanations or predictions, however, scientists use a model of atomic structure in which atoms are made of electrons, protons and neutrons. Protons and neutrons are made from quarks, and the electron is a member of the family of leptons.

Discovering the electron

Electrolysis – the effect of electric current in solutions

When electricity flows in an aqueous solution of silver nitrate, for example, silver metal appears at the negative electrode (cathode). This is an example of *electrolysis* and the best explanation is that:

- the silver exists in the solution as positively charged particles known as **ions** (Ag^+)
- one silver ion plus one unit of electricity gives one silver atom.

The name 'electron' was given to this unit of electricity by the Irish scientist George Johnstone Stoney in 1891.

Study of cathode rays

At normal pressures gases are usually very poor conductors of electricity, but at low pressures they conduct quite well. Scientists, such as William Crookes, first studied the effects of passing electricity through gases at low pressures. They saw that the glass of the containing vessel opposite the cathode (negative electrode) glowed when the applied potential difference (voltage) was sufficiently high.

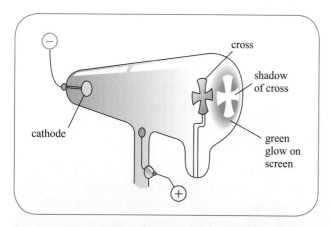

Figure 1.2 Cathode rays cause a glow on the screen opposite the cathode, and the 'Maltese Cross' casts a shadow. The shadow will move if a magnet is brought near to the screen. This shows that the cathode rays are deflected by a magnetic field. The term 'cathode ray' is still familiar today, as in 'cathode-ray oscilloscopes'.

A solid object, placed between the cathode and the glow, casts a shadow (Figure 1.2). They proposed that the glow was caused by rays coming from the cathode and called these *cathode rays*.

At the time there was some argument about whether cathode rays are waves, similar to visible light rays, or particles. The most important evidence is that they are strongly deflected by a magnetic field. This is best explained by assuming that they are streams of electrically charged particles. The direction of the deflection shows that the particles in cathode rays are negatively charged.

J. J. Thomson's *e/m* experiment

The great leap in understanding came in 1897, at the Cavendish Laboratory in Cambridge (Figure 1.3 and Figure 1.4). J. J. Thomson measured the deflection of a narrow beam of cathode rays in both magnetic and electric fields. His results allowed him to calculate the charge-to-mass ratio (*e/m*) of the particles. Their charge-to-mass ratio was found to be exactly the same, whatever gas or type of electrode was used in the experiment. The cathode-ray particles had a tiny mass, only approximately 1/2000th of the mass of a hydrogen atom. Thomson then decided to call them **electrons** – the name suggested earlier by Stoney for the 'units of electricity'.

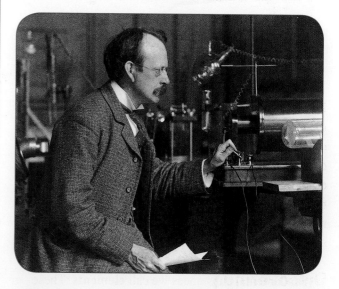

Figure 1.3 Joseph (J.J.) Thomson (1856–1940) using his cathode-ray tube.

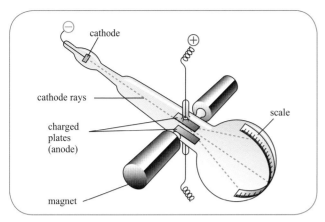

Figure 1.4 A drawing of Thomson's apparatus. The electrons move from the hot cathode (negative) through slits in the anode (positive).

Millikan's 'oil-drop' experiment

The electron charge was first measured accurately in 1909 by the American physicist Robert Millikan using his famous 'oil-drop' experiment (Figure 1.5). He found the charge to be 1.602×10^{-19} C (coulombs). The mass of an electron was calculated to be 9.109×10^{-31} kg, which is 1/1837th of the mass of a hydrogen atom.

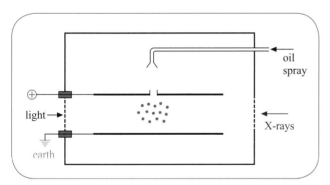

Figure 1.5 Oil drops which were sprayed into the container acquired a negative charge. The drops remained stationary when the upward force of attraction to the positive plates equalled the downward force due to gravity. Calculations on the forces allowed Millikan to find the charges on the drops. These were multiples of the charge on an electron.

Discovering protons and neutrons

New atomic models: 'plum-pudding' or 'nuclear' atom

Before electrons were discovered every atom was believed to be indivisible and to be made of the same 'material' all the way through – like a snooker ball. The discoveries about electrons demanded new models for atoms. If there are negatively charged electrons in all electrically neutral atoms, there must also be a positively charged part. For some time the most favoured atomic model was J. J. Thomson's 'plum-pudding', in which electrons (the 'plums') were embedded in a 'pudding' of positive charge (Figure 1.6).

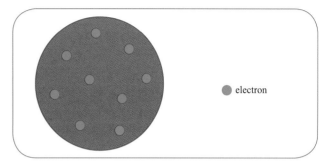

Figure 1.6 J. J. Thomson's 'plum-pudding' model of the atom. The electrons (plums) are embedded in a sphere of uniform positive charge.

Then, in 1909, came one of the experiments that changed everything. Two members of Ernest Rutherford's research team in Manchester University, Hans Geiger and Ernest Marsden, were investigating how α-particles (α is the Greek letter alpha) from a radioactive source were scattered when fired at very thin sheets of gold and other metals (Figure 1.7).

They detected the α-particles by the small flashes of light (called 'scintillations') that they caused on impact with a fluorescent screen. Since (in atomic terms) α-particles are heavy and energetic, Geiger and Marsden

Figure 1.7 Ernest Rutherford (right) and Hans Geiger using their apparatus for detecting α-particle deflections. Interpretation of the results led Rutherford to propose the nuclear model for atoms.

were not surprised that most particles passed through the metal with only slight deflections in their paths. These deflections could be explained, by the 'plum-pudding' model of the atom, as small scattering effects caused while the positive α-particles moved through the diffuse mixture of positive charge and electrons.

However, Geiger and Marsden also noticed some large deflections. A few (about one in 20 000) were so large that scintillations were seen on a screen placed on the same side of the gold sheet as the source of positively charged α-particles. This was unexpected. Rutherford said: 'it was almost as incredible as if you had fired a 15-inch shell at a piece of tissue paper and it came back and hit you!' (Figure 1.8).

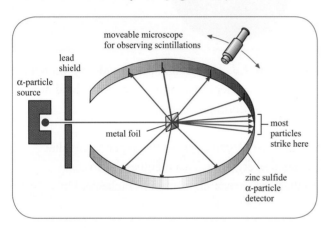

Figure 1.8 Geiger and Marsden's experiment, which investigated how α-particles are deflected by thin metal foils.

The plum-pudding model, with its diffuse positive charge, could not explain the surprising Geiger–Marsden observations. However, Rutherford soon proposed his convincing nuclear model of the atom. He suggested that every atom consists largely of empty space (where the electrons are) and that the mass is concentrated into a very small, positively charged, central core called the **nucleus**. The nucleus is about 10 000 times smaller than the atom itself – similar in scale to a marble placed at the centre of an athletics stadium.

Most α-particles will pass through the empty space in an atom with very little deflection. When an α-particle approaches on a path close to a nucleus, however, the positive charges strongly repel each other and the α-particle is deflected through a large angle (Figure 1.9).

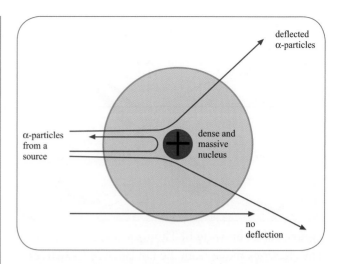

Figure 1.9 Ernest Rutherford's interpretation of the Geiger–Marsden observations. The positively charged α-particles are deflected by the tiny, dense, positively charged nucleus. Most of the atom is empty space.

Nuclear charge and 'atomic' number

In 1913, Henry Moseley, a member of Rutherford's research team in Manchester, found a way of comparing the positive charges of the nuclei of atoms of different elements. The charge increases by one unit from element to element in the Periodic Table. Moseley showed that the sequence of elements in the Table is related to the nuclear charges of their atoms, rather than to their relative atomic masses (see Chapter 7). The size of the nuclear charge was then called the **atomic number** of the element. Atomic number defined the position of the element in the Periodic Table.

Particles in the nucleus

The proton

After he proposed the nuclear atom, Rutherford reasoned that there must be particles in the nucleus which are responsible for the positive nuclear charge. He and Marsden fired α-particles through hydrogen, nitrogen and other materials. They detected new particles with positive charge and the approximate mass of a hydrogen atom. Rutherford eventually called these particles **protons**. A proton carries a positive charge of 1.602×10^{-19} C, equal in size but opposite in sign to the charge on an electron. It has a mass of 1.673×10^{-27} kg, about 2000 times as heavy as an electron.

Each electrically neutral atom has the same number of electrons outside the nucleus as there are protons within the nucleus.

The neutron

The mass of an atom, which is concentrated in its nucleus, cannot depend only on protons; usually the protons provide around half of the atomic mass. Rutherford proposed that there is a particle in the nucleus with a mass equal to that of a proton but with zero electrical charge. He thought of this particle as a proton and an electron bound together.

Without any charge to make it 'perform' in electrical fields, detection of this particle was very difficult. It was not until 12 years after Rutherford's suggestion that, in 1932, one of his co-workers, James Chadwick, produced sufficient evidence for the existence of a nuclear particle with a mass similar to that of the proton but with no electrical charge (Figure 1.10). The particle was named the **neutron**.

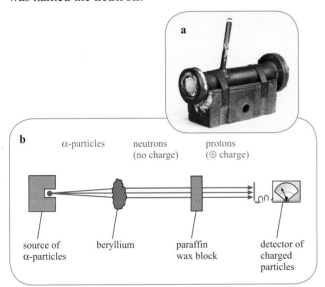

Figure 1.10 a Using this apparatus, James Chadwick discovered the neutron.
b Drawing of the inside of the apparatus. Chadwick bombarded a block of beryllium with α-particles ($^4_2$He). No charged particles were detected on the other side of the block. However, when a block of paraffin wax (a compound containing only carbon and hydrogen) was placed near the beryllium, charged particles were detected and identified as protons (H$^+$). Alpha-particles had knocked neutrons out of the beryllium, and in turn these had knocked protons out of the wax.

Atomic and mass numbers

Atomic number (Z)

The most important difference between atoms of different elements is in the number of protons in the nucleus of each atom. The number of protons in an atom determines the element to which the atom belongs. The atomic number of an element shows:
- the number of protons in the nucleus of an atom of that element
- the number of electrons in a neutral atom of that element
- the position of the element in the Periodic Table.

Mass number (A)

It is useful to have a measure for the total number of particles in the nucleus of an atom. This is called the **mass number**. For any atom:
- the mass number is the sum of the number of protons and the number of neutrons.

Summary table

Particle name	Relative mass	Relative charge
electron	negligible (about 1/2000th the mass of a proton)	−1
proton	1	+1
neutron	1	0

Isotopes

In Rutherford's model of the atom, the nucleus consists of protons and neutrons, each with a mass of one atomic unit. The relative atomic masses of elements should then be whole numbers. It was thus a puzzle why chlorine has a relative atomic mass of 35.5.

The answer is that atoms of the same element are not all identical. In 1913, Frederick Soddy proposed that atoms of the same element could have different atomic masses. He named such atoms **isotopes**. The word means 'equal place', i.e. occupying the same place in the Periodic Table and having the same atomic number.

The discovery of protons and neutrons explained the existence of isotopes of an element. In isotopes of one element, the number of protons must be the same, but the number of neutrons will be different.

Remember:

atomic number (Z) = number of protons

mass number (A) = number of protons + number of neutrons

Isotopes are atoms with the same atomic number, but different mass numbers. The symbol for isotopes is shown as:

$$^{\text{mass number}}_{\text{atomic number}}\text{X} \quad \text{or} \quad ^A_Z\text{X}$$

For example, hydrogen has three isotopes:

	Protium	Deuterium	Tritium
	^1_1H	^2_1H	^3_1H
protons	1	1	1
neutrons	0	1	2

It is also common practice to identify isotopes by name or symbol plus mass number only. For example, uranium, the heaviest naturally occurring element (Z = 92), has two particularly important isotopes of mass numbers 235 and 238. They are often shown as uranium-235 and uranium-238, as U-235 and U-238 or as ^{235}U and ^{238}U.

Numbers of protons, neutrons and electrons

It is easy to calculate the composition of a particular atom or ion:

number of protons = Z

number of neutrons = $A - Z$

number of electrons in neutral atom
= Z

number of electrons in positive ion
= Z − charge on ion

number of electrons in negative ion
= Z + charge on ion

For example, magnesium is element 12; it is in Group 2, so it tends to form doubly charged (2+) ions. The ionised isotope magnesium-25 thus has the full symbol:

$^{25}_{12}\text{Mg}^{2+}$,

and

number of protons = 12

number of neutrons = 13

number of electrons = 10

SAQ

1 a What is the composition (numbers of electrons, protons and neutrons) of neutral atoms of the two main uranium isotopes, U-235 and U-238?

    Hint

b What is the composition of the ions of potassium-40 (K^+) and chlorine-37 (Cl^-)? (Use the Periodic Table in the Appendix for the atomic numbers.)

    Hint

    Answer

Counting atoms and molecules

If you have ever had to sort and count coins, you will know that it is a very time-consuming business! Banks do not need to count sorted coins, as they can quickly check the amount by weighing. Chemists are also able to count atoms and molecules by weighing them. This is possible because atoms of different elements also have different masses.

We rely on tables of relative atomic masses for this purpose. *The relative atomic mass, A_r, of an element is the mass of the element relative to the mass of carbon-12; one atom of this isotope* (see page 5) *is given a relative isotopic mass of exactly 12.* The relative atomic masses, A_r, of the other elements are then found by comparing the average mass of their atoms with that of the carbon-12 isotope. Notice that we use the *average* mass of their atoms. This is because we take into account the abundance of their naturally occurring isotopes. Thus the precise relative atomic mass of hydrogen is 1.0079, whilst that of chlorine is 35.45. (Accepted relative atomic masses are shown on the Periodic Table in the Appendix.)

We use the term **relative isotopic mass** for the mass of an isotope of an element relative to carbon-12. For example, the relative isotopic mass of carbon-13 is 13.003. If the natural abundance of each isotope is known, together with their relative isotopic masses, we can calculate the relative atomic mass of the element as follows.

Chlorine, for example, occurs naturally as chlorine-35 and chlorine-37 with percentage natural abundances 75.5% and 24.5% respectively. So

$$\text{relative atomic mass} = \frac{(75.5 \times 35) + (24.5 \times 37)}{100}$$
$$= 35.5$$

SAQ

2 Naturally occurring neon is 90.9% neon-20, 0.3% neon-21 and 8.8% neon-22. Use these figures to calculate the relative atomic mass of naturally occurring neon.

Hint

Answer

The masses of different molecules are compared in a similar fashion. The **relative molecular mass**, M_r, of a compound is the mass of a molecule of the compound relative to the mass of an atom of carbon-12, which is given a mass of exactly 12.

To find the relative molecular mass of a molecule, we add up the relative atomic masses of all the atoms present in the molecule. For example, the relative molecular mass of methane, CH_4, is $12 + (4 \times 1) = 16$.

Where compounds contain ions, we use the term **relative formula mass**. Relative molecular mass refers to compounds containing molecules.

SAQ

3 Use the Periodic Table in the Appendix to calculate the relative formula mass of the following:
   a magnesium chloride, $MgCl_2$
   b copper sulfate, $CuSO_4$
   c sodium carbonate, $Na_2CO_3.10H_2O$
   (10H$_2$O means ten water molecules).

Answer

Building with atoms

Nanotechnology is the design and creation of objects that are so small we measure them in nanometres.

It is difficult for us to imagine a nanometre (nm). The tiny dimensions involved in nanotechnology are mind-boggling. The nanometre is the unit of measurement used on an atomic level:

$$1\,nm = 10^{-9}\,m$$

In other words, a nanometre is a billionth of a metre (or a millionth of a millimetre, check your ruler!).

Nanotechnologists are now making machines that are less than 100 nm in size. The skills and techniques to manipulate atoms, and position them where we want them, have only recently been developed.

The possibilities are very exciting. Imagine tiny machines patrolling your bloodstream. They could eventually check out cells in our bodies and deliver medication exactly to where it is needed. However, the use of nanoparticles, in for example cosmetics, is worrying some people. They believe the particles might pass through layers of skin, allowing substances designed for your skin to get inside the body. So they want tougher rules for the testing of these products.

Summary

Glossary

- Our current theory of the atom has developed from simpler theories; those early theories were changed or modified in the light of experimental results.

- Any atom has an internal structure with almost all of the mass in the nucleus, which has a diameter very much smaller than the diameter of the atom.

- The nucleus contains protons (positively charged) and neutrons (uncharged). Electrons (negatively charged) exist outside the nucleus.

- All atoms of the same element have the same atomic number (Z); that is, they have equal numbers of protons in their nuclei.

- The mass number (A) of an atom is the total number of protons and neutrons. Thus the number of neutrons $= A - Z$.

- The isotopes of an element are atoms with the same atomic number but different mass numbers. If neutral, they have the same number of protons and electrons but different numbers of neutrons.

- Atomic, isotopic and molecular masses are given relative to carbon-12, which has a mass of exactly 12.

Questions

1 Antimony, Sb, is a metal used in alloys to make lead harder. Bullets contain about 1% of antimony for this reason.

   a Antimony has two main isotopes.

     i What do you understand by the term isotopes? [1]

     ii Copy and complete the table below to show the properties of particles that make up isotopes. [2]

        Hint

	Proton	Neutron	Electron
Relative mass			
Relative charge			

   b Relative atomic mass, A_r, can be used to compare the masses of atoms of different elements.

        Hint

     i Explain what you understand by the term *relative atomic mass*. [3]

     ii The antimony in a bullet was analysed by a forensic scientist to help solve a crime. The antimony was found to have the following percentage composition by mass: ^{121}Sb, 57.21%; ^{123}Sb, 42.79%.

        Calculate a value for the relative atomic mass of the antimony. Give your answer to four significant figures. [2]

OCR Chemistry AS (2811) June 2006 [Total 8]

Answer

continued

2 Sulfur and sulfur compounds are common in the environment.

  a A sample of sulfur from a volcano contained 88.0% by mass of ^{32}S and 12.0% by mass of ^{34}S. Copy and complete the table below to show the atomic structure of each isotope of sulfur. [2]

Isotope	Number of		
	protons	neutrons	electrons
^{32}S			
^{34}S			

  b Calculate the relative atomic mass of the volcanic sulfur. Your answer should be given to three significant figures. [2]

OCR Chemistry AS (2811) June 2001 [Total 4]

Answer

3 Analysis of a sample of bromine in a mass spectrometer showed that it contained a mixture of the ^{79}Br and ^{81}Br isotopes in the proportions: ^{79}Br, 55.0%; ^{81}Br, 45.0%.
Calculate the relative atomic mass of bromine in this sample. [2]

OCR Chemistry AS (2811) Jan 2002 [Total 2]

Answer

4 A sample of element **B** was analysed in a mass spectrometer. The relative atomic mass of element **B** can be calculated from the results shown in the table below.

	Isotope 1	Isotope 2	Isotope 3
relative isotopic mass	58.0	60.0	62.0
percentage composition / %	68.2	27.3	4.5

Using the information in the table, calculate the relative atomic mass of this sample of **B**. Give your answer to three significant figures. [2]

OCR Chemistry AS (2811) June 2003 [Total 2]

Answer

5 Chemists use a model of an atom that consists of sub-atomic particles (protons, neutrons and electrons). The particles in each of the following pairs differ only in the number of protons or neutrons or electrons. Explain what the difference is within each pair.

  a ^{6}Li and ^{7}Li [1]

  b ^{32}S and ^{32}S^{2-} [1]

  c ^{39}K^{+} and ^{40}Ca^{2+} [1]

[Total 3]

Answer

continued

6 Here is one way in which scientists work:
- observations are made, usually from experiments
- a theory is suggested to explain these observations
- new observations are made – they either confirm or disprove the theory
- if the observations disprove the theory, a new theory is suggested.

Re-read the start of this chapter. Use the development of modern atomic theory to produce an essay that explains 'how science works'. The title of your essay is 'How scientists developed the modern theory of atomic structure'.

You should include:

– the original model of the atom

– the observations made by Crookes, Thomson and Millikan that led to the 'plum pudding' model

– the observations made by Geiger and Marsden that led to the 'nuclear atom' model

– Rutherford's ideas and Chadwick's observations that led to the 'protons, neutrons and electrons' model studied today at GCSE.

[Total max 10]

Answer

Chapter 2

Moles and equations

Background

e-Learning

Objectives

Counting chemical substances in bulk

The mole and Avogadro's constant

When chemists write a formula for a compound, it tells us how many atoms of each element are present in one molecule, or one formula unit, of the compound. For example, the formula of water is H_2O, and this tells us that two atoms of hydrogen are combined with one atom of oxygen in each water molecule. As the A_r of hydrogen is 1.0 and the A_r of oxygen is 16.0, the M_r is $2.0 + 16.0 = 18.0$ and the hydrogen and oxygen are combined in a mass ratio of $2:16$. Although atoms are too small to be weighed individually, any mass of water will have hydrogen and oxygen in this ratio.

For example, consider 18 g of water ($2 + 16 = 18$). This will actually contain 2 g of hydrogen and 16 g of oxygen. We can use any unit of mass as long as we keep to the same mass ratio. In 18 tonnes of water there will be 2 tonnes of hydrogen and 16 tonnes of oxygen. The actual number of atoms present will be very large indeed!

When we take the relative molecular mass or relative atomic mass of a substance in grams, we say that we have one mole of the substance. The mole is the chemist's unit of amount. A **mole** of substance is the mass of substance that has the same number of particles as there are atoms in exactly 12 g of carbon-12. The particles may be atoms, molecules, ions or even electrons.

The number of atoms or molecules in one mole is a constant known as Avogadro's constant. Avogadro's constant, N_A, is 6.02×10^{23} mol^{-1}.

We often refer to the mass of one mole of a substance as the molar mass, M. The units of molar mass are g mol^{-1}.

In Figure 2.1 a mole of some elements may be compared.

Units used by chemists

Mass is measured in g (kg in SI units).
Volume is measured in cm^3 or dm^3
   (1 dm^3 = 1000 cm^3 = 1 litre).
Amount of substance is measured in moles
   (abbreviation is mol).

You need to remember that amount has a specific meaning, as do mass and volume. Each has its own unit. SI is an abbreviation for the Système International d'Unités. In this internationally recognised system, kilogram, metre and mole are three of the seven base units from which all supplementary units are derived.

Figure 2.1 From left to right, one mole of each of copper, bromine, carbon, mercury and lead.

Moles are particularly helpful when we need to measure out reactants or calculate the mass of product from a reaction. Such information is very important when manufacturing chemicals. For example, if the manufacture of a drug requires a particularly expensive reagent, it is important to mix it with the correct amounts of the other reagents to ensure that it all reacts and none is wasted. You will need to be able to write formulae in order to calculate amounts in moles. (See page 14 for some help with writing formulae.)

To find the amount of substance present in a given mass, we must divide that mass by the molar mass, M, of the substance. For example, for NaCl, $M = 23.0 + 35.5 = 58.5\,\mathrm{g\,mol^{-1}}$; so in 585 g of sodium chloride (NaCl) there are 585/58.5 mol of NaCl, i.e. 10 mol NaCl.

SAQ

1 What amount of substance is there in:
   a 35.5 g of chlorine *atoms*?
   b 71 g of chlorine *molecules*, Cl_2?   ⟨Answer⟩

2 Use Avogadro's constant to calculate the total number of atoms of chlorine in:
   a 35.5 g of chlorine atoms
   b 71 g of chlorine molecules.   ⟨Answer⟩

To find the mass of a given amount of substance, we multiply the number of moles of the substance by the molar mass.

SAQ

3 Calculate the mass of the following:
   a 0.1 mol of carbon dioxide   ⟨Hint⟩
   b 10 mol of calcium carbonate, $CaCO_3$.   ⟨Answer⟩

Calculations involving reacting masses

If we are given the mass of reactants, we can find out the mass of products formed in a chemical reaction. To do this, a balanced equation is used. (See page 15 for revision on balancing equations.)

Consider the formation of water from hydrogen and oxygen:

	$2H_2$	$+ O_2$	$\rightarrow 2H_2O$
this reads	2	+ 1	→ 2
	molecules of hydrogen	molecule of oxygen	molecules of water
or	2 moles of hydrogen	+ 1 mole of oxygen	→ 2 moles of water
masses /g	$2 \times 2.0 = 4.0$	+ 32.0	→ $2 \times (2.0 + 16.0)$ = 36.0

If we mix 4 g of hydrogen with 32 g of oxygen we should produce 36 g of water on exploding the mixture. Notice that the number of moles of water does *not* equal the sum of the number of moles of hydrogen and oxygen, because they have chemically reacted to produce molecules with a different molecular mass.

SAQ

4 Hydrogen burns in chlorine to produce hydrogen chloride:
$$H_2 + Cl_2 \longrightarrow 2HCl$$
   a Calculate the ratio of the masses of reactants.   ⟨Hint⟩
   b What mass (in g) of hydrogen is needed to produce 36.5 g of hydrogen chloride?   ⟨Answer⟩

Suppose we wish to calculate the mass of iron that can be obtained from a given mass of iron oxide, Fe_2O_3. When iron ore is reduced by carbon monoxide in a blast furnace (Figure 2.2), the equation for the reaction is:
$$Fe_2O_3 + 3CO \longrightarrow 2Fe + 3CO_2$$
The molar mass of Fe_2O_3 is $(2 \times 55.8) + (3 \times 16.0) = 159.6\,\mathrm{g\,mol^{-1}}$. From the balanced equation, one mole of Fe_2O_3 produces two moles of iron.

Hence 159.6 g of Fe_2O_3 will produce $2 \times 55.8 = 111.6$ g iron so

1000 g of Fe_2O_3 will produce

$$\frac{111.6 \times 1000\,\mathrm{g}}{159.6} = 699.2\,\mathrm{g~iron}$$

Figure 2.2 Workers taking the slag from the top of the molten iron in an open-hearth blast furnace.

SAQ

5 Calculate the mass of iron produced from 1000 tonnes of Fe_2O_3. How many tonnes of Fe_2O_3 would be needed to produce 1 tonne of iron?
If the iron ore contains 12% of Fe_2O_3, how many tonnes of ore are needed to produce 1 tonne of iron?
(Note: 1 tonne = 1000 kg.)

[Hint]

[Answer]

Calculation of empirical and molecular formulae

The **empirical formula** of a compound shows the simplest whole-number ratio of the elements present. For many simple compounds it is the same as the molecular formula. The **molecular formula** shows the total number of atoms of each element present in a molecule of the compound. Some examples are shown below.

Compound	Empirical formula	Molecular formula
water	H_2O	H_2O
methane	CH_4	CH_4
butane	C_2H_5	C_4H_{10}
benzene	CH	C_6H_6

SAQ

6 Write down the empirical formulae of the following:
a hexane, C_6H_{14}
b hydrogen peroxide, H_2O_2.

[Hint]

[Answer]

The molecular formula is far more useful. It enables us to write balanced chemical equations and to calculate masses of compounds involved in a reaction. It is not possible to calculate the molecular formula from data that gives the percentage composition by mass of a compound. Such data alone does not tell us how atoms are arranged in a molecule. However, the empirical formula *can* be found from such data. To get the necessary data, experimental methods that determine the mass of each element present in a compound are needed.

For example, if magnesium is burned in oxygen, magnesium oxide is formed. Suppose that a piece of magnesium of known mass is burned completely and the magnesium oxide produced is weighed. The weighings enable us to calculate the empirical formula of magnesium oxide.

In such an experiment, 0.243 g of magnesium ribbon produced 0.403 g of magnesium oxide:

	Mg	O
mass /g	0.243	0.403 – 0.243
		= 0.160
amount /mol	$= \dfrac{mass}{M}$	$= \dfrac{mass}{M}$
	$= \dfrac{0.243}{24.3}$	$= \dfrac{0.160}{16.0}$
	= 0.0100	= 0.0100

Divide by the smallest amount to give whole numbers:

atoms /mol	1	1

Magnesium and oxygen atoms are present in the ratio 1 : 1. Hence the empirical formula of magnesium oxide is MgO. Notice that we convert the mass of each element to the amount in moles, as we need the ratio of the number of atoms of each element present.

SAQ

7 An oxide of copper has the following composition by mass: Cu, 0.635 g; O, 0.080 g. Calculate the empirical formula of the oxide.

[Hint]

[Answer]

Empirical formulae can also be calculated from percentage composition data.

For example, it is found that an oxide of iron consists of 69.9% iron and 30.1% oxygen. As before, the first step is to divide each percentage by the A_r of that element:

	Fe	O
Percentage	69.9	30.1
	$\dfrac{69.9}{55.8}$	$\dfrac{30.1}{16.0}$
	= 1.253	= 1.881

Divide by the smallest amount to give whole numbers:

1   1.5

Iron and oxygen atoms are present in the ratio 1:1.5. We convert this to the whole number ratio 2:3. Hence the empirical formula of this oxide of iron is Fe_2O_3.

SAQ

8 **a** A second oxide of iron consists of 77.7% iron and 22.3% oxygen. Calculate its empirical formula.

b A third oxide of iron consists of 72.3% iron and 27.7% oxygen. Calculate its empirical formula.

Answer

Writing chemical formulae

By this point in your study of chemistry, you will already know the formulae of some simple compounds. For advanced chemistry you will need to learn the formulae of a wide range of compounds. These formulae are determined by the electronic configurations of the elements involved and the ways in which they combine with other elements to form compounds. The chemical bonding of elements in compounds is studied in Chapter 6. It will help if you learn some generalisations about the names and formulae of compounds.

For an ionic compound, the total number of positive charges in one formula unit of the compound must exactly equal the total number of negative charges. For magnesium oxide, magnesium (in Group 2) forms Mg^{2+} ions. Oxygen (in Group 6) forms O^{2-} ions. Magnesium oxide is thus MgO ($+2-2 = 0$). Aluminium forms 3+ ions. Two Al^{3+} ions and three O^{2-} ions are needed in one formula unit of aluminium oxide ($2 \times (+3) + 3 \times (-2) = 0$). The formula is Al_2O_3. Note how the number of ions of each element in the formula is written as a small number following and below the element symbol.

Some compounds do not contain ions. These compounds consist of molecules held together by covalent bonds. The formulae of simple covalent compounds may be deduced from the numbers of electrons required to complete the outer electron shell

of each atom present. For example, in methane, carbon requires four more electrons to complete its outer shell. Hydrogen requires one. This means that one carbon atom will combine with four hydrogen atoms to form a CH_4 (methane) molecule. In Chapter 6, you will see that each hydrogen atom forms one bond to the carbon atom whilst the carbon atom forms four bonds to the four hydrogen atoms.

Table 2.1 summarises the charges on some of the common ions that you need to learn. The position of the elements in the Periodic Table is helpful. In many instances, the group in which an element lies gives you a clue to the charge on an ion of the element. Note that metals form positive ions, whilst non-metals form negative ions.

Charge	Examples
1+	hydrogen, H^+ Group 1, the alkali metal ions, e.g. Li^+, Na^+, K^+
2+	Group 2, the alkaline earth metal ions, e.g. Mg^{2+}, Ca^{2+}
3+	Al^{3+}
1−	Group 7, the halogens, e.g. F^-, Cl^-, Br^-, I^-
	nitrate, NO_3^-
2−	Group 6, O^{2-} and S^{2-}
	carbonate, CO_3^{2-}
	sulfate, SO_4^{2-}
3−	phosphate, PO_4^{3-}

Table 2.1 Charges on some common ions.

Metals do not usually change their names in compounds. However, non-metals change their name by becoming -ides. For example, chlorine becomes chloride in sodium chloride. Sodium has not changed its name, although its properties are now dramatically different! Many non-metals (and some metals) combine with other non-metals such as oxygen to form negative ions. These negative ions start with the name of the element and end in -ate, e.g. sulfate. Some of these ions are also included in Table 2.1.

SAQ

9 Write down the formula of each of the following compounds.
- **a** magnesium bromide
- **b** hydrogen iodide
- **c** calcium sulfide
- **d** sodium sulfate
- **e** potassium nitrate
- **f** nitrogen dioxide

Answer

10 Name each of the following compounds.
- **a** K_2CO_3
- **b** Al_2S_3
- **c** $LiNO_3$
- **d** $Ca_3(PO_4)_2$
- **e** SiO_2

Answer

Balancing chemical equations

Atoms are neither created nor destroyed in a chemical reaction. When we write a chemical equation we must, therefore, ensure we have the same number of atoms of each element on each side of the chemical equation. We do this by *balancing* the equation, as follows.

- Write down the formulae of all the reactants and all the products. It may help you to write these in words first.
- Now inspect the equation and count the atoms of each element on each side. As the atoms of each element present cannot be created or lost in the chemical reaction, we must balance the number of atoms of each element.
- Decide which numbers must be placed in front of each formula to ensure that the same number of each atom is present on each side of the equation.

It is most important that the formulae of the reactants and products are not altered; only the total number of each may be changed.

Worked example 1

Step 1 When iron(III) oxide is reduced to metallic iron by carbon monoxide, the carbon monoxide is oxidised to carbon dioxide. (The III in iron(III) oxide indicates that the iron has an oxidation state of +3. See Chapter 4 for more on oxidation states.)

iron(III) oxide + carbon monoxide $\longrightarrow$ iron + carbon dioxide

The formulae are:
$$Fe_2O_3 + CO \longrightarrow Fe + CO_2$$

Step 2 On inspection we note that there are two iron atoms in the oxide on the left-hand side but only one on the right-hand side. In order to balance the number of iron atoms we write:
$$Fe_2O_3 + CO \longrightarrow 2Fe + CO_2$$

Step 3 Next we count the oxygen atoms: three in the oxide plus one in carbon monoxide, on the left-hand side. As there are only two on the right-hand side in carbon dioxide, we must double the number of CO_2 molecules in order to balance the number of oxygen atoms:
$$Fe_2O_3 + CO \longrightarrow 2Fe + 2CO_2$$

Step 4 On checking we see we have solved one problem but created a new one. There are now two carbon atoms on the right but only one on the left. Doubling the number of CO molecules balances the carbon atoms but unbalances the oxygen atoms again!

If we examine the equation again, we see that in the reaction between Fe_2O_3 and CO, each CO molecule requires only one oxygen atom to form CO_2. Thus three CO molecules combine with the three oxygen atoms lost from Fe_2O_3. Three CO_2 molecules will be formed:
$$Fe_2O_3 + 3CO \longrightarrow 2Fe + 3CO_2$$

Step 5 Conclusion: the equation is now balanced.

We often need to specify the physical states of chemicals in an equation. This can be important when, for example, calculating enthalpy changes (see Chapter 16). The state symbols used are: (s) for solid; (l) for liquid; (g) for gas. A solution in water is described as aqueous, so (aq) is used. Addition of the state symbols to the equation for the reaction of iron(III) oxide with carbon monoxide produces:

$$Fe_2O_3(s) + 3CO(g) \longrightarrow 2Fe(s) + 3CO_2(g)$$

SAQ

11 Balance the equations for the following reactions.

   a The thermite reaction (used for chemical welding of lengths of rail):

$$Al + Fe_2O_3 \longrightarrow Al_2O_3 + Fe$$

   b Petrol contains octane, C_8H_{18}. Complete combustion in oxygen produces only carbon dioxide and water.

   c Lead nitrate, $Pb(NO_3)_2$, decomposes on heating to produce PbO, NO_2 and O_2.

[Answer]

Balancing ionic equations

In some situations, chemists prefer to use ionic equations. Such equations are simpler than the corresponding full equation (which show the full formulae of all compounds present). For example, when a granule of zinc is placed in aqueous copper(II) sulfate, copper metal is displaced, forming a red-brown deposit on the zinc. In the reaction, zinc reacts to form zinc sulfate. The full equation for the reaction is:

$$Zn(s) + CuSO_4(aq) \longrightarrow ZnSO_4(aq) + Cu(s)$$

During this reaction copper ions, Cu^{2+}, are converted to copper atoms, Cu, and zinc atoms, Zn, are converted to zinc ions, Zn^{2+}. The sulfate ion has remained unchanged. It is known as a spectator ion. The ionic equation does not show the ions that remain unchanged. It therefore provides a shorter equation which focusses our attention on the change taking place:

$$Zn(s) + Cu^{2+}(aq) \longrightarrow Zn^{2+}(aq) + Cu(s)$$

In an ionic equation we must balance the overall charge on the ions on each side of the equation. Notice that the charge on each side of this ionic equation is 2+. Ensure that the charges are balanced *before* balancing the number of atoms of each element.

The reaction of $Cu^{2+}(aq)$ with Zn(s) involves transfer of electrons. It is known as a **redox** reaction. You will learn more about such reactions in Chapter 4. Ionic equations are frequently used for redox reactions. Chemists also prefer to use ionic equations for **precipitation** reactions. For example, when sodium hydroxide is added dropwise to copper(II) sulfate a pale blue precipitate of copper(II) hydroxide, $Cu(OH)_2(s)$, is formed. The full equation is:

$$CuSO_4(aq) + 2NaOH(aq) \longrightarrow$$
$$Cu(OH)_2(s) + Na_2SO_4(aq)$$

The ionic equation is:

$$Cu^{2+}(aq) + 2OH^-(aq) \longrightarrow Cu(OH)_2(s)$$

Both sodium ions, $Na^+(aq)$, and sulfate ions, $SO_4^{2-}(aq)$, are spectator ions, meaning that they are unchanged, and can be omitted.

SAQ

12 Balance the following ionic equations.

   a $Cl_2(aq) + Br^-(aq) \longrightarrow$ [Hint]
$$Cl^-(aq) + Br_2(aq)$$

   b $Fe^{3+}(aq) + OH^-(aq) \longrightarrow$
$$Fe(OH)_3(s)$$ [Answer]

Calculations involving concentrations and gas volumes

Concentrations of solutions

When one mole of a compound is dissolved in a solvent to make one cubic decimetre ($1 \, dm^3$) of solution, the concentration is $1 \, mol \, dm^{-3}$. Usually the solvent is water and an aqueous solution is formed.

Traditionally, concentrations in $mol \, dm^{-3}$ have been expressed as molarities. For example, $2 \, mol \, dm^{-3}$ aqueous sodium hydroxide is 2M aqueous sodium hydroxide, where M is the molarity of the solution. Although this is still a convenient method for labelling bottles, etc., it is better to use the units of $mol \, dm^{-3}$ in your calculations. Although you may have used mol/dm^3 or mol/litre in the past, advanced chemistry requires you to use $mol \, dm^{-3}$.

A solution with a comparatively high concentration of solute is known as a concentrated solution. A solution with a comparatively low concentration of solute is known as a dilute solution.

An experimental technique in which it is essential to know the concentration of solutions is a **titration**. A titration (Figure 2.3) is a way of measuring quantities of reactants, and can be very useful in determining an unknown concentration or following the progress of a reaction. In titrations there are five things you need to know:

- the balanced equation for the reaction
- the volume of the solution of the first reagent
- the concentration of the solution of the first reagent
- the volume of the solution of the second reagent
- the concentration of the solution of the second reagent.

If we know four of these, we can calculate the fifth. Remember that concentrations may be in $mol\,dm^{-3}$ or $g\,dm^{-3}$.

In many titration calculations it is necessary to start by finding the amount of a reagent (in moles) from a given concentration and volume. For example, what amount of sodium hydroxide is present in $24.0\,cm^3$ of an aqueous $0.010\,mol\,dm^{-3}$ solution?

Figure 2.3 A titration enables the reacting volumes of two solutions to be accurately determined. One solution is measured with a graduated pipette into a conical flask, the other is added slowly from a burette. The point where complete reaction just occurs is usually shown using an indicator, which changes colour at this point (called the end-point).

Convert the volume to dm^3:

$$1\,dm^3 = 10 \times 10 \times 10\,cm^3 = 1000\,cm^3$$

$$24.0\,cm^3 = \frac{24.0}{1000}\,dm^3$$

number of moles $=$ volume $\times$ concentration

amount of NaOH in $24.0\,cm^3$
$$= \frac{24.0}{1000}\,dm^3 \times 0.010\,mol\,dm^{-3}$$

$$= 2.40 \times 10^{-4}\,mol$$

To check your calculations:

- notice how the units multiply and cancel: $dm^3 \times mol\,dm^{-3} = mol$ (use this as a check)
- ensure the units of the answer are those you would expect (e.g. mol for quantity or $mol\,dm^{-3}$ for concentration)
- think about the size of your answer (e.g. $24\,cm^3$ is much less than $1\,dm^3$, so we expect the quantity of sodium hydroxide in $24\,cm^3$ of $0.01\,mol\,dm^{-3}$ solution to be much less than $0.01\,mol$).

We also often need to find the concentration of a solution. We can do this if we know the amount of solute that is dissolved in a given volume of solvent. For example, what is the concentration of an aqueous solution containing $2 \times 10^{-4}\,mol$ of sulfuric acid in $10\,cm^3$ of solution?

As before, convert the volume to dm^3:

$$10\,cm^3 = \frac{10}{1000}\,dm^3 = 1 \times 10^{-2}\,dm^3$$

concentration $=$ number of moles/volume

concentration of sulfuric acid
$$= \frac{2 \times 10^{-4}\,mol}{1 \times 10^{-2}\,dm^3}$$

$$= 2 \times 10^{-2}\,mol\,dm^{-3}$$

Again check by looking at the units:

$$\frac{mol}{dm^3} = mol\,dm^{-3}$$

SAQ

13 a Calculate the amount in moles of nitric acid in $25.0\,cm^3$ of a $0.1\,mol\,dm^{-3}$ aqueous solution.

    b Calculate the concentration in $mol\,dm^{-3}$ of a solution in which there is $0.125\,mol$ of nitric acid dissolved in each $50\,cm^3$ of solution.

Hint

Hint

Answer

Changing concentrations expressed in $mol\,dm^{-3}$ to $g\,dm^{-3}$ and vice versa is straightforward. We multiply by the molar mass M to convert $mol\,dm^{-3}$ to $g\,dm^{-3}$. To convert $g\,dm^{-3}$ to $mol\,dm^{-3}$ we divide by M. (Notice how the units cancel correctly.)

SAQ

14 a What is the concentration in $g\,dm^{-3}$ of $0.50\,mol\,dm^{-3}$ aqueous ethanoic acid (CH_3CO_2H)?

    b What is the concentration in $mol\,dm^{-3}$ of an aqueous solution containing $4.00\,g\,dm^{-3}$ of sodium hydroxide?

Answer

Worked example 2

This example shows how such calculations are combined with a balanced chemical equation to interpret the result of a titration. Try to identify the 'five things to know' in this calculation. In the titration $20.0\,cm^3$ of $0.200\,mol\,dm^{-3}$ aqueous sodium hydroxide exactly neutralises a $25.0\,cm^3$ sample of sulfuric acid. The purpose of the titration is to calculate the concentration of the sulfuric acid, firstly in $mol\,dm^{-3}$, and then in $g\,dm^{-3}$.

Step 1

$$20.0\,cm^3 = \frac{20}{1000}\,dm^3 = 2.00 \times 10^{-2}\,dm^3$$

amount of
sodium hydroxide $= 2.00 \times 10^{-2}\,dm^3$
$\times 0.200\,mol\,dm^{-3}$
$= 4.00 \times 10^{-3}\,mol$

continued

Step 2 The balanced equation for the reaction is:
$$2NaOH(aq) + H_2SO_4(aq) \longrightarrow$$
$$Na_2SO_4(aq) + 2H_2O(l)$$

Step 3 Exact neutralisation requires $2\,mol$ of NaOH to $1\,mol$ of H_2SO_4. So

amount of H_2SO_4 neutralised in the titration

$$= \tfrac{1}{2} \times \text{amount of NaOH}$$

$$= \tfrac{1}{2} \times 4.00 \times 10^{-3}\,mol$$

$$= 2.00 \times 10^{-3}\,mol$$

Step 4
Volume of $H_2SO_4 = 25.0\,cm^3 = \dfrac{25.0}{1000}\,dm^3$

$$= 2.5 \times 10^{-2}\,dm^3$$

Step 5
a Concentration of H_2SO_4
$$= \frac{2.00 \times 10^{-3}\,mol}{2.5 \times 10^{-2}\,dm^3}$$
$$= 0.080\,mol\,dm^{-3}$$

b As $M(H_2SO_4) = 2.0 + 32.1 + (4 \times 16.0)$
$$= 98.1\,g\,mol^{-1}$$
concentration of $H_2SO_4 = 98.1\,g\,mol^{-1}$
$$\times 0.080\,mol\,dm^{-3}$$
$$= 7.85\,g\,dm^{-3}$$

SAQ

15 $20.0\,cm^3$ of $0.100\,mol\,dm^{-3}$ potassium hydroxide exactly neutralises a $25.0\,cm^3$ sample of hydrochloric acid. What is the concentration of the hydrochloric acid in

    a $mol\,dm^{-3}$

    b $g\,dm^{-3}$?

Hint

Answer

It is possible to use the results of a titration to arrive at the reacting mole ratio, called the *stoichiometric ratio*, and the balanced equation for a reaction. (You can read more about titration in Chapter 3.)

Worked example 3

This example illustrates how to use a titration result to find the *stoichiometric ratio*. A $25.0\,cm^3$ sample of $0.0400\,mol\,dm^{-3}$ aqueous metal hydroxide is titrated against $0.100\,mol\,dm^{-3}$ hydrochloric acid. $20.0\,cm^3$ of the acid were required for exact neutralisation of the alkali.

Step 1

Amount of metal hydroxide $= \dfrac{25.0}{1000}\,dm^3 \times 0.0400\,mol\,dm^{-3}$

$= 1.00 \times 10^{-3}\,mol$

Amount of hydrochloric acid $= \dfrac{20.0}{1000}\,dm^3 \times 0.100\,mol\,dm^{-3}$

$= 2.00 \times 10^{-3}\,mol$

Step 2 Hence the reacting (i.e. stoichiometric) mole ratio of metal hydroxide : hydrochloric acid is

$1.00 \times 10^{-3}:2.00 \times 10^{-3}$

or $\qquad\qquad$ 1:2

Step 3 Exactly one mole of the metal hydroxide neutralises exactly two moles of hydrochloric acid. One mole of HCl will neutralise one mole of hydroxide ions, so the metal hydroxide must contain two hydroxide ions in its formula. The balanced equation for the reaction is

$M(OH)_2(aq) + 2HCl(aq) \longrightarrow$
$MCl_2(aq) + 2H_2O(l)$

where M is the metal.

SAQ

16 Determine the stoichiometric ratio, [Hint] and hence the balanced equation, for the reaction of insoluble iron hydroxide with dilute nitric acid, HNO_3, in which $24.0\,cm^3$ of $0.05\,mol\,dm^{-3}$ nitric acid is exactly neutralised by $4.00 \times 10^{-4}\,mol$ of the iron hydroxide. [Answer]

Gas volumes

In 1811, Avogadro discovered that equal volumes of all gases contain the same number of molecules. (Note that the volumes must be measured under the same conditions of temperature and pressure.) This provides an easy way of calculating the amount of gas present in a given volume. At room temperature and pressure (r.t.p.), one mole of any gas occupies approximately $24.0\,dm^3$. For example, $24.0\,dm^3$ of carbon dioxide (CO_2) and $24.0\,dm^3$ of nitrogen (N_2) both contain one mole of molecules.

SAQ

17 a Calculate the number of moles of helium present in a balloon [Hint] with a volume of $2.4\,dm^3$. Assume that the pressure inside the balloon is the same as atmospheric pressure and that the balloon is at room temperature.

b Calculate the volume occupied by a mixture of 0.5 mol of propane and 1.5 mol of butane gases at room temperature and pressure. [Answer]

We can use reacting volumes of gases to determine the **stoichiometry** of a reaction. The experiments must be carried out under the same conditions of temperature and pressure. We can then assume that equal volumes of gases contain the same number of moles. For example, measurements show that $20\,cm^3$ of hydrogen react with exactly $10\,cm^3$ of oxygen to form water. The ratio of reacting volumes of hydrogen to oxygen is 20:10 or 2:1. Hence the reacting mole ratio for hydrogen : oxygen (the stoichiometry of the reaction) is also 2:1, and so the balanced equation is:

$2H_2(g) + O_2(g) \longrightarrow 2H_2O(l)$.

[Extension]

Summary

Glossary

- One mole of a substance is the amount of substance that has the same number of particles as there are atoms in exactly 12 g of carbon-12. This number is called Avogadro's constant.

- Empirical formulae show the simplest whole-number ratio of atoms in a compound whilst molecular formulae show the total number of atoms of each element present in one molecule or one formula unit of the compound. Empirical formulae may be determined from the composition by mass of a compound. The molecular formula may then be found if the molecular mass is known.

- Molar masses enable calculations to be made using moles and balanced chemical equations. Such calculations involve reacting masses, volumes and concentrations of solutions, and volumes of gases.

- Balanced chemical equations (which show the stoichiometry of a reaction) may also be derived by measuring reacting masses, volumes and concentrations of solutions, or volumes of gases.

Questions

1 Iridium reacts with fluorine to form a yellow solid **Y** with the percentage composition by mass: Ir, 62.75%; F, 37.25%.
   The empirical formula of **Y** can be calculated from this information.
   a Calculate the empirical formula of **Y**. [2]
   b Write a balanced equation for the reaction between iridium and fluorine. [1]
   OCR Chemistry AS (2811) May 2002 [Total 3]

Answer

2 Water, ammonia and sulfur dioxide react together to form a compound **A** which has the following percentage composition by mass:
   N   24.12%
   H   6.94%
   S   27.61%
   O   41.33%.
   a Calculate the empirical formula of compound **A**. [2]
   b Suggest a balanced equation for the formation of compound **A** from the reaction of water, ammonia and sulfur dioxide. [1]
   OCR Chemistry AS (2811) Jan 2006 [Total 3]

Answer

continued

3 In an experiment, a student bubbled chlorine through $120\,cm^3$ of an aqueous solution of $0.275\,mol\,dm^{-3}$ sodium hydroxide, NaOH(aq). The equation for the reaction that occurred is shown below.

$$Cl_2(g) + 2NaOH(aq) \longrightarrow NaCl(aq) + NaClO(aq) + H_2O(l)$$

Under the reaction conditions, 1 mole of $Cl_2(g)$ occupies $24.0\,dm^3$.

    a What is meant by the term *the mole*? [1]

    b How many moles of NaOH were in the $120\,cm^3$ volume of NaOH(aq)? [1]   Hint

    c Calculate the volume of $Cl_2(g)$ that was needed to react with the NaOH(aq) used. [2]   Hint

OCR Chemistry AS (2811) Jan 2006   [Total 4]

Answer

4 A student had a stomach-ache and needed to take something to neutralise excess stomach acid. He decided to take some Milk of Magnesia, which is an aqueous suspension of magnesium hydroxide, $Mg(OH)_2$.

    a The main acid in the stomach is hydrochloric acid, HCl(aq), and the unbalanced equation for the reaction that takes place with Milk of Magnesia is shown below.

        $\ldots Mg(OH)_2(s) + \ldots HCl(aq) \longrightarrow \ldots MgCl_2(aq) + \ldots H_2O(l)$

    Copy the equation and balance it by including numbers where necessary in the unbalanced equation. [1]

    b The student's stomach contained $500\,cm^3$ of stomach fluid with an acid concentration of $0.108\,mol\,dm^{-3}$. The student swallowed some Milk of Magnesia containing $2.42\,g\,Mg(OH)_2$.

    He wondered whether this dose was sufficient to neutralise the stomach acid.
Assume that all the acid in the stomach fluid was $0.108\,mol\,dm^{-3}$ hydrochloric acid.

      i How many moles of HCl were in the $500\,cm^3$ of stomach fluid? [1]   Hint

      ii Calculate the mass of $Mg(OH)_2$ necessary to neutralise this stomach fluid. [3]   Hint

      iii Determine whether the student swallowed too much, too little, or just the right amount of Milk of Magnesia to neutralise the stomach acid. [1]

OCR Chemistry AS (2811) June 2005   [Total 6]

Answer

5 A student reacted $2.74\,g$ of barium with water to form $250\,cm^3$ of aqueous barium hydroxide.

    a Calculate how many moles of Ba reacted. [1]

    b Calculate the concentration, in $mol\,dm^{-3}$ of $Ba(OH)_2$ that was formed. [1]

    c Calculate the volume of H_2 that would be produced at room temperature and pressure (r.t.p.). [1 mol of gas molecules occupies $24.0\,dm^3$ at r.t.p.] [1]   Hint

OCR Chemistry AS (2811) June 2001   [Total 3]

Answer

Chapter 3

Acids

Objectives

Acids and bases

An **acid** is a compound that releases H^+ ions (protons) in aqueous solution. Three common acids are sulfuric acid, H_2SO_4, hydrochloric acid, HCl, and nitric acid, HNO_3. These acids all release H^+ ions when they dissolve in water. The release of H^+ ions can be written as an equation. The equation for HCl dissolving in water is:

$$HCl(g) \xrightarrow{water} H^+(aq) + Cl^-(aq)$$

SAQ

1 Write a similar equation for nitric acid dissolving in water. (Pure nitric acid is a liquid.)

Answer

2 When sulfuric acid dissolves in water the sulfate ion (SO_4^{2-}) is formed.
   Write an equation for sulfuric acid dissolving in water. (Pure sulfuric acid is a liquid.)

Answer

A **base** is a substance that accepts (or 'receives') H^+ ions from an acid. Bases can neutralise acids. Metal oxides and metal hydroxides are bases, as is ammonia. Some common bases and their formulae are listed in Table 3.1.

Name of base	Formula
calcium oxide	CaO
magnesium oxide	MgO
copper(II) oxide	CuO
calcium hydroxide	Ca(OH)$_2$
sodium hydroxide	NaOH
potassium hydroxide	KOH
ammonia	NH$_3$

Table 3.1 Some common bases.

Alkalis

Sodium hydroxide and potassium hydroxide are bases. They are also **alkalis**. An alkali is a soluble base. Alkalis release OH^- ions in aqueous solution. OH^- ions are called hydroxide ions. The equation for NaOH dissolving in water is:

$$NaOH(s) \xrightarrow{water} Na^+(aq) + OH^-(aq)$$

SAQ

3 Write a similar equation for potassium hydroxide dissolving in water.

Answer

When ammonia gas dissolves in water some of the ammonia molecules react with water molecules. The product of this reaction is ammonium hydroxide solution:

$$NH_3(g) + H_2O(l) \rightleftharpoons NH_4^+(aq) + OH^-(aq)$$

Ammonia solution is therefore an alkali.

Neutralisation

When a base neutralises an acid, the base accepts the H^+ ions donated by the acid. At the same time the H^+ ions of the acid are replaced by metal ions or NH_4^+ (ammonium) ions, forming a **salt**.

If the base is a metal hydroxide, the OH^- ions accept the H^+ ions, forming water. For example, when sodium hydroxide neutralises hydrochloric acid:

$$Na^+OH^-(aq) + H^+Cl^-(aq) \longrightarrow Na^+Cl^-(aq) + H_2O(l)$$

Sodium chloride (Na^+Cl^-) is a salt. It forms when the H^+ ions of the hydrochloric acid are replaced by Na^+ ions (Figure 3.1).

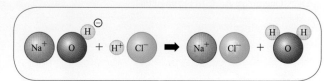

Figure 3.1 When sodium hydroxide neutralises hydrochloric acid the OH^- ions accept the H^+ ions, forming water.

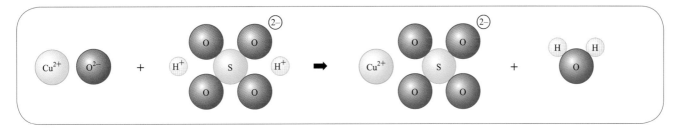

Figure 3.2 When copper(II) oxide neutralises sulfuric acid the O^{2-} ions accept the H^+ ions, forming water.

If the base is a metal oxide, the O^{2-} ions accept the H^+ ions, forming water. For example, when copper(II) oxide neutralises sulfuric acid:

$$Cu^{2+}O^{2-}(s) + H^+_2SO_4^{2-}(aq) \longrightarrow$$
$$Cu^{2+}SO_4^{2-}(aq) + H_2O(l)$$

Copper(II) sulfate ($Cu^{2+}SO_4^{2-}$) is a salt. It forms when the H^+ ions of the sulfuric acid are replaced by Cu^{2+} ions.

Ammonia solution contains $NH_3(aq)$ as well as $NH_4^+OH^-(aq)$. When ammonia neutralises an acid the NH_3 molecules accept the H^+ ions, forming NH_4^+ ions. For example, when ammonia neutralises nitric acid:

$$NH_3(aq) + H^+NO_3^-(aq) \longrightarrow NH_4^+NO_3^-(aq)$$

Ammonium nitrate ($NH_4^+NO_3^-$) is a salt. It forms when the H^+ ions of the nitric acid are replaced by NH_4^+ ions.

SAQ

4 Write a balanced chemical equation for the reaction between calcium hydroxide and nitric acid, showing the charges on the ions. Explain how the acid is neutralised and name the salt formed.

Answer

Making salts

Many salts are very useful. Salts can be made by neutralising acids. For example, the reaction illustrated in Figure 3.2 can be used to make copper(II) sulfate. Excess copper(II) oxide has to be added to the sulfuric acid. Adding excess copper(II) oxide ensures that all of the sulfuric acid is neutralised.

Since copper(II) oxide is an insoluble base, the excess can be removed by filtration.

Sugar-free gum

People today tend to eat too much sugar – and that is not just in sweets and biscuits. Many processed foods, such as baked beans, often have sugar added.

This sugar is converted into acid by bacteria in your mouth. The acid then attacks the enamel on your teeth. However, by chewing a sugar-free gum, more saliva is produced which can be slightly alkaline. This will neutralise the acid that would otherwise attack your teeth.

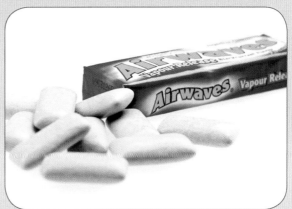

Figure 3.3 Neutralising acid in your mouth helps prevent tooth decay.

SAQ

5 Look at Figure 3.4 below.

a

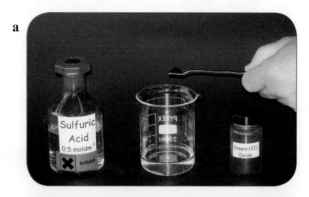

b

c   **d**

e

Figure 3.4 Method for preparing copper sulfate.

Write a caption for each of the five stages.

Answer

Salts can also be made by neutralising acids with metal carbonate compounds. For example, if hydrochloric acid is neutralised by calcium carbonate, the salt made is calcium chloride.

$$Ca^{2+}CO_3{}^{2-}(s) + 2H^+Cl^-(aq) \longrightarrow$$
$$Ca^{2+}Cl^-{}_2(aq) + CO_2(g) + H_2O(l)$$

Effervescence (fizzing) is seen as the calcium carbonate is added to the hydrochloric acid. This is because carbon dioxide gas is given off. When this reaction is used as a way of making calcium chloride, the method used is the same as the one in Figure 3.4.

SAQ

6 Write balanced chemical equations for reactions **a** to **d** below, showing the charges on any ions present. Name the salt formed in each case.
   a copper(II) carbonate with hydrochloric acid
   b potassium hydroxide with sulfuric acid
   c magnesium oxide with nitric acid
   d ammonia solution with sulfuric acid

Answer

Titration

When an acid is neutralised by an alkali the products are a salt and water – see Figure 3.1. However, the method shown in Figure 3.4 cannot be used with this reaction to make a sample of the salt. This is because alkalis are soluble in water. Excess alkali cannot be filtered out. Instead the method used is **titration**. Figure 3.5 shows titration being used to make ammonium sulfate

The acid is measured out, indicator is added, then the acid is neutralised by alkali from a burette. When the volume of alkali needed to neutralise the volume of acid is known, the experiment is repeated without the indicator. The product obtained is a pure salt solution.

SAQ

7 Potassium chloride is a salt that can be made from potassium hydroxide solution and hydrochloric acid. Describe how you would do this in a lab, ending with a good crystalline sample of the salt. Explain your choice of method.

Answer

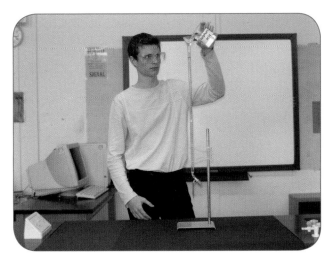

Figure 3.5 a A funnel is used to fill the burette with sulfuric acid.

Figure 3.5 b A graduated pipette is used to measure 25 cm³ of ammonia solution into a conical flask.

Figure 3.5 c An indicator called litmus solution is added to the ammonia solution which turns blue.

Figure 3.5 d 12.5 cm³ of sulfuric acid from the burette have been added to the 25 cm³ of alkali in the conical flask. The litmus has gone purple, showing that this volume of acid was just enough to neutralise the alkali.

Figure 3.5 e When this is repeated without using indicator, ammonium sulfate solution is produced.

Figure 3.5 f When the water is evaporated off, solid ammonium sulfate is obtained.

Anhydrous and hydrated salts

Some salts can have water molecules incorporated in their structure. Such salts are said to be **hydrated**. The water is known as **water of crystallisation**. A familiar example of this is blue copper(II) sulfate. Its formula is $CuSO_4.5H_2O$. The '$.5H_2O$' in the formula means there are five water molecules associated with each $CuSO_4$ formula unit. Copper(II) sulfate can also exist without any water of crystallisation. It is then said to be **anhydrous**, and its formula is simply $CuSO_4$.

For a hydrated salt (such as blue copper(II) sulfate) composition data enables us to calculate the number of water molecules per formula unit.

Figure 3.6 Hydrated copper(II) sulfate (left) and anhydrous copper(II) sulfate (right).

Worked example

How to calculate the number of water molecules per formula unit

0.720 g of hydrated cobalt(II) chloride consists of 0.178 g of cobalt, 0.215 g of chlorine, 0.036 g of hydrogen and 0.291 g of oxygen. The hydrogen and oxygen are present as water of crystallisation.

Step 1 Calculate the number of moles of each element present, using a table:

Step 2 Divide by the smallest amount to change the number of moles of each element into a whole-number ratio:

Step 3 Conclusion: the empirical formula of hydrated cobalt(II) chloride is $CoCl_2H_{12}O_6$, which is better written as $CoCl_2.6H_2O$ since hydrogen and oxygen are present as water of crystallisation.

	Co	Cl	H	O
mass /g	0.178	0.215	0.036	0.291
amount /mol	$= \dfrac{mass}{M}$	$= \dfrac{mass}{M}$	$= \dfrac{mass}{M}$	$= \dfrac{mass}{M}$
	$= \dfrac{0.178}{58.9}$	$= \dfrac{0.215}{35.5}$	$= \dfrac{0.036}{1.0}$	$= \dfrac{0.291}{16.0}$
	$= 0.0030$	$= 0.0061$	$= 0.0360$	$= 0.0182$
atoms /mol	1	2	12	6

SAQ

8 0.859 g of hydrated copper(II) chloride consists of 0.320 g of copper, 0.358 g of chlorine, 0.020 g of hydrogen and 0.161 g of oxygen. The hydrogen and oxygen are present as water of crystallisation. Calculate the empirical formula of hydrated copper(II) chloride.

Hint

Answer

9 The empirical formula of anhydrous manganese(II) sulfate is $MnSO_4$. It also exists in a hydrated form.
When 2.480 g of hydrated manganese(II) sulfate was heated strongly until all the water of crystallisation had been driven off, 1.680 g of anhydrous manganese(II) sulfate remained.
  a Calculate the number of grams of water that was driven off.
  b Calculate the number of moles of water that was driven off.
  c Calculate the number of moles of anhydrous manganese(II) sulfate that was left at the end.
  d What is the empirical formula of hydrated manganese(II) sulfate?

Hint

Answer

SAQ

10 Calcium nitrate exists in a hydrated form which can be made by reacting calcium carbonate with nitric acid. The composition by mass of hydrated calcium nitrate is 16.98% calcium, 11.86% nitrogen, 67.77% oxygen and 3.39% hydrogen. Hydrated calcium nitrate can be converted to anhydrous calcium nitrate. The composition by mass of anhydrous calcium nitrate is 24.44% calcium, 17.06% nitrogen and 58.50% oxygen.

a Describe in detail how you would make crystals of hydrated calcium nitrate by reacting calcium carbonate with nitric acid.

b How would you convert hydrated calcium nitrate to anhydrous calcium nitrate?

c Calculate the empirical formula of anhydrous calcium nitrate.

d Calculate the empirical formula of hydrated calcium nitrate.

Answer

Summary

Glossary

- Acids are proton (H^+) donors.

- Bases are proton (H^+) acceptors.

- Acids release H^+ ions in aqueous solution.

- Alkalis are soluble bases that release OH^- ions in aqueous solution.

- A salt is produced when the H^+ ion of an acid is replaced by a metal ion or an ammonium ion.

- Acids react with many metals to form salts and hydrogen.

- Acids are neutralised by:
  – basic metal oxides to form salts and water
  – metal carbonates to form salts, water and carbon dioxide
  – alkalis to form salts and water.

- Ammonia behaves as a base. For example, it neutralises sulfuric acid to form the salt ammonium sulfate.

- Some salts can have water molecules incorporated in their structure. Such salts are said to be *hydrated*. The water is known as *water of crystallisation*.

Questions

1 A small amount of magnesium oxide, MgO, was reacted with excess dilute hydrochloric acid.
  a Describe what would be seen as the reaction occurs. [1]
  b Write a balanced equation for this reaction. [1]
OCR Chemistry AS (2813) June 2006 [Total 2]

Hint

Answer

2 a Hydrochloric acid is a strong acid. What is meant by the term *acid*? [1]
  b Hydrochloric acid reacts with a solution of sodium carbonate.
    i Write appropriate state symbols in the equation for the reaction shown below.
    $$2HCl\ldots + Na_2CO_3\ldots \longrightarrow 2NaCl\ldots + CO_2\ldots + H_2O\ldots$$ [1]
    ii State what you would see to indicate that the reaction was taking place. [1]
OCR Chemistry AS (2813) June 2004 [Total 3]

Hint

Answer

3 Hydrogen chloride, HCl, is a colourless gas which dissolves very readily in water forming hydrochloric acid.
  (1 mol of gas occupies $24.0\,dm^3$ at room temperature and pressure, r.t.p.)
  a At r.t.p., $1\,dm^3$ of water dissolved $432\,dm^3$ of hydrogen chloride gas.
    i How many moles of hydrogen chloride dissolved in the water? [1]
    ii The hydrochloric acid formed has a volume of $1.40\,dm^3$. What is the concentration, in $mol\,dm^{-3}$, of the hydrochloric acid? [1]
  b Hydrochloric acid reacts with magnesium carbonate, $MgCO_3$. For this reaction, state what you would see and write a balanced equation. [2]
OCR Chemistry AS (2811) June 2001 [Total 4]

Hint

Answer

4 The concentration of sulfuric acid can be checked by titration. A sample of sulfuric acid was analysed as follows.
  $10.0\,cm^3$ of sulfuric acid was diluted with water to make $1.00\,dm^3$ of solution.
  The diluted sulfuric acid was then titrated with aqueous sodium hydroxide, NaOH.
  $$H_2SO_4(aq) + 2NaOH(aq) \longrightarrow Na_2SO_4(aq) + 2H_2O(l)$$
  In the titration, $25.0\,cm^3$ of $0.100\,mol\,dm^{-3}$ aqueous sodium hydroxide required $20.0\,cm^3$ of the diluted sulfuric acid for neutralisation.
  a Calculate how many moles of NaOH were used. [1]
  b Calculate the concentration, in $mol\,dm^{-3}$, of the diluted sulfuric acid, H_2SO_4. [2]
  c Calculate the concentration, in $mol\,dm^{-3}$, of the original sample of sulfuric acid. [1]
OCR Chemistry AS (2811) Jan 2001 [Total 4]

Hint

Answer

Chapter 4

Redox

Objectives

Oxidation numbers and oxidation states

Oxidation states

When you look at the formulae of many compounds you see that there are differences in the ratios of the atoms that combine with each other – MgO and Al_2O_3, for example. Chemists have devised various ways for comparing the 'combining ability' of individual elements. One term, much used in the past, but less so nowadays, is 'valency' meaning 'strength'. A more useful measure is **oxidation state**. This is a numerical value associated with atoms of each element in a compound or ion. Some chemists prefer the term **oxidation number** (abbreviated ox. no.). The only difference between this and oxidation state is that we say an atom *'has* an oxidation number of +2' but '*is in* an oxidation state of +2'.

There are rules for determining the values of oxidation states.

- Oxidation states are usually calculated as the number of electrons that atoms lose, gain or share when they form ionic or covalent bonds in compounds.
- The oxidation state of uncombined elements (that is, not in compounds) is always zero. For example, each atom in $H_2(g)$ or $O_2(g)$ or $Na(s)$ or $S_8(s)$ has an oxidation state of zero; otherwise, in a compound the numbers are always given a sign, + or –.
- For a monatomic ion, the oxidation state of the element is simply the same as the charge on the ion. For example:

ion	Na^+	Ca^{2+}	Cl^-	O^{2-}
ox. state	+1	+2	–1	–2

- In a chemical species (compound or ion), with atoms of more than one element, the most electronegative element is given the negative oxidation state. Other elements are given positive oxidation states. (For an explanation of the term 'electronegative', see Chapter 6.) For example,

in the compound disulfur dichloride, S_2Cl_2, chlorine is more electronegative than sulfur. The two chlorine atoms each have an oxidation state of –1, and thus the two sulfur atoms each have an oxidation state of +1.

- The oxidation state of hydrogen in compounds is +1, except in metal hydrides (e.g. NaH), when it is –1.
- The oxidation state of oxygen in compounds is –2, except in peroxides (e.g. H_2O_2), when it is –1, or in OF_2, when it is +2.
- The sum of all the oxidation states in a neutral compound is zero. In an ion, the sum equals the overall charge. For example, the sum of the oxidation states in $CaCl_2$ is 0; the sum of the oxidation states in OH^- is –1.

Some examples of determining oxidation states are shown below:

In CO_2	the ox. state of each O atom is –2 giving a total of –4 CO_2 is neutral therefore the ox. state of C is +4
In $MgCl_2$	the ox. state of Mg is +2 the ox. state of each Cl is –1
In NO_3^-	the ox. state of each O is –2 total for O_3 is –6 the overall charge on the ion is –1 therefore ox. state of N in NO_3^- is +5

SAQ

1 What is the oxidation state of:
   a C in CO_3^{2-}
   b Al in Al_2Cl_6?

Answer

Redox: oxidation and reduction

The term **redox** is used for the simultaneous processes of *red*uction and *ox*idation. Originally oxidation and reduction were related only to reactions of oxygen and hydrogen. They now include any reactions in which electrons are transferred.

For example, consider what happens when iron reacts with oxygen and with chlorine.

i with oxygen:
$$4Fe(s) + 3O_2(g) \longrightarrow 2Fe_2O_3(s)$$

ii with chlorine:
$$2Fe(s) + 3Cl_2(g) \longrightarrow 2FeCl_3(s)$$

In both of these reactions, each iron atom has lost three electrons and changed oxidation state from 0 in $Fe(s)$ to +3 in $Fe_2O_3(s)$ and $FeCl_3(s)$.

$$Fe \longrightarrow Fe^{3+} + 3e^-$$
ox. state   0   +3

This is **oxidation**. In all oxidation reactions, atoms of an element in a chemical species *lose electrons* and increase their oxidation states.

In reaction **i** above, the oxygen atoms each gain two electrons and change oxidation state from 0 in $O_2(g)$ to –2 in $Fe_2O_3(s)$.

$$O_2 + 4e^- \longrightarrow 2O^{2-}$$
ox. state of atoms   0   –2

Similarly, in reaction **ii**, chlorine atoms each gain one electron and change oxidation state from 0 to –1.

$$Cl_2 + 2e^- \longrightarrow 2Cl^-$$
ox. state of atoms   0   –1

These are processes of **reduction**. In all reduction reactions, atoms of an element in a chemical species *gain electrons* and decrease their oxidation states.

We call reactions, such as **i** and **ii** above, redox reactions, as both oxidation and reduction take place at the same time. Any chemical reaction which involves the oxidised and reduced forms of the same element, compound or ion is called a *redox reaction*. The element, compound or ion that *gains* electrons acts as an oxidising agent; the one that *loses* electrons acts as a reducing agent.

SAQ

2   **a**   Calcium carbonate reacts with hydrochloric acid. A balanced chemical equation for this reaction is:
$$Ca^{2+}CO_3{}^{2-}(s) + 2H^+Cl^-(aq) \longrightarrow$$
$$Ca^{2+}Cl^-{}_2(aq) + CO_2(g) + H_2O(l)$$
Is this a redox reaction? Justify your answer by giving the oxidation state of each element before and after the reaction.

b   Potassium manganate(VII) reacts with hydrochloric acid. A balanced chemical equation for this reaction is:
$$2K^+MnO_4{}^-(s) + 16H^+Cl^-(aq) \longrightarrow$$
$$2K^+Cl^-(aq) + 2Mn^{2+}Cl^-{}_2(aq) + 5Cl_2(g) + 8H_2O(l)$$
Is this a redox reaction? Justify your answer by giving the oxidation state of each element before and after the reaction.

Answer

Extraction and corrosion of metals

Whenever we extract metals from their compounds found in ores, we use redox reactions. The metal ions are always reduced to form metal atoms.

For example, in a blast furnace used to extract iron, the main redox reaction taking place is:
$$2Fe_2O_3 + 3CO \longrightarrow 4Fe + 3CO_2$$
The oxidation number of Fe in Fe_2O_3 is +3. This gets reduced to 0 in the atoms of the element iron that are formed in the reaction. In effect:
$$Fe^{3+} + 3e^- \longrightarrow Fe$$
This is called a half-equation. It tells us that Fe^{3+} ions are reduced to Fe atoms. At the same time, carbon monoxide is oxidised to carbon dioxide.

Figure 4.1 The iron in the hull of this ship is protected by bars of magnesium metal. The magnesium atoms (oxidation number = 0) are oxidised to Mg^{2+} ions in preference to iron atoms changing to Fe^{3+}. This is called sacrificial protection.

continued

The overall reaction is a redox reaction.

Unfortunately, iron will rust when exposed to air and water. In this reaction the iron atoms (oxidation number = 0) return to their +3 oxidation state, forming hydrated iron(III) oxide. This redox reaction costs society millions of pounds each year in replacing corroded parts or trying to prevent the rusting taking place.

SAQ

3 Write a half-equation to show the oxidation of magnesium atoms to magnesium ions. [Answer]

Oxidation numbers in compound names

The names of some compounds include the oxidation state of one of the elements involved. The oxidation state is shown as a Roman numeral in brackets. An example of this is the way we name the two different iron chlorides: $Fe^{2+}Cl^-_2$ is called iron(II) chloride, $Fe^{3+}Cl^-_3$ is called iron(III) chloride.

Figure 4.2 Iron(II) chloride and iron(III) chloride.

SAQ

4 What are the names of
  a Cu^+Cl^-
  b $Cu^{2+}Cl^-_2$? [Answer]

Sodium, nitrogen and oxygen can form two different compounds. Their formulae are $Na^+NO_2^-$ and $Na^+NO_3^-$. Nitrogen has a different oxidation state in each compound. $Na^+NO_2^-$ is called 'sodium nitrate(III)'. $Na^+NO_3^-$ is called 'sodium nitrate(V)'. The III and the V tell us the oxidation state of nitrogen in each compound. Sodium is Na^+ in both compounds; its oxidation state is therefore +1 in both compounds. The oxidation state of oxygen is –2 in both compounds. The oxidation state of nitrogen is +3 in sodium nitrate(III), and +5 in sodium nitrate(V).

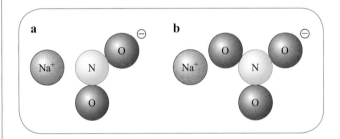

Figure 4.3 a One formula unit of 'sodium nitrate(III)' and b one formula unit of 'sodium nitrate(V)'.

SAQ

5 What are the names of: [Hint]
  a $Na^+_2SO_3^{2-}$
  b $Na^+_2SO_4^{2-}$? [Answer]

Many formulae can be stated unambiguously without reference to oxidation states. For example, Na^+Cl^- is just called 'sodium chloride'. This is because sodium exists as Na^+ in *all* of its common compounds, so its oxidation state is always +1 in compounds. There is no need to call it 'sodium(I) chloride'. Similarly 'chlor*ide*' *always* refers to the Cl^- ion, so the oxidation state of chlorine in every compound called chloride is –1.

Care is still needed, however. There are four different compounds called sodium chlor*ate*. The 'ate' means that each formula unit also contains oxygen. One of these compounds is called sodium chlorate(I), the Roman numeral gives the oxidation state of the chlorine. Its formula is Na^+ClO^-. The other three compounds are called sodium chlorate(III), sodium chlorate(V) and sodium chlorate(VII). They differ from each other by the amount of oxygen in one formula unit of the compound.

SAQ

6 What are the formulae of sodium chlorate(III), sodium chlorate(V) and sodium chlorate(VII)?

[Hint]

[Answer]

Metals and non-metals

Often a metal and a non-metal will react together to form an ionic compound. In such a reaction each metal atom loses one or more electrons. The metal atoms become positive ions. Each non-metal atom gains one or more electrons. The non-metal atoms become negative ions. For example, when magnesium reacts with chlorine:

$$Mg(s) + Cl_2(g) \longrightarrow Mg^{2+}Cl^-_2(s)$$

Each magnesium atom loses two electrons. The oxidation number of magnesium increases from 0 to +2. Magnesium is oxidised in this reaction. Each chlorine atom gains one electron. The oxidation number of chlorine decreases from 0 to −1. Chlorine is reduced in this reaction.

When a metal reacts with a non-metal a redox reaction takes place. The metal is oxidised and the non-metal is reduced.

SAQ

7 Calcium reacts with oxygen to form calcium oxide, $Ca^{2+}O^{2-}$.

  a Write a balanced chemical equation for this reaction.

  b Which element is oxidised in this reaction? Justify your answer in two ways.

  c Which element is reduced in this reaction? Justify your answer in two ways.

[Answer]

Metals react with acids

Magnesium metal reacts with dilute hydrochloric acid to produce a salt (magnesium chloride) and hydrogen:

$$Mg(s) + 2H^+Cl^-(aq) \longrightarrow Mg^{2+}Cl^-_2(aq) + H_2(g)$$

This is a redox reaction. The magnesium atoms lose electrons and become magnesium ions. Magnesium is oxidised. The hydrogen ions gain electrons and become hydrogen atoms. Hydrogen is reduced. The chloride ions are unchanged at the end of the reaction. The chloride ions are neither oxidised nor reduced.

Figure 4.4 Magnesium metal reacting with dilute hydrochloric acid.

Other metals react with dilute acids in this way too. Only very unreactive metals like copper, silver and gold fail to do so.

SAQ

8 Zinc metal reacts with dilute sulfuric acid, forming hydrogen and zinc sulfate, $Zn^{2+}SO_4^{2-}$.

[Hint]

  a Write a balanced chemical equation for this reaction, showing the charge on any ions present.

  b Which element is oxidised in this reaction? Justify your answer in two ways.

  c Which element is reduced in this reaction? Justify your answer in two ways.

[Answer]

Summary

Glossary

- Oxidation is loss of electrons.

- Reduction is gain of electrons.

- When an atom or ion is oxidised its oxidation number increases.

- When an atom or ion is reduced its oxidation number decreases.

- The oxidation state of an element can be shown in the compound's name (if more than one possible oxidation state is possible) e.g. iron(III) bromide.

- Metal atoms tend to be reducing agents. When they react they lose electrons (i.e. are oxidised themselves) and the oxidation number increases as positive ions are formed. The opposite happens with non-metals.

Questions

1 Domestic tap water has been chlorinated. Chlorine reacts with water as shown below.

$$Cl_2(g) + H_2O(l) \longrightarrow HOCl(aq) + HCl(aq)$$

State the oxidation number of chlorine in

a Cl_2 [1]   Hint

b HOCl [1]

c HCl. [1]   Hint

OCR Chemistry AS (2811) June 2006   [Total 3]

Answer

2 The reaction between barium and water is a redox reaction.

$$Ba(s) + 2H_2O(l) \longrightarrow Ba(OH)_2(aq) + H_2(g)$$

a Explain, in terms of electrons, what is meant by

   i oxidation [1]

   ii reduction. [1]

b Which element has been oxidised in this reaction? [1]   Hint

   Deduce the change in its oxidation state. [1]

OCR Chemistry AS (2811) June 2001   [Total 4]

Answer

3 A student prepared an aqueous solution of calcium chloride by reacting calcium with hydrochloric acid. Calcium chloride contains Ca^{2+} and Cl^- ions.

a Complete and balance the following equation for this reaction.

$$\ldots Ca(s) + \ldots HCl(aq) \longrightarrow \ldots CaCl_2(aq) + \ldots\ldots$$ [2]

continued

b This is a redox reaction.

Use oxidation states to show that calcium has been oxidised. [2]

OCR Chemistry AS (2811) Jan 2005 [Total 4]

Answer

4 The formation of magnesium oxide, MgO, from its elements involves both oxidation and reduction in a redox reaction.

a Write a full equation, including state symbols, for the formation of MgO from its elements. [2]

b MgO reacts when heated with acids such as nitric acid, HNO_3.

$$MgO(s) + 2HNO_3(aq) \longrightarrow Mg(NO_3)_2(aq) + H_2O(l)$$

Using oxidation numbers, explain whether or not the reaction between MgO and HNO_3 is a redox reaction. [2]

OCR Chemistry AS (2811) Jan 2002 [Total 4]

Answer

Chapter 5

Electron structure

e-Learning

Objectives

Electrons in atoms

Electrons hold the key to almost the whole of chemistry. Protons and neutrons give atoms their mass, but electrons are the outer part of the atom and only electrons are involved in the changes that happen during chemical reactions. If we knew everything about the arrangements of electrons in atoms and molecules, we could predict most of the ways that chemicals behave purely from mathematics. So far this has proved very difficult, even with the most advanced computers – but it may yet happen.

SAQ

1　Suggest why the isotopes of an element have the same chemical properties, though they have different relative isotopic masses.

Answer

Which models are currently accepted about how electrons are arranged around the nucleus? One of the first simple ideas – that they just orbit randomly around the nucleus – was soon rejected.

A model you may have used at GCSE level considers the electrons to be arranged in shells. These 'shells' correspond to different energy levels occupied by the electrons.

Arrangements of electrons: energy levels and 'shells'

There was a great advance in atomic theory when, in 1913, the Danish physicist Niels Bohr proposed his ideas about arrangements of electrons in atoms.

Earlier the German physicist Max Planck had proposed, in his 'Quantum Theory' of 1901, that energy can only be transferred in fixed amounts or 'packets' of energy which he called *quanta*; a single packet of energy is a *quantum*. Bohr applied this idea to the energy of electrons. He suggested that, as electrons could only possess energy in quanta, they

would not exist in a stable way anywhere outside the nucleus unless they were in fixed or 'quantised' energy levels. If an electron gained or lost energy, it could move to higher or lower energy levels but not somewhere in between. It is a bit like climbing a ladder; you can only stay in a stable state on one of the rungs. You will find that, as you read more widely, there are several names given to these energy levels. The most common name is *shells*.

Shells are numbered 1, 2, 3, 4, etc. These numbers are known as *principal quantum numbers* (symbol n). Such numbers correspond to the numbers of rows (or periods) in the Periodic Table.

We can now write the simple electronic configurations as shown in Table 5.1.

	Atomic number	Number of electrons in shell		
		$n = 1$	$n = 2$	$n = 3$
H	1	1		
He	2	2		
Li	3	2	1	
Be	4	2	2	
B	5	2	3	
C	6	2	4	
N	7	2	5	
O	8	2	6	
F	9	2	7	
Ne	10	2	8	
Na	11	2	8	1

Table 5.1 Simple electronic configurations of the first eleven elements in the Periodic Table.

Remember that the atomic number tells us the number of electrons present in an atom of the element. For a given element, electrons are added to the shells as follows:

- up to 2 electrons in shell 1
- up to 8 electrons in shell 2
- up to 18 electrons in shell 3
- up to 32 electrons in shell 4.

Some of the best evidence for the existence of electron shells comes from ionisation energies.

Ionisation energy

When an atom loses an electron it becomes a positive ion. We say that it has been *ionised*. Energy is needed to remove electrons and this is generally called **ionisation energy**.

> The *first ionisation energy* of an element is the amount of energy needed to remove one electron from each atom in a mole of atoms of an element in the gaseous state.

The general symbol for ionisation energy is ΔH_i and for a first ionisation energy it is ΔH_{i1}.
The process may be shown by the example of calcium as:
$$Ca(g) \longrightarrow Ca^+(g) + e^-; \quad \Delta H_{i1} = +590\,kJ\,mol^{-1}$$

(If the symbols seem unfamiliar at this stage, see Chapter 16.)

The energy needed to remove a second electron from each ion in a mole of gaseous ions is the *second ionisation energy*. For calcium:
$$Ca^+(g) \longrightarrow Ca^{2+}(g) + e^-; \quad \Delta H_{i2} = +1150\,kJ\,mol^{-1}$$
Note that the second ionisation energy is much larger than the first. The reasons for this are discussed in the bullet pointed list below.

We can continue removing electrons until only the nucleus of an atom is left. The sequence of first, second, third, fourth, etc. ionisation energies (or 'successive ionisation energies') for the first eleven elements in the Periodic Table are shown in Table 5.2.

We see the following for any one element.

- Successive ionisation energies increase. As each electron is removed from an atom, the remaining ion becomes more positively charged. Moving the next electron away from the increased positive charge is more difficult and the next ionisation energy is even larger.
- There are one or more particularly large rises within the set of ionisation energies of each element (except hydrogen and helium).

		\multicolumn{11}{c|}{Electrons removed}										
		1	2	3	4	5	6	7	8	9	10	11
1	H	1310										
2	He	2370	5250									
3	Li	520	7300	11 800								
4	Be	900	1760	14 850	21 000							
5	B	800	2420	3660	25 000	32 800						
6	C	1090	2350	4620	6220	37 800	47 300					
7	N	1400	2860	4580	7480	9450	53 300	64 400				
8	O	1310	3390	5320	7450	11 000	13 300	71 300	84 100			
9	F	1680	3470	6040	8410	11 000	15 200	17 900	92 000	106 000		
10	Ne	2080	3950	6120	9370	12 200	15 200	20 000	23 000	117 000	131 400	
11	Na	510	4560	6940	9540	13 400	16 600	20 100	25 500	28 900	141 000	158 700

Table 5.2 Successive ionisation energies for the first eleven elements in the Periodic Table (to nearest $10\,kJ\,mol^{-1}$).

Ionisation energies of elements are measured mainly by two techniques:

- using calculations based on the light emitted by the element when it is heated strongly, or when an electric current is passed through it in the gaseous state.
- using electron bombardment of the element in the gaseous state.

These data may be interpreted in terms of the atomic numbers of elements and their simple electronic configurations.

Before doing so, we must consider the factors which influence ionisation energies.

Factors influencing ionisation energies

The three strongest influences on ionisation energies of elements are the following.

- *The size of the positive nuclear charge* This charge affects all the electrons in an atom. The increase in nuclear charge with atomic number will tend to cause an increase in ionisation energies.
- *The distance of the electron from the nucleus* All forces of attraction decrease rapidly as the distance between the attracted bodies increases. Thus the attraction between the nucleus of an atom and an electron is less for electrons further from the nucleus, and greater for electrons nearer the nucleus. The further a shell is from the nucleus, the lower are the ionisation energies for electrons in that shell.
- *The 'shielding' effect by electrons in filled inner shells* All electrons are negatively charged and repel each other. Electrons in the filled inner shells repel electrons in the outer shell and reduce the effect of the positive nuclear charge. This is called the *shielding effect*. The greater the shielding effect upon an electron, the lower is the energy required to remove it and thus the lower the ionisation energy.

Consider the example of the successive ionisation energies of lithium in Table 5.2. We see a low first ionisation energy, followed by much larger second and third ionisation energies. This confirms that lithium has one electron in its outer shell $n = 2$, which is easier to remove than either of the two electrons in the inner shell $n = 1$. The electron in the $n = 2$ shell is easier to remove because it is *further from the nucleus* and *more shielded* than the electrons in the $n = 1$ shell. The large increase in ionisation energy indicates where there is a change from shell $n = 2$ to shell $n = 1$.

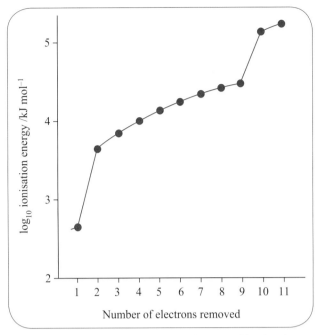

Figure 5.1 Graph of logarithm ($\log_{10}$) of ionisation energy of sodium against the number of electrons removed.

The pattern is seen even more clearly if we plot a graph of ionisation energies (y axis) against number of electrons removed (x axis). As the ionisation energies are so large, we must use logarithms to base 10 ($\log_{10}$) to make the numbers fit on a reasonable scale. The graph for sodium is shown in Figure 5.1.

SAQ

2 a In Figure 5.1 why are there large increases between the first and second ionisation energies of sodium and again between the ninth and tenth ionisation energies?

  b How does this graph confirm the suggested simple electronic configuration for sodium of (2,8,1)?

Successive ionisation energies are thus helpful for predicting or confirming the simple electronic configurations of elements. In particular, they confirm the number of electrons in the outer shell. This also leads to confirmation of the position of the element in the Periodic Table. Elements with one electron in their outer shell are in Group 1, elements with two electrons in their outer shell are in Group 2, and so on.

SAQ

3 The first four ionisation energies of an element are: 590, 1150, 4940 and 6480 kJ mol^{-1}. Suggest the Group in the Periodic Table to which this element belongs.

Hint

Answer

Need for a more complex model

Table 5.1 tells us correctly how many electrons there are in each of the shells 1, 2 and 3 for the first eleven elements. Scientists now understand electron configurations to be more complicated than this. You will see in Chapter 7 that the first ionisation energies of the elements 3 (lithium) to 10 (neon) do not increase evenly. This, and other, variations show the need for a more complex model of electron configurations than the Bohr model.

Here is a summary of this more complex model.

- The energy levels (shells) of principal quantum numbers $n = 1, 2, 3, 4$, etc. do not have precise energy values. Instead, they each consist of a set of subshells, which contain *orbitals* with different energy values.
- The subshells are of different types labelled **s**, **p**, **d** and **f**. An s subshell contains one orbital; a p subshell contains three orbitals; a d subshell contains five orbitals; and an f subshell contains seven orbitals.
- An electron orbital represents a region of space around the nucleus of an atom, capable of accommodating two electrons. Therefore an s subshell can accommodate two electrons, a p subshell can accommodate six electrons, a d subshell can accommodate ten electrons and an f subshell can accommodate fourteen electrons.
- Each orbital has its own approximate, three-dimensional shape. It is not possible to draw the shape of orbitals precisely. They do not have exact boundaries but are fuzzy, like clouds; indeed, they are often called 'charge-clouds'.

Approximate representations of orbitals are shown in Figure 5.2. Note that there is only one type of s orbital but three different p orbitals (p_x, p_y, p_z). There are five different d orbitals and seven f orbitals.

If you wish to learn more about this we suggest you research into quantum mechanics and, in particular, the Schrödinger equation and Heisenberg's uncertainty principle.

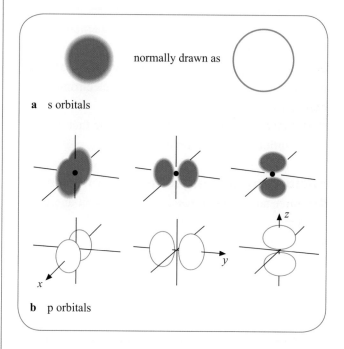

Figure 5.2 Representations of orbitals (the position of the nucleus is shown by the black dot): **a** s orbitals are spherical; **b** p orbitals, p_x, p_y and p_z, have 'lobes' along x, y and z axes.

Orbitals: spin-pairing

The shell $n = 1$ consists of just one 'subshell' containing a single s orbital called 1s; $n = 2$ consists of s and p orbitals in subshells called 2s and 2p; $n = 3$ consists of s, p and d orbitals in subshells called 3s, 3p and 3d.

There is an important principle concerning orbitals that affects all electronic configurations. This is the theory that any individual orbital can hold *one* or *two* electrons but *not more*.

You may wonder how an orbital can hold two electrons with negative charges that repel each other strongly. It is explained by the idea of 'spin-pairing'. Along with charge, we say that electrons have a property called 'spin'. We can visualise spin as an electron rotating at a fixed rate. Two electrons can exist as a pair in an orbital by each having opposite spin (Figure 5.3); this reduces the effect of repulsion (see also later in this chapter). Clockwise spin is shown as ↑, anticlockwise spin as ↓.

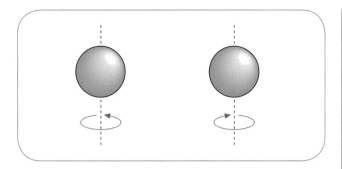

Figure 5.3 Representation of opposite spins of electrons.

Taking all the known evidence into consideration, scientists now believe:

- shell $n = 1$ contains up to two electrons in an s orbital
- shell $n = 2$ contains up to eight electrons, two in an s orbital and six in the p subshell, with two in each of the p_x, p_y and p_z orbitals
- shell $n = 3$ contains up to 18 electrons, two in an s orbital, six in the p subshell and ten in the d subshell, with two in each of the five orbitals.

Order of filling shells and orbitals

In each successive element of the Periodic Table, the order of filling the shells and orbitals is the order of their relative energy. The electronic configuration of each atom is the one that gives as low an energy state as possible to the atom as a whole. This means that the lowest-energy orbitals are filled first. The order of filling is:

first 1s, then 2s, 2p, 3s, 3p, 4s, 3d, 4p, …

As you see, the order (shown diagrammatically in Figure 5.4) is not quite what we might have predicted!

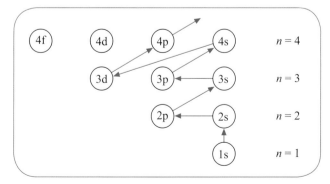

Figure 5.4 Diagram to show the order in which orbitals are filled up to shell $n = 4$.

An expected order is followed up to the 3p subshell, but then there is a variation, as the 4s is filled before the 3d. This variation and other variations further along in the order are caused by the increasingly complex influences of nuclear attractions and electron repulsions upon individual electrons.

Electronic configurations

Representing electronic configurations

The most common way of representing the electronic configuration of an atom is shown below. For example, hydrogen has one electron in an s orbital in the shell with principal quantum number $n = 1$. We show this as:

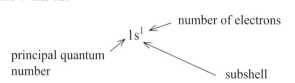

Helium has two electrons, both in the 1s orbital, and is shown as $1s^2$.

The electronic configurations for the first eighteen elements (H to Ar) are shown in Table 5.3.

1	H	$1s^1$
2	He	$1s^2$
3	Li	$1s^2\ 2s^1$
4	Be	$1s^2\ 2s^2$
5	B	$1s^2\ 2s^2\ 2p^1$
6	C	$1s^2\ 2s^2\ 2p^2$
7	N	$1s^2\ 2s^2\ 2p^3$
8	O	$1s^2\ 2s^2\ 2p^4$
9	F	$1s^2\ 2s^2\ 2p^5$
10	Ne	$1s^2\ 2s^2\ 2p^6$
11	Na	$1s^2\ 2s^2\ 2p^6\ 3s^1$
12	Mg	$1s^2\ 2s^2\ 2p^6\ 3s^2$
13	Al	$1s^2\ 2s^2\ 2p^6\ 3s^2\ 3p^1$
14	Si	$1s^2\ 2s^2\ 2p^6\ 3s^2\ 3p^2$
15	P	$1s^2\ 2s^2\ 2p^6\ 3s^2\ 3p^3$
16	S	$1s^2\ 2s^2\ 2p^6\ 3s^2\ 3p^4$
17	Cl	$1s^2\ 2s^2\ 2p^6\ 3s^2\ 3p^5$
18	Ar	$1s^2\ 2s^2\ 2p^6\ 3s^2\ 3p^6$

Table 5.3 Electronic configurations for the first eighteen elements in the Periodic Table.

For the next set of elements 19 (potassium) to 36 (krypton), it is more convenient to represent part of the configuration as a 'noble-gas core'. In this case the core is the configuration of argon. For convenience we sometimes represent $1s^2\,2s^2\,2p^6\,3s^2\,3p^6$ as [Ar] rather than write it out each time. Some examples are shown in Table 5.4.

19	Potassium (K)	[Ar] $4s^1$
20	Calcium (Ca)	[Ar] $4s^2$
21	Scandium (Sc)	[Ar] $3d^1\ 4s^2$
⋮		
24	Chromium (Cr)	[Ar] $3d^5\ 4s^1$
25	Manganese (Mn)	[Ar] $3d^5\ 4s^2$
⋮		
29	Copper (Cu)	[Ar] $3d^{10}\ 4s^1$
30	Zinc (Zn)	[Ar] $3d^{10}\ 4s^2$
31	Gallium (Ga)	[Ar] $3d^{10}\ 4s^2\ 4p^1$
⋮		
35	Bromine (Br)	[Ar] $3d^{10}\ 4s^2\ 4p^5$
36	Krypton (Kr)	[Ar] $3d^{10}\ 4s^2\ 4p^6$

Table 5.4 Electronic configurations for some of the elements 19 to 36, where [Ar] is the electronic configuration of argon, $1s^2\,2s^2\,2p^6\,3s^2\,3p^6$.

The following points should be noted.
- When the 4s orbital is filled, the next electron goes into a 3d orbital (see scandium). This begins a pattern of filling up the 3d subshell, which finishes at zinc. The elements that add electrons to the d subshells are called the **d-block elements**; a subset of these is called **transition elements.**
- There are variations in the pattern of filling the d subshell at elements 24 (chromium) and 29 (copper). These elements have only one electron in their 4s orbital. Chromium has five d electrons, rather than the expected four; copper has ten d electrons rather than nine. This is the outcome of the complex interactions of attractions and repulsions in their atoms.
- From element 31 (gallium) to 36 (krypton) the electrons add to the 4p subshell. This is similar to the pattern of filling the 3p subshell from elements 13 (aluminium) to 18 (argon) in Period 3.

The Periodic Table

The elements in the s-block have their outer electron in an s subshell. Similarly the elements in the p-, d- and f-blocks have their outer electrons in p, d and f subshells respectively.

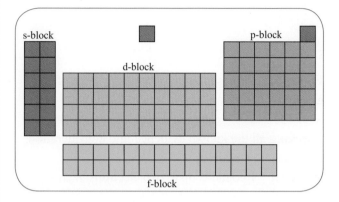

Figure 5.5 There are four blocks of elements in the Periodic Table.

SAQ

4 **a** How many electrons will fit into each of an s subshell, a p subshell, a d subshell and an f subshell?

b How many electrons will fit into each of shell 1, shell 2, shell 3 and shell 4?

c Where in the Periodic Table are: the s-block, the p-block, the d-block and the f-block?

Answer

Filling of orbitals

Whenever possible, electrons will occupy orbitals singly. This is due to the repulsion of their negative charges. Electrons remain unpaired until the available orbitals of *equal energy* have one electron each. When there are more electrons than the orbitals can hold as singles, they pair up by spin-pairing. This means that, if there are three electrons occupying a p subshell, one each will go to the p_x, p_y and p_z orbitals, rather than two in p_x, one in p_y and none in p_z. When there are four electrons occupying a p subshell, two will spin-pair in one orbital, leaving a single electron in each of the other two orbitals. Similarly, five electrons in a d subshell will remain unpaired in the five orbitals (see Mn in Figure 5.6).

	Sodium atom	**Sodium ion**	**Fluorine atom**	**Fluoride ion**
Symbol	Na	Na$^+$	F	F$^-$
Atomic number	11	11	9	9
Electrons	11	10	9	10
Configuration	$1s^2\, 2s^2\, 2p^6\, 3s^1$	$1s^2\, 2s^2\, 2p^6$	$1s^2\, 2s^2\, 2p^5$	$1s^2\, 2s^2\, 2p^6$

Table 5.5 Electronic configurations of some ions

As an example, we can show how orbitals are occupied in atoms of carbon, nitrogen and oxygen as:

carbon (six electrons) $1s^2\ 2s^2\ 2p_x^1\ 2p_y^1\ 2p_z^0$
nitrogen (seven electrons) $1s^2\ 2s^2\ 2p_x^1\ 2p_y^1\ 2p_z^1$
oxygen (eight electrons) $1s^2\ 2s^2\ 2p_x^2\ 2p_y^1\ 2p_z^1$

(Normally electronic configurations are shown in less detail – as in Table 5.3 and Table 5.4.)

Electronic configurations of ions

The number of electrons in an ion is found from the atomic number of the element and the charge of the ion. Some examples are shown in Table 5.5.

Note that both the sodium ion, Na$^+$, and the fluoride ion, F$^-$, have the same electronic configuration as the noble gas neon. This has implications for the formation of, and bonding in, the compound sodium fluoride (see Chapter 6 for a discussion of ionic bonding).

Electronic configurations in boxes

Another useful way of representing electronic configurations is in box form. We can show the electrons as arrows with their clockwise or anticlockwise spin as ↑ or ↓.

Figure 5.6 shows the electronic configurations of some of the first 36 elements represented in this way.

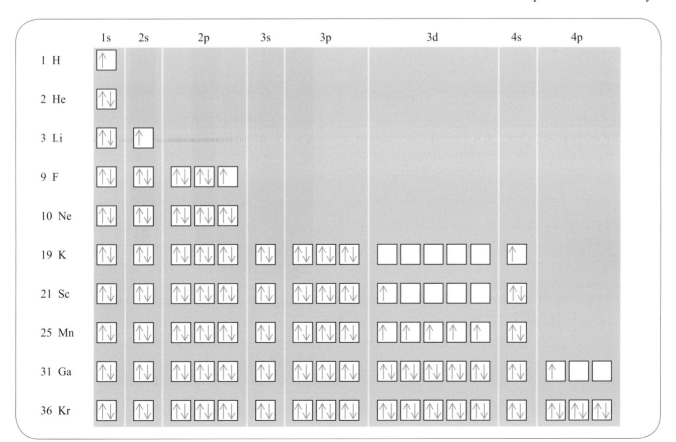

Figure 5.6 Electronic configurations of some elements in box form.

SAQ

5 Draw box-form electronic configurations for: boron, oxygen, argon, nickel and bromine.

Hint

Answer

Luminescence – exciting electrons!

By absorbing energy, electrons can jump into higher energy levels. When these electrons drop back to lower energy levels they give out energy, sometimes as light. This is called 'luminescence'.

You might have seen this in action inside the glowing safety tubes used to attract the attention of rescuers in dark remote locations. Party-goers use them too – then they are known as 'glow-sticks'.

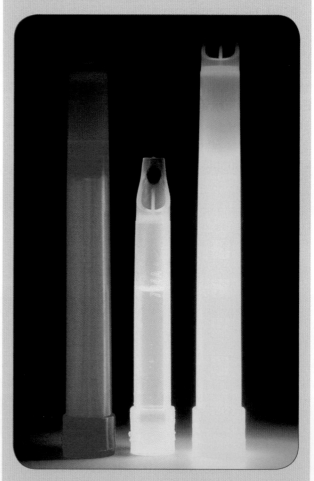

Figure 5.7 Electrons moving between energy levels are the source of this light energy. Chemicals in the tube are said to be 'chemiluminescent'.

Summary

Glossary

- Electrons can exist only at certain energy levels and gain or lose 'quanta' of energy when they move between the levels.

- The main energy levels or 'shells' are given principal quantum numbers $n = 1, 2, 3, 4$, etc. Shell $n = 1$ is the closest to the nucleus.

- The shells consist of subshells known as s, p, d or f and each subshell consists of orbitals. Subshells s, p, d and f have one, three, five and seven orbitals respectively. Orbitals s, p, d and f have different, distinctive shapes; we have looked at the shapes of s and p orbitals.

- s orbitals are spherical. p orbitals have two 'lobes'.

- Each orbital holds a maximum of two electrons, so that full subshells of s, p, d and f orbitals contain two, six, ten and fourteen electrons respectively. The two electrons in any single orbital are spin-paired.

- Electrons remain unpaired among orbitals of equal energy until numbers require them to spin-pair.

- The first ionisation energy of an element is the energy required to remove one electron from each atom in a mole of atoms of the element in the gaseous state.

- Successive ionisation energies are the energies required to remove first, second, third, fourth etc. electrons from each atom in a mole of gaseous atoms of an element.

- Large changes in the values of successive ionisation energies of an element indicate that the electrons are being removed from different shells. This gives evidence for the electronic configuration of atoms of the element and helps to confirm the position of the element in the Periodic Table.

Questions

1 **a** State what is meant by an orbital. [1]

   b Draw diagrams to show the shape of an s orbital and of a p orbital. [2]

   c Complete the table below to show how many electrons completely fill each of the following:

	number of electrons
a p *orbital*	
a d *subshell*	
the third *shell* ($n = 3$)	

[3]

OCR Chemistry AS (2811) June 2001 [Total 6]

Answer

continued

2 a Successive ionisation energies provide evidence for the arrangement of electrons in atoms. The graph below shows the eight successive ionisation energies of oxygen.

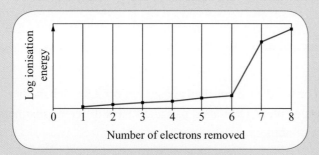

Number of electrons removed

> **i** Write an equation, including state symbols, to represent the second ionisation energy of oxygen. [2]
> **ii** How does this graph provide evidence for the existence of two electron shells in oxygen? [2]

b i Complete the electronic configuration for an aluminium atom: $1s^2$... [1]

> **ii** Using axes as shown below, sketch a graph to show the 13 successive ionisation energies of aluminium. [2]

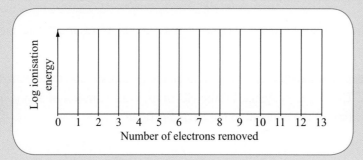

Number of electrons removed

OCR Chemistry AS (2811) May 2002 [Total 7]

Hint

Answer

3 The first six successive ionisation energies of an element D are shown in the table below.

element	ionisation energy / kJ mol⁻¹					
	1st	2nd	3rd	4th	5th	6th
D	1086	2353	4621	6223	37832	47278

a Define the term *first ionisation energy*. [3]
b Write an equation, with state symbols, to represent the *third* ionisation energy of element D. [2]
c Use the table to deduce which group of the Periodic Table contains element D. Explain your answer. [3]

Hint

OCR Chemistry AS (2811) Jan 2003 [Total 8]

continued

4 This question refers to calcium chloride, made up of Ca^{2+} and Cl^- ions.

   a Copy and complete the table below.   [2]

species	number of	
	protons	electrons
Ca^{2+}		
Cl^-		

   b Complete the electronic configuration of Ca^{2+} : $1s^2...$ [1]   Hint

   c What is the formula of calcium chloride?   [1]

OCR Chemistry AS (2811) Jan 2002   [Total 4]

Answer

5 Electrons are arranged in energy levels. The diagram below for the seven electrons in a nitrogen atom is incomplete. It shows two electrons in the 1s level.

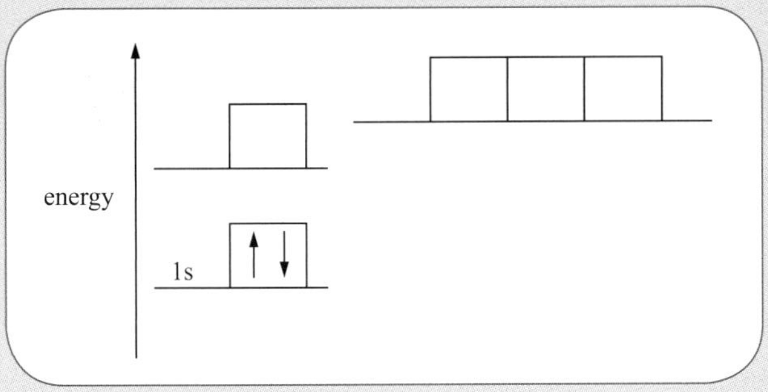

   a Copy and complete the diagram for the seven electrons in a nitrogen atom by
     i adding labels for the other subshell levels.   [1]
     ii showing how the electrons are arranged.   [2]   Hint

   b Magnesium reacts with nitrogen forming magnesium nitride, which is an ionic compound.
     i Complete the electronic configuration for the 12 electrons in a magnesium atom: $1s^2...$ [1]
     ii What is the charge on each ion in magnesium nitride?   [2]
     iii Complete the electronic configuration of each *ion* in magnesium nitride.   [2]
     iv Deduce the formula of magnesium nitride.   [1]

OCR AS Chemistry 2811 Jan 2001   [Total 9]

Answer

Chapter 6

Bonding and structure

Background

e-Learning

Objectives

Figure 6.1 A salt mountain with a salt pan in the foreground, Sardinia.

Ionic bonding

Many familiar substances are ionic compounds. An example is common salt (sodium chloride). Sodium chloride and many other ionic compounds are present in sea water. Crystals of salt are readily obtained by the partial evaporation of sea water in a salt pan (Figure 6.1).

We need to understand the bonding in compounds in order to explain their structure and physical properties. Ionic compounds have the following physical properties:

- they are crystalline solids with high melting points
- they conduct electricity in aqueous solution or when they are molten – the passing of electric current makes them decompose, which is called electrolysis
- they are hard and brittle with crystals that cleave easily
- they are often soluble in water.

Ionic bonding results from the *electrostatic attraction* between the *oppositely charged ions*. In sodium chloride, the ions are arranged in a giant lattice which determines the shape of the crystals grown from sea water. Most other minerals are also

found as well-formed crystals. The shapes of these crystals arise from the way in which the ions are packed together in the lattice. Some crystals are shown in Figure 6.2.

Figure 6.2 A selection of minerals.

The high melting points of ionic compounds can make them particularly useful, e.g. aluminium oxide (melting point 2072 °C). A fibrous form of aluminium oxide is used in tiles on the Space Shuttle. They provide protection from the high temperatures experienced on re-entry into the atmosphere (Figure 6.3) and are also used in the lining of portable gas forges.

Figure 6.3 The space shuttle *Columbia*, seen during the fitting of the thermal insulation tiles.

Electrolysis

Another important characteristic of ionic compounds is their ability to conduct electricity, with decomposition, when in aqueous solution or when they are molten. The passing of electric current makes them decompose. This process is called **electrolysis**. Electrolysis is used to produce chlorine from brine (concentrated aqueous sodium chloride) (Figure 6.4) and aluminium from molten aluminium oxide.

Figure 6.4 Industrial electrolysis: chlorine cell.

Ions are free to move through the aqueous solution or molten compound and are attracted to the oppositely charged electrode. Positive ions (cations) are attracted to the negative electrode (cathode), and negative ions (anions) to the positive electrode (anode). At the electrode ions are discharged: e.g. chloride ions change to atoms of chlorine which combine to form Cl_2 gas, aluminium ions change to atoms of aluminium metal. When it is discharged, an ion either gains or loses electrons. *Cations* (positively charged ions) gain electrons. *Anions* (negatively charged ions) lose electrons. The number of electrons gained or lost depends on the magnitude of the charge on the ion. A chloride ion loses one electron; an aluminium ion gains three electrons.

These changes may be represented as follows:
- at the positive electrode (anode):
  $Cl^- \longrightarrow Cl + e^-$ then $2Cl \rightarrow Cl_2$
  This can be summarised as: $2Cl^- \rightarrow Cl_2 + 2e^-$
  (or $Cl^- \longrightarrow \frac{1}{2}Cl_2 + e^-$).
- at the negative electrode (cathode):
  $Al^{3+} + 3e^- \longrightarrow Al$.

SAQ

1 Write similar equations, including electrons, for the discharge of copper and bromide ions during the electrolysis of copper bromide, $CuBr_2$, using carbon electrodes. Indicate the electrode at which each reaction will occur.

Hint

Answer

Recycling aluminium

Aluminium is extracted from an ore called bauxite which contains aluminium oxide. The bauxite is dug from open-cast mines, creating huge craters. Then the aluminium oxide is separated from the ore. The waste from the process is stored in large lagoons of water which become brown from iron impurities in the ore.

The aluminium oxide is then shipped to a processing plant where the aluminium is extracted by electrolysis. Before electrolysis can take place the aluminium oxide must be melted. This

continued

requires a large amount of energy even though some cryolite (sodium aluminium fluoride) is added to lower the temperature needed to melt the oxide. Then the electrolysis takes place. A typical plant will use as much electricity as a small town!

This is why we should recycle as much aluminium as possible. There is a 95% saving in energy if you compare recycled aluminium with aluminium extracted from bauxite. This helps tackle global warming as we burn less fossil fuels, so less carbon dioxide is released into the air. We also reduce other pollutant gases produced in the extraction process. Then there is the benefit of preserving the ore supplies, and less open-cast mining means fewer scars on the landscape. The aluminium that would have been thrown away would also take up precious space in our landfill sites.

Figure 6.5 Remember to recycle your cans. As well as aluminium cans, we can also save energy and resources by recycling steel cans.

continued

Formation of ions from atoms

Positive ions are formed when electrons are removed from atoms. This happens most easily with metallic elements. Atoms of non-metallic elements tend to gain electrons to form negative ions. Hence when metals combine with non-metals, electrons are transferred from the metal atoms to the non-metal atoms. Usually each metal atom will lose all of its outer-shell electrons and each non-metal atom will gain enough electrons to fill its outer shell. After this electron transfer, the metal atoms and non-metal atoms have filled outer shells (noble gas electronic configurations). The ionic bonding results from the electrostatic attraction between the oppositely charged ions.

Dot-and-cross diagrams are used to show the electronic configurations of atoms and ions. The electrons in atoms of one element in the compound are shown by dots, and the electrons of the other element are shown by crosses. Table 6.1 shows some examples.

Atom	Electronic configuration	Dot-and-cross diagram	Ion	Electronic configuration	Dot-and-cross diagram
Na	2,8,1		Na^+	2,8	
Cl	2,8,7		Cl^-	2,8,8	

Table 6.1 Examples of dot-and-cross diagrams.

Usually when we draw a dot-and-cross diagram, the filled inner electron shells are omitted. Only the outer shells are drawn. In the case of a sodium ion, Na^+, this shell no longer contains any electrons. The nucleus of the element is shown by the symbol for the element. The dot-and-cross diagram for an ion is placed in square brackets with the charge outside the brackets. Electrons are placed in pairs for clarity. (See page 53 for more about 'lone pairs' of electrons.)

The outer shell dot-and-cross diagram for sodium chloride is:

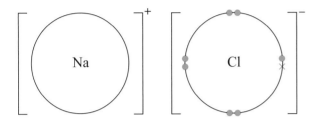

It is acceptable to leave out the circle representing the outer shell from dot-and-cross diagrams, showing only the outer electrons.

SAQ

2   Draw dot-and-cross diagrams for the following ionic compounds.

   a KF   **b** Na_2O **c** MgO
   d $CaCl_2$

   [Answer]

Compound ions

Ions such as Na^+, Cl^- and Mg^{2+} are known as simple ions. A simple ion has formed from a single atom that has gained or lost one or more electrons. Some ions consist of a small group of atoms, held together by covalent bonds, with an overall electric charge. These ions are called compound ions. Five different compound ions are shown in Figure 6.6.

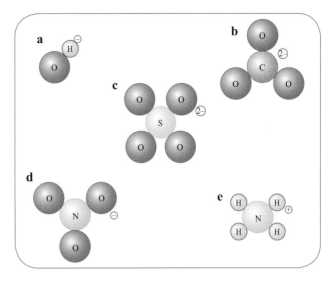

Figure 6.6 a A hydroxide ion. Its formula is OH^-.
b A carbonate ion. Its formula is CO_3^{2-}. **c** A sulfate ion. Its formula is SO_4^{2-}. **d** A nitrate ion. Its formula is NO_3^-; **e** An ammonium ion. Its formula is NH_4^+.

SAQ

3   What is the formula of
   a magnesium hydroxide
   b sodium sulfate
   c ammonium carbonate
   d calcium nitrate?

   [Hint]

   [Answer]

The properties of ionic compounds

The typical properties of ionic compounds can be explained by their structure. An ionic compound is made of ions which are arranged in a giant ionic lattice. In the ionic lattice, positive and negative ions alternate in a three-dimensional arrangement. The way in which the ions are arranged depends on their relative sizes. Sodium chloride has a cubic ionic lattice, which is shown in Figure 6.7.

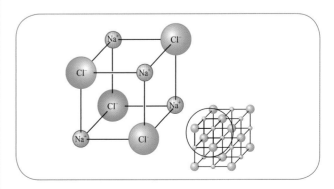

Figure 6.7 The sodium chloride (NaCl) lattice.

Magnesium oxide has the same cubic structure with magnesium ions in place of sodium ions and oxide ions in place of chloride ions. In the lattice, the attraction between oppositely charged ions binds them together. These attractions greatly outweigh repulsions between similarly charged ions since each ion is surrounded by six oppositely charged ions. So the melting points of ionic compounds are very high. The melting point usually increases as the charges on the ions increase. Sodium chloride with its singly charged ions has a melting point of 801 °C, and magnesium oxide with its doubly charged ions has a melting point of 2852 °C.

Ionic compounds are hard and brittle. The cleavage of gemstones and other ionic crystals occurs between planes of ions in the ionic lattice. If an ionic crystal is tapped sharply in the direction of one of the crystal planes with a sharp-edged knife, it will split cleanly. As a plane of ions is displaced by the force of the knife, ions of similar charge come together and the repulsions between them cause the crystal to split apart. The natural shape of ionic crystals is the same as the arrangement of the ions in the lattice. This is because the crystal grows as ions are placed in the lattice and this basic shape continues to the edge of the crystal. Hence sodium chloride crystals are cubic. The smallest repeating unit in the lattice is known as the *unit cell*.

Gemstones and other semi-precious stones, such as emeralds, sapphires and rubies (Figure 6.8), are ionic compounds valued for their colour and hardness. Gemstones are crystalline and are cut so

that they sparkle in light. They are cut by exploiting the cleavage planes between layers of ions in the crystal structure.

Ionic compounds may dissolve in water. As a general rule all metal nitrates and most metal chlorides are soluble, so are almost all of the salts of the Group 1 metals. Ionic compounds that carry higher charges on the ions tend to be less soluble or insoluble. For example, whilst Group 1 hydroxides are soluble, Group 2 and 3 hydroxides are sparingly soluble or insoluble in water (a *sparingly soluble* compound has only a very low solubility, e.g. calcium hydroxide as lime water). When ionic compounds dissolve, energy must be provided to overcome the strong attractive forces between the ions in the lattice. This energy is provided by the strong attractive forces between the water molecules and the ions. Water molecules are attracted to ions by strong electrostatic forces. The oxygen atoms of the water molecules are attracted to the positive ions. The hydrogen atoms of the water molecules are attracted to the negative ions. This happens because water molecules are polar (see page 54). An ion surrounded by water molecules is called a hydrated ion. Figure 6.9 shows one hydrated sodium ion and one hydrated chloride ion. The water

Figure 6.8 Sapphires in the form of both rough crystals and cut gemstones.

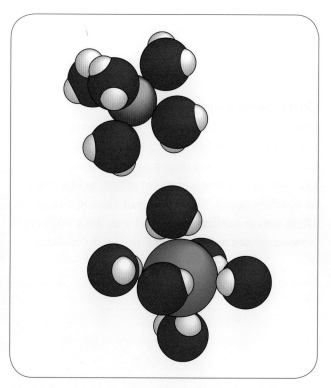

Figure 6.9 A hydrated sodium ion (grey) and a hydrated chloride ion (green).

molecules are shown with red oxygen atoms and white hydrogen atoms.

Electrolysis of ionic compounds can only occur when the ions are free to move. In the lattice the ions are in fixed positions, and so ionic solids will not conduct electricity. On melting, or dissolving in water, the ions are no longer in fixed positions so they are free to move towards electrodes (Figure 6.10).

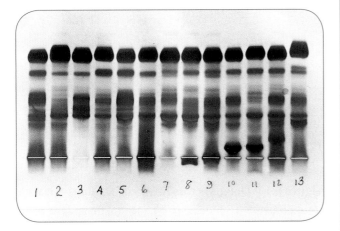

Figure 6.10 In this diagnostic test for leukaemia, negatively charged polypeptide ions (proteins) move towards a positive electrode (towards the top of the paper).

Covalent bonding

Many familiar compounds are liquids or gases or solids with low melting points, e.g. water, ammonia, methane, ethanol, sucrose and poly(ethene). Such compounds have very different properties to ionic compounds. They all contain molecules – groups of atoms which are held together by covalent bonds. Such compounds have a *simple molecular structure*. They are non-conductors of electricity and may be insoluble in water. They may dissolve in organic solvents such as ethanol or cyclohexane.

However, some crystalline covalent compounds are very hard, have high melting points and are more difficult to cleave than ionic compounds. Such compounds do not consist of small molecules. Instead the covalent bonds extend throughout the crystal in a giant lattice structure. An example of this is silicon dioxide, which forms quartz crystals (Figure 6.11). Silicon dioxide has a *giant covalent structure*.

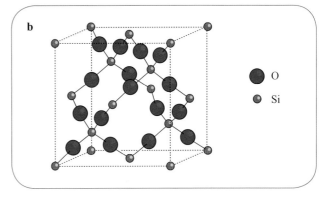

Figure 6.11 a Quartz crystals. **b** Model of the arrangement of atoms in a quartz lattice.

Single covalent bonds

A **single covalent bond** is a shared pair of electrons. The eight compounds in Figure 6.12 are made up of molecules held together by single covalent bonds.

You should note that the boron atom in a boron trifluoride molecule does not have a noble gas configuration. Neither does the sulfur atom in a sulfur hexafluoride molecule. An explanation of this requires a more sophisticated model of bonding that is not needed at this level.

SAQ

4 Close this book and try to remember how to draw the dot-and-cross diagrams for hydrogen, chlorine, sulfur hexafluoride, methane, water, ammonia, hydrogen chloride and boron trifluoride. Use the Periodic Table to work out the electronic configurations of the atoms involved.

Answer

51

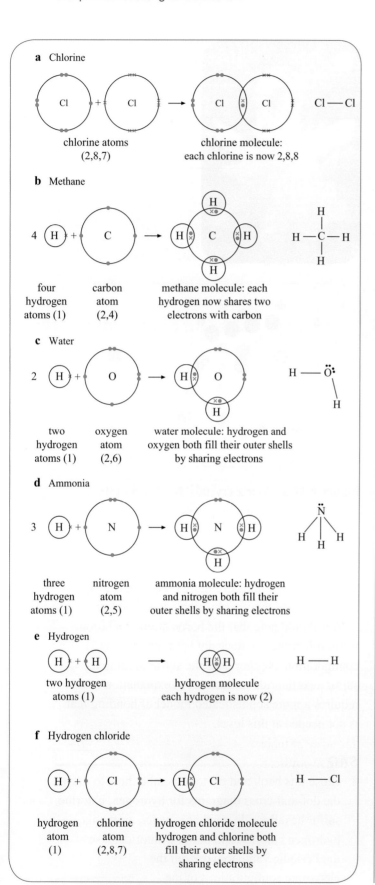

a Chlorine

chlorine atoms
(2,8,7)

chlorine molecule:
each chlorine is now 2,8,8

b Methane

four
hydrogen
atoms (1)

carbon
atom
(2,4)

methane molecule: each
hydrogen now shares two
electrons with carbon

c Water

two
hydrogen
atoms (1)

oxygen
atom
(2,6)

water molecule: hydrogen and
oxygen both fill their outer shells
by sharing electrons

d Ammonia

three
hydrogen
atoms (1)

nitrogen
atom
(2,5)

ammonia molecule: hydrogen
and nitrogen both fill their
outer shells by sharing electrons

e Hydrogen

two hydrogen
atoms (1)

hydrogen molecule
each hydrogen is now (2)

f Hydrogen chloride

hydrogen
atom
(1)

chlorine
atom
(2,8,7)

hydrogen chloride molecule
hydrogen and chlorine both
fill their outer shells by
sharing electrons

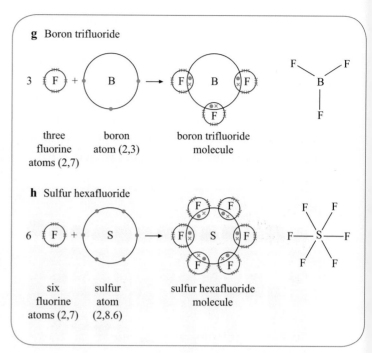

g Boron trifluoride

three
fluorine
atoms (2,7)

boron
atom (2,3)

boron trifluoride
molecule

h Sulfur hexafluoride

six
fluorine
atoms (2,7)

sulfur
atom
(2,8.6)

sulfur hexafluoride
molecule

Figure 6.12 Dot-and-cross diagrams for some covalent compounds: **a** chlorine (Cl_2);   **b** methane (CH_4);   **c** water (H_2O);   **d** ammonia (NH_3);   **e** hydrogen (H_2);   **f** hydrogen chloride (HCl);   **g** boron trifluoride (BF_3);   **h** sulfur hexafluoride (SF_6).

Multiple covalent bonds

Most covalent bonds are single bonds, consisting of one shared pair of electrons. Atoms can also bond together by sharing two pairs of electrons. This bond is called a **double covalent bond**. Figure 6.13 shows two molecules that have one or more double bonds.

Atoms can also bond together by sharing three pairs of electrons. This bond is called a **triple covalent bond**. Figure 6.14 shows a molecule that has a triple bond.

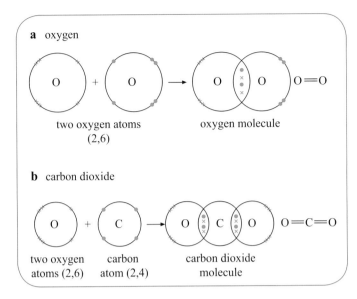

Figure 6.13 Dot-and-cross diagrams for compounds showing a double covalent bond: **a** oxygen; **b** carbon dioxide.

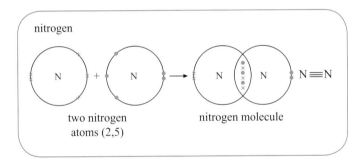

Figure 6.14 Dot-and-cross diagram for nitrogen, which shows a triple covalent bond.

SAQ ——————————————

5 **a** Close this book and try to remember how to draw the dot-and-cross diagrams for carbon dioxide, nitrogen and oxygen. Use the Periodic Table to work out the electronic configurations of the atoms involved.

  b Which of the three molecules has one double bond?

  c Which of the three molecules has two double bonds?

  d Which of the three molecules has one triple bond?

  Answer

Lone pairs

Atoms in molecules frequently have pairs of electrons in their outer shells that are not involved in covalent bonds. These non-bonding electron pairs are called lone pairs. In an ammonia molecule the nitrogen atom has one lone pair of electrons. In a water molecule the oxygen atom has two lone pairs of electrons.

Dative covalent bonding

A lone pair can sometimes be used to form a covalent bond to an atom that can accommodate two more electrons in its outer shell. An example of this is when an ammonia molecule and an H^+ ion combine to form an ammonium ion, NH_4^+ (shown in Figure 6.15).

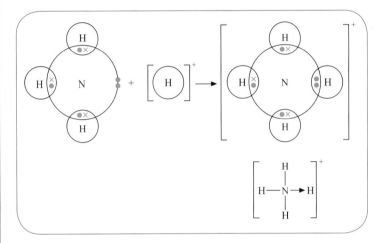

Figure 6.15 The NH_4^+ ion contains a dative covalent bond.

The bond between the NH_3 molecule and the H^+ ion consists of two electrons that both came from the nitrogen atom. The lone pair of the NH_3 molecule has formed this bond. A covalent bond like this is called a **dative covalent bond** or a coordinate bond. A dative covalent bond is represented by an arrow in displayed formulae of molecules. Another molecule which has a dative covalent bond is carbon monoxide (see Figure 6.16).

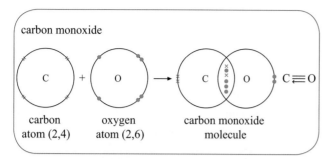

Figure 6.16 Dot-and-cross diagram for carbon monoxide, which shows a dative covalent bond.

SAQ

6 a A water molecule can form a dative bond to an H^+ ion, forming the oxonium ion H_3O^+. Draw a dot-and-cross diagram and the displayed formula of the oxonium ion.

> Hint

b BF_3 and gaseous ammonia (NH_3) combine to form a white solid with the formula F_3BNH_3. There is a bond between the boron and the nitrogen. Draw a dot-and-cross diagram and the displayed formula of F_3BNH_3.

> Hint

> Answer

Polar molecules

The electrons in a covalent bond may, on average, spend their time nearer to one of the two bonded atoms. That atom gains a small negative charge (written δ–) and the other bonded atom gains a small positive charge (written δ+). A covalent bond that is like this is said to be *polar*. **Polar covalent bonds** are caused by a difference in electronegativity between the elements. **Electronegativity** is the ability of a bonded atom to attract the electrons in the bond towards itself.

The electronegativity of the elements increases from Group 1 to Group 7 across the Periodic Table. Electronegativity also increases up each group of elements as the atomic number decreases. Several attempts have been made to put numerical values on electronegativity. For our purposes, it is sufficient to recognise that electronegativities increase:

a moving from left to right across a period in the Periodic Table, and

b vertically up groups.

The electronegativity of hydrogen is lower than that of most non-metallic elements. In the few cases where it is not lower, it is of a very similar magnitude to that of the non-metal.

increasing electronegativity

→

$Cl < N < O < F$

These electronegativity differences between atoms introduce a degree of polarity in covalent bonds between different atoms. A bigger difference in electronegativity will cause a greater degree of bond polarity. This accounts for the polarity of many simple diatomic molecules such as hydrogen chloride, HCl.

The situation is more complicated in molecules that consist of more than two atoms. Here the shape of the molecule must be taken into account. A symmetrical distribution of polar covalent bonds produces a non-polar molecule. The dipoles of the bonds exert equal and opposite effects on each other. An example is tetrachloromethane, CCl_4. This tetrahedral molecule has four polar C–Cl bonds. The four dipoles point towards the corners of the tetrahedral molecule, cancelling each other out. So CCl_4 is not polar.

In the closely related trichloromethane, $CHCl_3$, the three C–Cl dipoles point in a similar direction. Their combined effect is not cancelled out by the C–H bond. (The C–H bond is virtually non-polar.) Hence trichloromethane is a very polar molecule. (See Figure 6.17.)

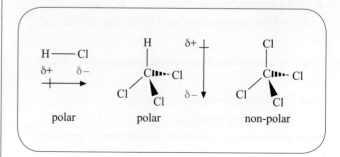

Figure 6.17 The polarity of molecules.

SAQ

7 Predict the polarity of the following molecules.
   a O_2 **b** HF   **c** CH_3Br
   d SCl_2 (a non-linear molecule)

Answer

Bond polarity can be a helpful indication of the reactivity of a molecule. This is clearly illustrated by a comparison of nitrogen and carbon monoxide. Both molecules contain triple bonds, which require a similar amount of energy to break them. (The CO bond actually requires more energy than the N_2 bond!) However, carbon monoxide is a very reactive molecule, whereas nitrogen is very unreactive. Non-polar nitrogen will only undergo reactions at high temperatures or in the presence of a catalyst. Carbon monoxide, which is a polar molecule, may be burned in air and it combines more strongly with the iron in haemoglobin than oxygen does. Many chemical reactions are started by a reagent attacking one of the electrically charged ends of a polar bond. Non-polar molecules are consequently much less reactive towards ionic or polar reagents. Other important polar molecules include water and ammonia.

As a knowledge of molecular shape is needed to predict the polarity of some molecules, the next section shows how you can predict the shapes of simple molecules.

Shapes of simple molecules

Molecules vary in shape, as shown by the six examples in Figure 6.18.

Electron-pair repulsion theory

As electrons are negatively charged, they repel each other. In Chapter 5, you saw that electrons may pair up with opposite spins in orbitals. This is also true in molecules. An electron pair in the outer shell of the central atom in a simple molecule will exert a repulsion on the other electron pairs. Each pair will repel each of the other pairs. The effect of these repulsions will cause the electron pairs to move as far apart as possible within the confines of the bonds between the atoms in the molecule. This will determine the three-dimensional shape of the molecule.

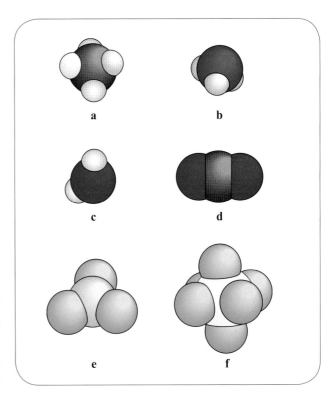

Figure 6.18 Shapes of molecules. These space-filling models show the molecular shapes of: **a** methane (CH_4);   **b** ammonia (NH_3);   **c** water (H_2O);   **d** carbon dioxide (CO_2);   **e** boron trifluoride (BF_3) and   **f** sulfur hexafluoride (SF_6).

The concept of electron-pair repulsion is a powerful theory. It successfully predicts shapes, which have been confirmed by modern experimental techniques.

In order to predict the shape of a molecule, the number of pairs of outer-shell electrons on the central atom is needed. It is best to start with a dot-and-cross diagram and then to count the electron pairs, as shown in the following examples.

● **Methane**
   As there are four bonding pairs of electrons, these repel each other towards the corners of a regular tetrahedron. The molecule thus has a tetrahedral shape.

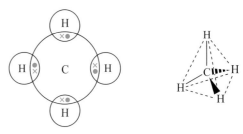

- **Ammonia**
  This has three bonding pairs and one lone pair on the central atom, nitrogen. The four electron pairs repel each other and occupy the corners of a tetrahedron, as in methane. However, the nitrogen and three hydrogen atoms form a triangular pyramidal molecule.

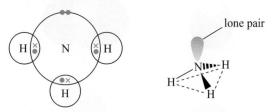

- **Water**
  There are two bonding pairs and two lone pairs. Again, these repel each other towards the corners of a tetrahedron, leaving the oxygen and two hydrogen atoms as a non-linear (or bent) molecule.

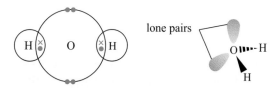

- **Carbon dioxide**
  This has two carbon–oxygen double bonds. Multiple bonds are best considered in the same way as single electron pairs. The two double bond pairs repel each other as far as possible. The molecule is linear (i.e. the OCO angle is 180°):

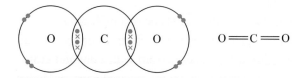

- **Boron trifluoride**
  This is an interesting molecule, as boron only has six electrons on its outer shell, distributed between three bonding pairs. The three bonding pairs repel each other equally, forming a trigonal planar molecule with bond angles of 120°. Boron trifluoride is very reactive and will accept a non-bonding (lone) pair of electrons. For example, with ammonia, $H_3N \longrightarrow BF_3$ is formed (note the dative covalent bond indicated by the arrow).

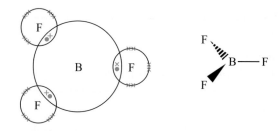

- **Sulfur hexafluoride**
  There are six bonding pairs and no lone pairs. Repulsion between six electron pairs produces the structure shown. All angles are 90°. The shape produced is an octahedron (i.e. eight faces).

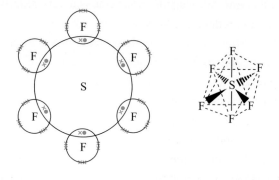

SAQ

8 a Draw dot-and-cross diagrams for **i** PCl_3, **ii** NH_4^+, **iii** H_2S and **iv** SCl_2.

b Consider the bonding and lone pairs you have drawn in part **a**. Predict the shape of each molecule and illustrate each shape with a diagram.

Answer

Lone pairs, bonding pairs and bond angles

Lone pairs of electrons are attracted by only one nucleus, unlike bonding pairs, which are shared between two nuclei. As a result, lone pairs are pulled closer to the nucleus than bonding pairs. The electron charge-cloud in a lone pair has a greater width than a bonding pair. The diagram below shows the repulsions between lone pairs (pink) and bonding pairs (white) in a water molecule.

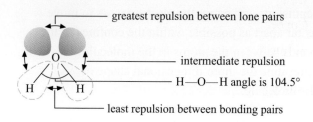

greatest repulsion between lone pairs

intermediate repulsion

H—O—H angle is 104.5°

least repulsion between bonding pairs

The repulsion between lone pairs is thus greater than that between a lone pair and a bonding pair. The repulsion between a lone pair (LP) and a bonding pair (BP) is greater than that between two bonding pairs. To summarise:

LP–LP repulsion > LP–BP repulsion >
BP–BP repulsion

This variation in repulsion produces small but measurable effects on the bond angles in molecules. In methane, all the HCH angles are the same at 109.5°. In ammonia, the slightly greater repulsion of the lone pair pushes the bonding pairs slightly closer together and the angle reduces to 107°. In water, two lone pairs reduce the HOH angle to 104.5°.

Bond enthalpy and bond length

In general, double bonds are shorter than single bonds. In addition, the energy required to break a double bond is greater than that needed to break a single bond. The **bond enthalpy** is the energy required to break one mole of the given bond in the gaseous molecule (see also Chapter 16). Table 6.2 shows some examples of bond enthalpies and bond lengths.

Bond	Bond enthalpy /kJ mol^{-1}	Bond length /nm
C–C	347	0.154
C=C	612	0.134
C–O	358	0.143
C=O	805	0.116

Table 6.2 Some bond enthalpies and bond lengths. See Chapter 16 for an explanation of enthalpy.

Metallic bonding

Metals have very different properties to both ionic and covalent compounds. In appearance they are usually shiny (Figure 6.19).

Figure 6.19 Metals. Clockwise from top left: sodium; gold; mercury, magnesium and copper.

They are good conductors of both heat and electricity. They conduct electricity in the solid state (or when molten) and are not decomposed by the passing current, unlike ionic compounds. They are easily worked and may be drawn into wires or hammered into a different shape, i.e. they are ductile and malleable. They often possess high tensile strengths and they are usually hard. Table 6.3 provides information on some of the properties of aluminium, iron and copper, with the non-metal sulfur for comparison.

	Density /g cm^{-3}	Tensile strength /10^{10} Pa	Thermal conductivity /W m^{-1} K^{-1}	Electrical conductivity /10^8 S m^{-1}
Aluminium	2.70	7.0	238	0.38
Iron	7.86	21.1	82	0.10
Copper	8.92	13.0	400	0.59
Sulfur	2.07		0.029	1×10^{-23}

Table 6.3 Properties of three metals and sulfur.

It is this range of properties that has led humans to use metals to make tools, weapons and jewellery. Two major periods in our history are named after the metals in use at the time (Figure 6.20). The change from bronze (an alloy of tin and copper) to iron reflected the discovery of methods for extracting different metals.

A simplified model of metallic bonding is adequate for our purposes. In a metallic lattice, the atoms lose their outer shell electrons to become positive ions. The outer shell electrons occupy new energy levels, which extend throughout the metal lattice.

The bonding is often described as a 'sea' of mobile electrons surrounding a lattice of positive ions. This is shown in Figure 6.21. The lattice is held together by the strong attractive forces between the mobile electrons and the positive ions. The mobile electrons are described as 'delocalised' as they are free to move through the structure and are not attached to any particular metal ion.

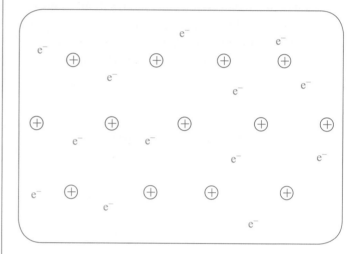

Figure 6.21 Metallic bonding. There are strong attractive forces between positively charged ions and a 'sea' of delocalised electrons.

The properties of metals can be explained using this bonding model. Electrical conduction can take place in any direction, as electrons are free to move throughout the lattice. Conduction of heat occurs by vibration of the positive ions as well as via the mobile electrons.

Metals are both ductile and malleable because the bonding in the metallic lattice is not broken when they are physically deformed. As a metal is hammered or drawn into a wire, the metal ions slide over each other to new lattice positions. The mobile electrons continue to hold the lattice together. Some metals will even flow under their own weight. Lead has a problem in this respect. Lead used on roofs suffers from 'creep' as the metal slowly flows downwards under the influence of gravity.

The transition elements (see Chapter 5) are metals that possess both hardness and high tensile strength. Hardness and high tensile strength are also due to the strong attractive forces between the metal ions and the mobile electrons in the lattice.

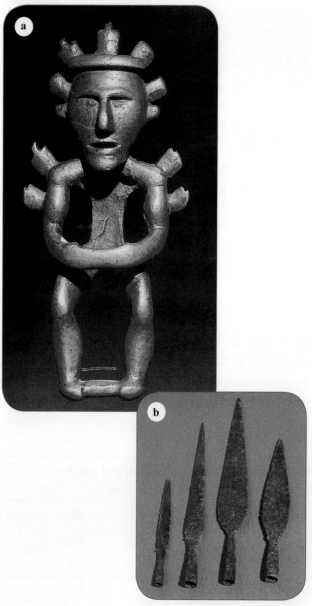

Figure 6.20 **a** Bronze age statue and **b** iron age spears.

9 Use Table 6.3 to answer the following questions and give full explanations in terms of metallic bonding. (Assume steel and stainless steel have similar properties to iron.)

a Why do some stainless steel saucepans have a copper base?

b Aluminium with a steel core is used for overhead power cables in preference to copper. Why is aluminium preferred? What is the function of the steel core?

c Apart from overhead power cables, copper is chosen for almost all other electrical uses. Suggest reasons for the choice of copper.

Answer

Intermolecular forces

Substances with a simple molecular structure often have a low melting point and a low boiling point. However weak attractive forces, called intermolecular forces, *do* exist between the molecules. You should note the difference between the *strong* forces within each molecule, which are covalent bonds, and the *weak* forces between neighbouring molecules, which are intermolecular forces.

Before we discuss the attractive forces that exist between molecules, it may be helpful to review the *kinetic theory of matter*. Matter exists in solid, liquid and gaseous states. In the solid state, the particles are packed together in a regular ordered way. This order breaks down when a substance melts. In the liquid state, there may be small groups of particles with some degree of order, but, overall in the liquid, particles are free to move past each other. In order to do this, many of the forces between the particles must be overcome on melting. In the gaseous state, the particles are widely separated. They are free to move independently, and all the forces between the particles in the solid or liquid state have been overcome on vaporisation. In the gaseous state, the particles move randomly in any direction. As they do so, they exert a force which causes pressure (vapour pressure) on the walls of their container.

The properties of all substances made up of small molecules are dependent on weak intermolecular forces. It is the properties of these substances that provide evidence for the existence of intermolecular forces and help us to understand the nature of these forces. If a gas is able to condense to a liquid, which can then be frozen to a solid, there must be an attraction between the molecules of the gas. When a solid melts or a liquid boils, energy is needed to overcome this attraction. For example, water in a kettle will stop boiling if the electricity is switched off. The temperature of the water is constant whilst the water is boiling, and the heating effect of energy from the electricity is separating the water molecules from each other to produce water vapour.

There are three types of intermolecular forces: van der Waals' forces, dipole–dipole forces and hydrogen bonds.

Van der Waals' forces

Even the atoms of noble gases must exert an attraction on each other. Figure 6.22 shows the enthalpy change of vaporisation of the noble gases plotted against the number of electrons present. (*Enthalpy change of vaporisation* is the energy required to convert the liquid to a gas.)

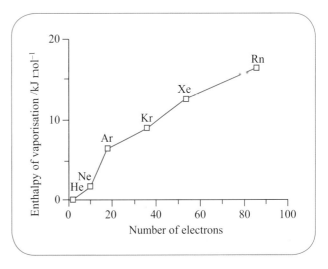

Figure 6.22 Enthalpy change of vaporisation of the noble gases plotted against the number of electrons present.

The trend in the enthalpy change of vaporisation shows an increase from helium to xenon as the number of electrons increases. Alkanes (Chapter 11) show a similar trend; their enthalpies of vaporisation also increase with increasing numbers of atoms in the molecules (and hence with increasing numbers of electrons). Both the noble gases and the alkanes have attractive forces between atoms and molecules, which are now known to depend on the number of electrons and protons present.

The forces arise because electrons in atoms or molecules can be thought of as moving at very high speeds in orbitals. At any instant in time it is possible for more electrons to lie to one side of the atom or molecule than the other. When this happens, an instantaneous electric dipole occurs. The momentary imbalance of electrons provides the negative end of a dipole, with the atomic nucleus providing the positive end of the dipole. This instantaneous dipole produces an induced dipole in a neighbouring atom or molecule, which is hence attracted (Figure 6.23).

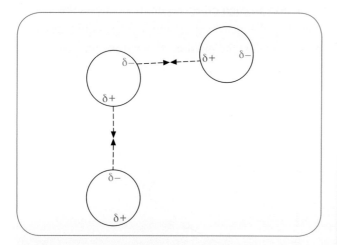

Figure 6.23 Induced dipole attractions.

This is rather like the effect of a magnet (magnetic dipole) on a pin. The pin becomes temporarily magnetised and is attracted to the magnet. Intermolecular forces of this type are called *instantaneous dipole–induced dipole forces* or **van der Waals' forces**. The strength of van der Waals' forces increases with the number of electrons and protons present.

Van der Waals' forces are the weakest type of attractive force found between atoms or molecules. They are responsible for the slippery nature of graphite (Figure 6.24a) in contrast to the great hardness of diamond (Figure 6.24b), and for the volatility of bromine and iodine (Figure 6.25).

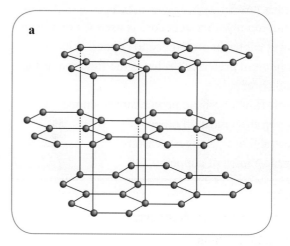

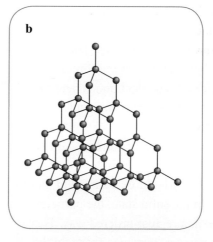

Figure 6.24 a The structure of graphite. In the planar sheets of carbon atoms, all the bonding electrons are involved in covalent bonds. The attraction between sheets are much weaker van der Waals' forces. These forces are easily overcome, allowing the sheets to slide over each other (rather like a pack of cards). Graphite is often used as a lubricant.

b The structure of diamond. In contrast to graphite, each carbon atom forms four covalent bonds to four other carbon atoms. The resulting network of covalent bonds requires considerable energy to separate the atoms. The strength of the bonding in diamond is responsible for its great hardness.

Figure 6.25 Bromine **a** and iodine **b** exist as covalent molecules. They are both volatile, as only very weak van der Waals' forces need to be overcome to vaporise them.

A polymer is a molecule built up from a large number of small molecules (called monomers). Low-density poly(ethene), LDPE, and high-density poly(ethene), HDPE, have differing properties because of the way the polymer molecules are packed (Figure 6.26). The HDPE molecules can pack much more closely as they are not branched. LDPE molecules are branched at intervals, which prevents them packing as closely. As a result, the van der Waals' forces are not as strong.

Teflon is poly(tetrafluoroethene), PTFE. A model of part of a PTFE molecule is shown in Figure 6.27. The van der Waals' forces between oil or grease and PTFE are much weaker than those present in the oil or grease itself. This gives rise to the polymer's non-stick properties.

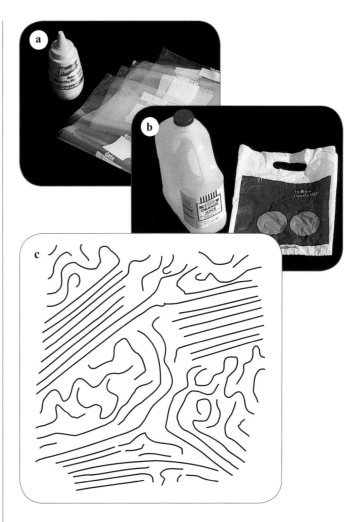

Figure 6.26 Low-density and high-density poly(ethene). **a** LDPE is made under high pressure with a trace of oxygen as catalyst. The product consists primarily of a tangled mass of polymer chains with some regions where the chains have some alignment. **b** HDPE is made using catalysts developed by the Swiss chemist Ziegler and the Italian chemist Natta. (They received a Nobel Prize for their discoveries.) In HDPE, the polymer chains are arranged in a much more regular fashion. This increases the density of the material and makes it more opaque to light. As the molecules are closer together in HDPE, the van der Waals' forces between the non-polar poly(ethene) molecules are greater and the tensile strength of the material is higher. **c** Diagram of crystalline and non-crystalline regions in poly(ethene). LDPE has fewer crystalline regions than HDPE. In the crystalline regions, polymer chains (shown as lines in the diagram) lie parallel to each other.

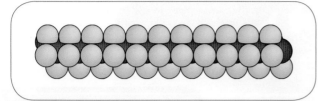

Figure 6.27 Model of PTFE. Note how the fluorine atoms (yellow-green) surround the carbon atoms to produce a non-polar polymer.

Permanent dipole–dipole forces

A nylon rod may be given a charge of static electricity by rubbing it with a dry sheet of thin poly(ethene). If this is brought near a fine jet of water, the stream of water is attracted by the charge on the nylon rod. You can try this for yourself. Use a nylon comb and as fine a trickle of water from a tap as possible (see Figure 6.28).

Figure 6.28 Deflection of water by an electrically charged nylon comb.

The water molecules are attracted to the charged nylon rod or comb because they have a *permanent* electric dipole. A force of this type is called a *dipole–dipole force*. The dipole of every water molecule arises because of the bent shape of the molecule and the greater electron charge around the oxygen atom. As we saw earlier in the case of the hydrogen chloride molecule, a molecule is often polar if its atoms have different electronegativities. The diagram

below shows the lone pairs and electric dipole of a water molecule (note that the arrow head shows the negative end of the dipole).

SAQ

10 The nylon rod carries a positive charge. Which end of the water molecule is attracted to the rod? Why are no water molecules repelled by the rod? A poly(ethene) rod may be given a negative charge when rubbed with a nylon cloth. Will the charge on the poly(ethene) rod attract or repel a thin stream of water?

Answer

Dipole–dipole forces make poly(ester) strong. Many fabrics are made using poly(ester) fibres because of its strength. Production of poly(ester) fibres together with a section of a poly(ester) molecule are shown in Figure 6.29.

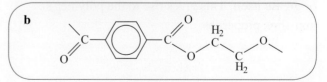

Figure 6.29 **a** The production of poly(ester) fibre. **b** A section of the poly(ester) chain. The strength of poly(ester) fibre is due to the strong dipole–dipole forces between ester groups of adjacent molecules.

SAQ

11 Copy the section of the poly(ester) chain shown in Figure 6.29 and mark on your copy the polar groups, showing the δ+ and δ– charges. Draw a second section of poly(ester) chain alongside your first section and mark in the dipole–dipole forces with dotted lines.

Answer

Water is peculiar

Figure 6.30 shows the enthalpy changes of vaporisation of water and other hydrides of Group 6 elements.

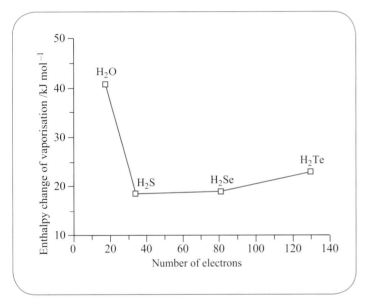

Figure 6.30 Enthalpy changes of vaporisation of Group 6 hydrides, including water, plotted against number of electrons present.

SAQ

12 a Explain the underlying increase in the enthalpy change of vaporisation from H_2S to H_2Se to H_2Te.

b Estimate a value for the enthalpy change of vaporisation of water based on this trend.

c What is the cause of the much higher value observed for water?

Answer

Water has a much higher enthalpy change of vaporisation than expected. Water has several other peculiar properties.

● The boiling point of water is much higher than predicted by the trend in boiling points for the hydrides of other Group 6 elements.
● Water has a very high surface tension and a high viscosity.
● The density of ice is less than the density of water (Figure 6.31). Most solids are denser than their liquids, as molecules usually pack closer in solids than in liquids.

Figure 6.31 Ice floats on water.

Surface tension of water

You can demonstrate the high surface tension of water for yourself by floating a needle on water. Rinse a bowl several times with water. Fill the bowl with water. Place a small piece of tissue paper on the surface of the water. Now place a needle on the tissue. Leave it undisturbed. The paper will sink, leaving the needle floating. Now carefully add a few drops of washing-up liquid (which lowers the surface tension of water) and observe the effect.

Hydrogen bonds

The peculiar nature of water is explained by the presence of the strongest type of intermolecular force – the **hydrogen bond** (indicated on diagrams by dotted lines). Water is highly polar owing to the large difference in electronegativity between hydrogen and oxygen. The resulting intermolecular attraction between oxygen atoms and hydrogen atoms on neighbouring water molecules is a very strong dipole–dipole attraction called a hydrogen bond (Figure 6.32). Each water molecule can form two hydrogen bonds to other water molecules. These form in the directions of the lone pairs.

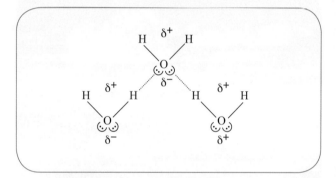

Figure 6.32 The hydrogen bonds between these water molecules are shown as dotted lines.

In the liquid state, water molecules collect in groups. On boiling, the hydrogen bonds must be broken. This raises the boiling point significantly as the hydrogen bonds are stronger than the other intermolecular forces. Similarly, the enthalpy change of vaporisation is much higher than it would be if no hydrogen bonds were present.

In ice, a three-dimensional hydrogen-bonded lattice is produced. In this lattice, each oxygen atom is surrounded by a tetrahedron of hydrogen atoms bonded to further oxygen atoms. The structure is shown in Figure 6.33. The extensive network of hydrogen bonds raises the freezing point significantly above that predicted by the trend for other Group 6 hydrides.

Figure 6.33 Model of ice. Oxygen atoms are red, hydrogen atoms are white, hydrogen bonds are lilac. This hydrogen-bonded arrangement makes ice less dense than water.

SAQ

13 A diamond-type lattice is present in ice. The O····H hydrogen bond length is 0.159 nm and the O–H covalent bond length is 0.096 nm. When ice melts, some hydrogen bonds break and the density rises. Use Figure 6.33 and these values to explain why ice has a lower density than water.

Answer

The high surface tension of water is explained by the presence of a hydrogen-bonded network of water molecules at the surface. This network is sufficiently strong to enable a needle to be floated on the surface of water.

Within the bulk of water, small groups of molecules are attracted by hydrogen bonds. The hydrogen bonds are constantly breaking and reforming at room temperature. As the temperature of water is raised towards the boiling point, the number of hydrogen bonds reduces. On boiling, the remaining hydrogen bonds are broken. Water vapour consists of widely separated water molecules.

SAQ

14 Why does a needle floating on water sink on the addition of washing-up liquid to the water?

15 The boiling points of the hydrides of the Group 5 element hydrides are as follows.

Hydride	Boiling point /°C
ammonia, NH_3	−33
phosphine, PH_3	−88
arsine, AsH_3	−55
stibine, SbH_3	−17
bismuthine, BiH_3	+22

Plot a graph of these boiling points against the relative molecular mass of the hydrides.

a Explain the steadily rising trend in the boiling points from phosphine to bismuthine.

b Explain why the boiling point of ammonia does not follow this trend.

Answer

Conserving our water supplies

Across the world, water is used as shown in Figure 6.34.

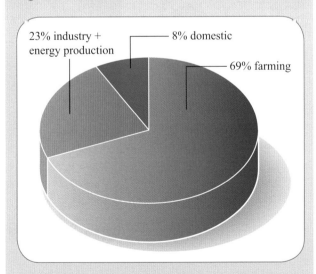

Figure 6.34 Global water usage: farming 69%, industry and energy production 23%, domestic 8%.

continued

The actual proportions differ from country to country. For example, some African nations can use almost 90% of their water consumption in farming as they need to irrigate their crops (see Figure 6.35), whereas in European countries, the proportion used in industry and in producing energy is likely to be over 50%. The percentage of water used in industry can be used as a measure of how developed a country is.

The individual usage of water (average per person) also varies widely. A person in the USA is likely to use at least 20 times as much water as a person in Africa each day.

Figure 6.35 Irrigation in Sudan.

You might wonder why we should bother trying to save water when it all gets recycled naturally in the water cycle. However, once we have used water, there is always a price to pay before putting it back into the environment. We can't just dump dirty water directly into rivers and seas. Cleaning up our waste water costs us money.

In the future, an ever-growing population will be competing for the planet's water. Therefore we need to conserve our water and do even more to restore its quality after use. By saving energy, we are also helping to conserve water because power stations use a significant proportion of our water.

Extension

Hydrogen bonds play a very important part in the structures and properties of biochemical polymers. For example, protein chains often produce a helical structure, and the ability of DNA molecules to replicate themselves depends primarily on the hydrogen bonds, which hold the two parts of the molecules together in a double helix (Figure 6.36).

Relative bond strengths

Table 6.4 shows the relative strengths of intermolecular forces and other bonds.

	Energy /kJ mol^{-1}
van der Waals' forces, e.g. in xenon	15
hydrogen bond, e.g. in water	22
O–H covalent bond in water	464
ionic bonding, sodium chloride	760

Table 6.4 Relative strengths of intermolecular forces and bonds.

Note that the intermolecular forces are much weaker than the forces of attraction found in typical covalent bonds or in ionic bonding. Van der Waals' forces are weaker than dipole–dipole forces. Hydrogen bonds are the strongest of the intermolecular forces.

SAQ

16 Ammonia is a gas which liquifies easily under pressure due to the formation of hydrogen bonds. Draw a diagram to show hydrogen bond formation between two adjacent molecules.

> Hint

> Answer

Table 6.5 provides a summary of the pattern and variety of structures and bonding found among elements and compounds.

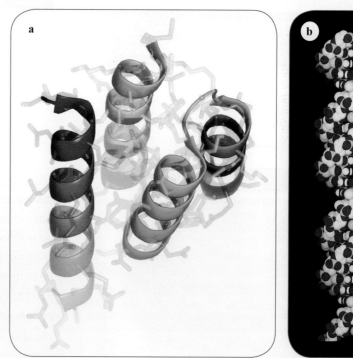

Figure 6.36 Photographs of models of biochemical polymers. **a** A molecular model of a protein found in the measles virus. The regions where the protein chain forms a helical structure can be seen clearly. **b** A section of the DNA molecule.

	Type of structure				
	Giant lattices			Simple molecular	Atomic
	Ionic	Covalent	Metallic		
Where this type of structure is found	compounds formed between metals and non-metals	Group 4 elements and some of their compounds	metals	some elements and some compounds formed between non-metals	noble gases
Some examples	sodium chloride, magnesium oxide	diamond, graphite, silicon(IV) oxide	aluminium, copper	hydrogen H_2, chlorine Cl_2, methane CH_4, ammonia NH_3	helium, neon
Particles present	ions	atoms	positive ions and electrons	small molecules	atoms
Attractions that hold particles together	between oppositely charged ions	electrons in covalent bonds attract nuclei	delocalised sea of electrons attracts positive ions	various intermolecular forces between molecules, covalent bonds within molecule	intermolecular forces between atoms: instantaneous dipole–induced dipole (van der Waals' forces) only
Common physical state(s) at room temperature and pressure	solid	solid	solid	solids, liquids and gases	gases
Melting and boiling points, enthalpy change of vaporisation	high	very high	moderately high to high	low	very low
Hardness	hard, brittle	very hard	hard, malleable	solids, usually soft	
Electrical conductivity	conduct when molten or in aqueous solution	usually non-conductors	conduct when solid or molten	non-conductors	non-conductors
Solubility in water	many ionic compounds are soluble	insoluble	insoluble, some react liberating hydrogen	usually insoluble unless very polar and capable of forming hydrogen bonds to water	sparingly soluble

Table 6.5 Summary of structure and bonding.

Summary

- All bonding involves electrostatic attractive forces.

- In ionic bonding, the attractive forces are between oppositely charged ions.

- In a covalent bond (one electron from each atom) or a dative covalent bond (both electrons from one atom), the forces are between two atomic nuclei and pairs of electrons situated between them.

- In metallic bonding, the forces are between delocalised electrons and positive ions.

- Intermolecular attractive forces also involve electrostatic forces.

- Intermolecular forces (hydrogen bonds, dipole–dipole and van der Waals' forces) are much weaker than ionic, covalent or metallic bonding forces.

- Dot-and-cross diagrams enable ionic and covalent bonds to be described. Use of these diagrams with electron-pair repulsion theory enables molecular shapes to be predicted.

- Physical properties and structures of elements and compounds may be explained in terms of kinetic theory and bonding (Table 6.5).

Questions

1 In this question, one mark is available for the quality of use and organisation of scientific terms. Nitrogen and oxygen are elements in Period 2 of the Periodic Table. The hydrogen compounds of oxygen and nitrogen, H_2O and NH_3, both form hydrogen bonds.

   a **i** Draw a diagram containing two H_2O molecules to show what is meant by *hydrogen bonding*. On your diagram, show any lone pairs present and relevant dipoles.   [3]

      ii State and explain <u>two</u> anomalous properties of water resulting from hydrogen bonding.   [4]

   b The dot-and-cross diagram of an ammonia molecule is shown below.

Hint

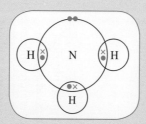

      Predict, with reasons, the bond angle in an ammonia molecule.   [4]

OCR Chemistry AS (2811) June 2005   [Total 11]

Answer

continued

2 The compounds NH_3, BF_3 and HI all have covalent bonding and simple molecular structures. The Pauling electronegativity values shown in the table can be used to predict polarity in these compounds.

H
2.1

Li	Be	B	C	N	O	F
1.0	1.5	2.0	2.5	3.0	3.5	4.0
Na						Cl
0.9						3.0
K						Br
0.8						2.8
						I
						2.5

a Explain the term *electronegativity*.   [2]

b The electronegativity values in the table can be used to predict the polarity of a bond. Copy the boxes below and show the polarity of each bond by adding δ+ or δ– to each bond.   **Hint**
   The first box has been completed for you.

$^{\delta-}$O–H$^{\delta+}$	H–N	F–B	H–I

                                                                            [2]

c Using outer electron shells only, draw dot-and-cross diagrams for molecules of NH_3 and BF_3.   [2]

d The diagrams below show the shapes of molecules of NH_3 and BF_3. State the bond angle in each molecule and state the name of each shape.

                                                                            [4]

e Explain why NH_3 has polar molecules whereas molecules of BF_3 are non-polar.   [2]

f Polar molecules of NH_3 form hydrogen bonds. Draw a diagram to show this hydrogen bonding.   [1]

g NH_3 reacts with HI to form the ionic compound NH_4I, made up of NH_4^+ and I^- ions.

   $$NH_3 + HI \rightarrow NH_4I$$

   Explain why the H–N–H bond angle in NH_3 is less than that in NH_4^+.   [2]

OCR Chemistry AS (2811) Jan 2002   [Total 15]

Answer

continued

3 Water and carbon dioxide both consist of covalent molecules.

   a Draw dot-and-cross diagrams for a molecule of water and a molecule of carbon dioxide. Show outer electron shells only. [3]

   b The shape of a water molecule is different from the shape of a carbon dioxide molecule.

     i Draw the shapes of these molecules and state the bond angles. [4]

     ii Explain why a water molecule has a different shape from a carbon dioxide molecule. [2]

   c Water and carbon dioxide both have polar bonds. Explain why water has polar molecules but carbon dioxide has non-polar molecules. [2]

Hint

OCR Chemistry AS (2811) Jan 2004 [Total 11]

Answer

4 This question is about the simple molecular compounds water, ammonia and sulfur dioxide.

   a Pairs of electrons in molecules may be present as *bonding pairs* or as *lone pairs*.

     i Copy and complete the table below for water, ammonia and sulfur dioxide.

molecule	H_2O	NH_3	SO_2
number of bonding pairs of electrons			4 (2 double bonds)
number of lone pairs of electrons around central atom			1

[2]

     ii Use your answers to **a i** to help you draw the shape of a molecule of NH_3 and a molecule of SO_2.

      Clearly show values of the bond angles in your diagrams. [4]

   b The O–H bonds in water and the N–H bonds in ammonia have dipoles. Why do these bonds have dipoles? [1]

   c Describe and explain the density of ice compared with water. [2]

OCR Chemistry AS (2811) Jan 2006 [Total 9]

Answer

continued

5 Chemicals show a range of different structures. The table below shows four types of structure.

structure	example
giant metallic	i
giant ionic	ii
giant molecular	iii
simple molecular	iv

a Copy and complete the table by giving an example of each type of structure. [4]

b A giant metallic structure has metallic bonding.

   i Draw a labelled diagram to show metallic bonding. [2]

   ii How does a substance with a giant metallic structure conduct electricity? [1]

c Explain why a substance with a giant ionic lattice conducts electricity when molten but not when solid. [2]

d Explain why a substance with a giant molecular structure has a higher boiling point than a substance with a simple molecular structure. [3]

OCR Chemistry AS (2811) June 2003   [Total 12]

Hint

Answer

Chapter 7

Periodicity

e-Learning

Objectives

Finding periodic patterns

Patterns of chemical properties and atomic masses

If you were given samples of all the elements (some are shown in Figure 7.1) and the time to observe their properties, you would probably find many ways of arranging them.

You could classify them by their states at a particular temperature (solids, liquids or gases) or as metals and non-metals; you might find patterns in their reactions with oxygen or water or other chemicals. Would you consider trying to link these properties to the relative atomic masses of the elements?

If you have studied the metallic elements lithium, sodium and potassium, you will know that they have similar reactions with oxygen, water and chlorine, and form similar compounds. The rates of their reactions show that sodium comes between lithium and potassium in reactivity. Now look at their relative atomic masses:

Li	Na	K
6.9	23	39.1

The relative atomic mass of sodium is the mean of the relative atomic masses of lithium and potassium. There is a pattern here which is also shown by other groups of elements in threes – chlorine, bromine and iodine, for example. The 'middle' element has the mean relative atomic mass and other properties in between those of the other two. This pattern was first recorded by the German chemist Johann Döbereiner (1780–1849) as his 'Law of Triads' (Figure 7.2).

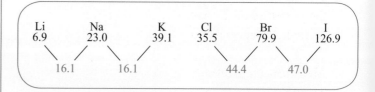

Figure 7.2 Two of Döbereiner's 'triads'.

At the time, however, it was little more than a curiosity, as too few elements were known and values for atomic masses were uncertain.

Later in the century, more elements were known and atomic masses could be measured more accurately. A British chemist, John Newlands (1837–98), suggested that, when the elements were arranged in order of increasing atomic mass, 'the eighth element, starting from a given one, is a kind of

Figure 7.1 A few of the 115 or so known elements.

b CHEMICAL NEWS,
March 9, 1866.

Table II.—Elements arranged in Octaves.

No.	No.	No.	No.	No.	No.	No.	No.
H 1	F 8	Cl 15	Co & Ni 22	Br 29	Pd 36	I 42	Pt & Ir 50
Li 2	Na 9	K 16	Cu 23	Rb 30	Ag 37	Cs 44	Os 51
G 3	Mg 10	Ca 17	Zn 24	Sr 31	Cd 38	Ba & V 45	Hg 52
Bo 4	Al 11	Cr 19	Y 25	Ce & La 33	U 40	Ta 46	Tl 53
C 5	Si 12	Ti 18	In 26	Zr 32	Sn 39	W 47	Pb 54
N 6	P 13	Mn 20	As 27	Di & Mo 34	Sb 41	Nb 48	Bi 55
O 7	S 14	Fe 21	Se 28	Ro & Ru 35	Te 43	Au 49	Th 56

Figure 7.3 a John Newlands. **b** This is the table Newlands presented to the Chemical Society in 1866 in a paper entitled 'The Law of Octaves, and the Causes of Numerical Relations among the Atomic Weights'. Note that some elements have symbols that we do not use today, e.g. G and Bo. What are these elements?

repetition of the first, like the eighth note in an octave of music'.

Newlands presented his ideas for a 'Law of Octaves' to a meeting of the Chemical Society in 1866 (Figure 4.3). They were not well received. Unfortunately, his 'octaves' seemed to apply only to the first 16 elements. After these, his 'Law of Octaves' did not seem to be obeyed. He had not allowed space in his table for the possibility of new elements to be discovered.

Despite the Chemical Society's sceptical reception of Newlands' ideas, we now know that he had found the important pattern of periodicity. This means that the properties of elements have a regularly recurring or 'periodic' relationship with their relative atomic masses.

Mendeleev's periodic table

The greatest credit for producing chemistry's most famous organisation of elements – the Periodic Table – is always given to Dmitri Mendeleev (1834–1907) from Russia.

Mendeleev arranged the elements, just as Newlands had, in order of increasing relative atomic mass (Figure 7.4). At the time (late 1860s) over 60 elements were known, and he saw that there was some form of regularly repeating pattern of properties.

Mendeleev made several crucial decisions that ensured the success of his first Periodic Table. The most important decisions were the following.

- He left spaces in the table so that similar elements could always appear in the same group.
- He said that the spaces would be filled by elements not then known. Furthermore, he predicted what the properties of these elements might be, based on the properties of known elements in the same group. He made predictions, for example, about the element between silicon and tin in Group 4. (This element was only discovered about 15 years later.) Mendeleev had called it 'eka-silicon'; it is now known as germanium (Table 7.1).

Property	Mendeleev's predictions for 'eka-silicon'	Germanium
Appearance	light-grey solid	dark-grey solid
Atomic mass	72	72.59
Density /g cm^{-3}	5.5	5.35
Oxide formula	eka-SiO_2	GeO_2
Oxide density	4.7	4.2
Chloride density	1.9 (liquid)	1.84 (liquid)
Chloride b.p. /°C	<100	84

Table 7.1 Comparison of Mendeleev's predictions for eka-silicon with known properties of germanium.

Figure 7.4 a Mendeleev's first published periodic table in the *Zeitschrift für Chemie* in 1869. Note that the elements with similar properties (e.g. Li, Na, K) are in horizontal rows in this table.

b This photograph shows a late version of Mendeleev's Periodic Table on the building where he worked in St Petersburg. Elements with similar properties are now arranged vertically in groups.

SAQ

1 From Table 7.1, how well do you think Mendeleev's predicted properties for eka-silicon compare with the known properties of germanium?

Answer

A theory or model is most valuable when it is used to explain and predict. Mendeleev's Periodic Table was immensely successful. By linking the observed periodicity in the properties of elements with the atomic theory of matter, the table helped to organise and unify the science of chemistry and led to much further research. It has been greatly admired ever since. It was even able to cope with the discovery of a whole new group of elements, now called 'the noble gases' (helium to radon), though these had not been predicted by Mendeleev.

Making new elements

Wouldn't it be great to have an element named after you! Not surprisingly, Dmitri Mendeleev has that honour – but it came many years after his death. Element 101 is called mendelevium. It was made in 1955 by chemists at the Lawrence Berkeley National Laboratory in California. Glenn Seaborg led teams that created several new elements and he was able to develop the f-block of the Periodic Table. He could use the Periodic Table to make predictions about the chemistry of the new elements discovered (just as Mendeleev had done in the previous century).

Elements after uranium (atomic number 92) have had to be made artificially. Scientists use powerful particle accelerators to smash ions into each other at incredible speeds. Then they detect the particles made in the aftermath of the collisions (see Figure 7.5).

Most of these new heavy elements are very unstable, decomposing in nuclear reactions to form smaller atoms, giving off radiation in the process. This research is very costly and the elements made have few real uses, but calculations predict that certain isotopes should be relatively stable. For example, Russian scientists made an isotope of element 114, containing 175 neutrons which existed for a matter of seconds instead of the usual microseconds.

What are your views on such research?

continued

Figure 7.5 Particle accelerators cost many millions of pounds to build.

Atomic structure and periodicity

In 1913 the British scientist Henry Moseley was able to show that the real sequence in the Periodic Table is not the order of relative atomic masses. The sequence is the order of atomic numbers – the numbers of protons in the nuclei of atoms of the elements (see Chapter 1). This sequence of elements by atomic numbers is close to the sequence by relative atomic masses, but is not exactly the same.

SAP

2 a What do we mean by 'isotopes', 'atomic number' and 'relative atomic mass'?

b Why are the relative atomic mass values for tellurium (Te) and iodine (I, given the symbol J in Mendeleev's table) the 'wrong' way around in the Periodic Table, whereas their atomic numbers fit the table?

Answer

Moseley's work was about the nature of the nucleus and led to the correct sequence of elements. It did not, however, answer questions about the periodic variations in physical and chemical properties. This is because these properties depend much more upon the numbers and distributions of electrons in atoms.

Versions of the Periodic Table

Chemists have enjoyed displaying the Periodic Table in many different ways (for two versions, see Figure 7.6). You may be able to invent some new versions.

The Periodic Table most often seen is shown in Figure 7.7 (and also in the appendix). Its main features are:

● the vertical groups of elements, labelled 1, 2, 3 up to 7 (the noble gases are not called Group 8 but Group 0)

● the horizontal periods labelled 1, 2, 3, etc.

Blocks of elements in the Periodic Table

Chemists find it helpful to identify 'blocks' of elements by the type of electron orbital most affecting the properties. These are shown on the Periodic Table in Figure 7.7.

● Groups 1 and 2 elements are in the *s-block*.

● Groups 3 to 7 and Group 0 (except He) are in the *p-block*.

● The transition elements are included in the *d-block*.

● The lanthanide and actinide elements are included in the *f-block*.

Periodic patterns of physical properties of elements

We shall now look in more detail at the physical properties of elements and their relationships with the electronic configurations of atoms.

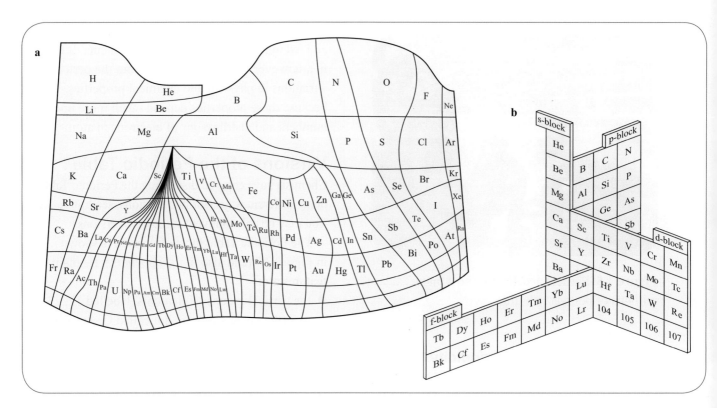

Figure 7.6 Two rather unusual versions of the Periodic Table:
a the elements according to relative abundance and **b** a three-dimensional, four-vaned model.

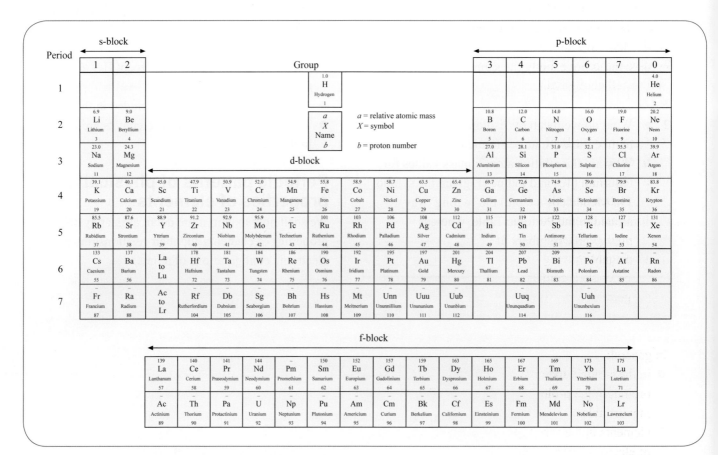

Figure 7.7 The Periodic Table of the elements.

H₂(g) mols																	He(g) atoms
Li(s) metal	Be(s) metal											B(s) giant cov	C(s) giant cov	N₂(g) mols	O₂(g) mols	F₂(g) mols	Ne(g) atoms
Na(s) metal	Mg(s) metal											Al(s) metal	Si(s) giant cov	P₄(s) mols	S₈(s) mols	Cl₂(g) mols	Ar(g) atoms
K(s) metal	Ca(s) metal	Sc(s) metal	Ti(s) metal	V(s) metal	Cr(s) metal	Mn(s) metal	Fe(s) metal	Co(s) metal	Ni(s) metal	Cu(s) metal	Zn(s) metal	Ga(s) metal	Ge(s) metal	As(s) giant cov	Se(s) mols	Br₂(l) mols	Kr(g) atoms

Figure 7.8 Structures of elements 1 (hydrogen, H) to 36 (krypton, Kr) (mols = simple molecules; giant cov = giant covalent).

Summary of structure of the first 36 elements

Figure 7.8 shows details of the structure and state of elements 1 to 36, hydrogen (H) to krypton (Kr).

Periodic patterns of electronic configurations

Magnesium and calcium are both s-block elements in Group 2. Their electronic configurations are:

Mg $1s^2\ 2s^2\ 2p^6\ 3s^2$
Ca $1s^2\ 2s^2\ 2p^6\ 3s^2\ 3p^6\ 4s^2$

Notice that they have the same outer-shell configuration, s^2. All the Group 2 elements have an outer-shell configuration of s^2.

SAQ

3 Carbon and silicon are both p-block elements and in Group 4.
   a Write down the electronic configurations of carbon and silicon.
   b What do you notice about their outer-shell configurations?
   c Suggest, giving a reason, the outer-shell configuration for germanium.

   [Answer]

In general:
- in the s-block, the outermost electrons are in an s orbital; in the p-block, the outermost electrons are in p orbitals
- elements in the same group have the same number of electrons in their outer shell

- for the elements in Groups 1 to 7, the number of outer-shell electrons is the same as the group number; for example, chlorine in Group 7 has seven outer-shell electrons: $1s^2\ 2s^2\ 2p^6\ 3s^2\ 3p^5$, a total of seven electrons in the third shell
- Group 0 elements, the noble gases, have a full outer shell of eight electrons: $s^2\ p^6$; for example, neon has the electronic configuration $1s^2\ 2s^2\ 2p^6$.

Periodic patterns of atomic radii

The size of an atom cannot be measured precisely as the electron shells do not define a clear outer limit. However, one measure of the size of an atom is its 'atomic radius'. This can be either the 'covalent radius' or the 'metallic radius' (Figure 7.9). *Covalent radius* is half the distance between the nuclei of neighbouring atoms in molecules. *Metallic radius* is half the distance between the nuclei of neighbouring atoms in metallic crystals. The covalent radius can be measured for most elements and is usually what is meant when we use the term **atomic radius**.

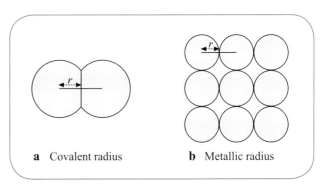

a Covalent radius   b Metallic radius

Figure 7.9 Covalent and metallic radii.

H 0.037																	He –
Li 0.123	Be 0.089											B 0.080	C 0.077	N 0.074	O 0.074	F 0.072	Ne –
Na 0.157	Mg 0.136											Al 0.125	Si 0.117	P 0.110	S 0.104	Cl 0.099	Ar –
K 0.203	Ca 0.174	Sc 0.144	Ti 0.132	V 0.122	Cr 0.117	Mn 0.117	Fe 0.116	Co 0.116	Ni 0.115	Cu 0.117	Zn 0.125	Ga 0.125	Ge 0.122	As 0.121	Se 0.117	Br 0.114	Kr –

Figure 7.10 The atomic (covalent) radii of elements 1 to 36, measured in nanometres (nm).

Atomic radii of elements 1 to 36 are shown in Figure 7.10.

When atomic (covalent) radii are plotted against atomic numbers for the first 36 elements, the graph appears as in Figure 7.11. The noble gases are not included as they do not have covalent radii.

SAQ

4 Why do the noble gases not have any measured covalent radii?

Answer

Note the relative positions of the elements in any one group, such as Group 1 (alkali metals) or Group 7 (halogens), and across Periods 2 and 3. The trends (Figure 7.12) show that atomic radii:
- increase down a group
- decrease across a period

- after some decrease, are relatively constant across the transition elements, titanium to copper.

Note that the trends in atomic radii are generally in the opposite direction to the trends in first ionisation energies (see page 80). As atomic radii become larger, first ionisation energies become smaller. In any one atom, both trends are due to the same combined effects of:
- the size of the nuclear charge
- the distance of the outer electron shell from the nucleus
- the shielding effect of filled inner electron shells upon the outer shell.

Down any one group, the nuclear charges increase, but the distance and shielding effects increase even more as extra electron shells are added. The overall result is an increase in atomic radii.

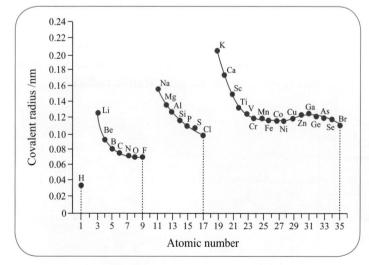

Figure 7.11 Plot of atomic (covalent) radii against atomic number of elements. The noble gases (He–Kr) are not included.

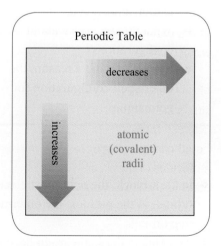

Figure 7.12 Trends of atomic (covalent) radii.

H −253																	He −269
Li 1331	Be 2477											B 3927	C 4827	N −196	O −183	F −188	Ne −246
Na 890	Mg 1117											Al 2447	Si 2677	P 281	S 445	Cl −34	Ar −186
K 766	Ca 1492	Sc 2477	Ti 3277	V 3377	Cr 2642	Mn 2041	Fe 2887	Co 2877	Ni 2837	Cu 2582	Zn 908	Ga 2237	Ge 2827	As 613	Se 685	Br 58	Kr −153

Figure 7.13 Boiling points of elements 1 to 36, measured in degrees Celsius (°C).

Across Periods 2 and 3, the nuclear charges increase from element to element. The distance and shielding effects remain fairly constant, because electrons are added to the same outer shell. As the increasing attraction pulls the electrons closer to the nuclei, the radii of the atoms decrease.

Periodic patterns of boiling points for elements 1 to 36

The variation in boiling points is shown in Figure 7.13 and Figure 7.14. In any boiling liquid, particles are entering the vapour phase in large numbers. If the forces of attraction between the particles are strong, the boiling point is high; if the forces are weak, the boiling point is low.

Note that the 'peaks' of the graph are occupied by elements from the same group – carbon and silicon from Group 4. Their extremely high boiling points are due to the strong covalent bonds between atoms of these elements, which exist in the giant covalent lattice structure and persist, to some extent, in the liquid phase. These bonds are broken when the elements boil.

The 'troughs' are occupied by elements that consist of diatomic molecules (H_2, N_2, O_2, F_2, Cl_2, Br_2) or single atoms (He, Ne, Ar, Kr). The forces of attraction between the particles are very weak, even in the liquid phase, and are readily broken. The diatomic molecules or atoms are easily separated from each other as the temperature rises.

Elements in Groups 1, 2 and 3 occupy similar positions on the rising parts of the curve. Most of

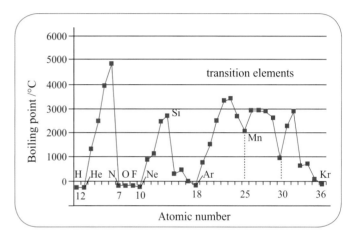

Figure 7.14 Plot of boiling points of elements 1 to 36 against atomic numbers.

the elements in these groups are metals, and the rising boiling points show that their metallic bonding persists into the liquid phase. The metallic bonding is stronger on moving from Group 1 to Group 2 to Group 3, as there are more outer-shell electrons available to be mobile and take part in the bonding.

SAQ

5 Why do both phosphorus (P_4) and sulfur (S_8) occupy low positions on the boiling-point curve but are higher than chlorine?

Hint

Answer

6 Figure 7.15 shows the melting points of the elements from Period 3. Plot these melting points against the atomic numbers for the elements.

continued

79

Na	Mg	Al	Si	P(white)	S	Cl	Ar
98	649	660	1410	44	113	−101	−189

Figure 7.15 Melting points of the elements from Period 3, measured in degrees Celsius (^{o}C).

Explain the following in terms of structure and bonding of the elements:

a the increase in the melting points between sodium and aluminium

b the very high melting point of silicon

c the low melting points of sulfur to argon.

Answer

Extension

Periodic patterns of first ionisation energies

The first ionisation energies of the first 36 elements in the Periodic Table are shown in Figure 7.16. Their variation with atomic number is displayed in Figure 7.17.

The most significant features of the graph are:

● the 'peaks' are all occupied by elements of the same group (Group 0, the noble gases)

● the 'troughs' are all occupied by elements of the same group (Group 1, the alkali metals)

● there is a general increase in ionisation energy across a period, from the Group 1 elements to the Group 0 elements, although the trend is uneven

● the first ionisation energies of the elements 21 (scandium) to 29 (copper) (the d-block elements of Period 4) show very little variation.

How are these periodic variations in first ionisation energies to be explained in terms of the model of atomic structure and electronic configurations outlined in Chapter 5?

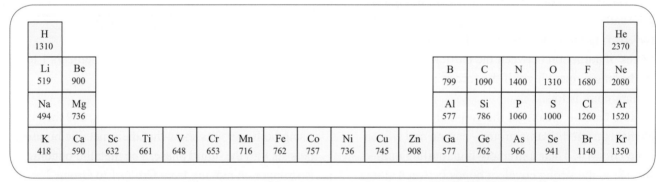

H 1310																	He 2370
Li 519	Be 900											B 799	C 1090	N 1400	O 1310	F 1680	Ne 2080
Na 494	Mg 736											Al 577	Si 786	P 1060	S 1000	Cl 1260	Ar 1520
K 418	Ca 590	Sc 632	Ti 661	V 648	Cr 653	Mn 716	Fe 762	Co 757	Ni 736	Cu 745	Zn 908	Ga 577	Ge 762	As 966	Se 941	Br 1140	Kr 1350

Figure 7.16 The first ionisation energies of elements 1 to 36, measured in kilojoules per mole ($kJ\,mol^{-1}$).

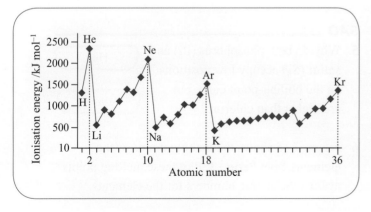

Figure 7.17 The first ionisation energies of elements 1 to 36, plotted against atomic number.

Consider some examples:

● fluorine, element 9, has the configuration $1s^2\,2s^2\,2p^5$

● neon, element 10, has the configuration $1s^2\,2s^2\,2p^6$

● sodium, element 11, has the configuration $1s^2\,2s^2\,2p^6\,3s^1$

As you see in Figure 7.17, the ionisation energy of neon is higher than that of fluorine; sodium's ionisation energy is much lower. The main differences in the atoms of these elements are:

● the numbers of protons in their nuclei, and hence their positive nuclear charges, are different

- the outer occupied orbital in both fluorine and neon is in the 2p subshell, but sodium has an electron in the next shell, in its 3s orbital.

These differences between the atoms may be explained by the factors that influence their first ionisation energies (see Chapter 5).

- An increase in positive nuclear charge will tend to cause an increase in first ionisation energies.
- The forces of attraction between the positive nuclear charge and the negatively charged electrons decreases as the quantum number of the shells increases. The further the shell is from the nucleus the lower the first ionisation energy is.
- Filled inner electron shells shield outer electrons. The outer electrons are repelled by the electrons in the filled inner shells so the first ionisation energy falls.

We shall now apply these ideas to the three elements, 9, 10 and 11 – fluorine, neon and sodium.

The outer electrons in both fluorine and neon atoms are in the 2p orbitals. This means that the 'distance' effect and the 'shielding' effect are similar. However, the nuclear charge in a neon atom is larger and attracts the 2p electrons more strongly. This causes the first ionisation energy of neon to be higher than that of fluorine.

The outer electron of sodium is in the 3s orbital, as the 2p orbitals are full. The ionisation energy of sodium is much lower than that of neon, even though a sodium atom has a larger nuclear charge. This shows how the combined effects of increased distance and of shielding reduce the effective nuclear charge. The 3s electron in a sodium atom is further from the nucleus than any 2p electrons. It is also shielded from the attractions of the nuclear charge by two complete inner shells ($n = 1$ and $n = 2$). The electrons in shell $n = 2$ are shielded only by the electrons in one shell ($n = 1$).

First ionisation energies across a period

From Figure 7.17 you will see that there is a general trend of increasing ionisation energies across a period. However, the trend is uneven. Look, for example, at elements 3, 4 and 5, lithium ($1s^2\ 2s^1$), beryllium ($1s^2\ 2s^2$) and boron ($1s^2\ 2s^2\ 2p^1$). We might have predicted that boron would have the highest ionisation energy of the three; in fact, it is beryllium. Experimental evidence such as this

leads to a further assumption about electronic configurations: it is easier to remove electrons from p orbitals than from s orbitals in the same shell. Our modern theories for electronic structure show that the p orbitals are of slightly higher energy levels than the s orbital for a given quantum number. Hence our theories predict that an electron is more easily removed from the p orbital than the s orbital. Thus the 2p electron in boron is easier to remove than one of the 2s electrons. Though the nuclear charge in boron is larger than in beryllium, boron has the lower first ionisation energy.

Now look at the other elements in Period 2.

	Carbon	*Nitrogen*	*Oxygen*
atomic number	6	7	8
electronic config.	$1s^22s^22p^2$	$1s^22s^22p^3$	$1s^22s^22p^4$
box config. for 2p	↑ ↑ ☐	↑ ↑ ↑	↑↓ ↑ ↑

	Fluorine	*Neon*
atomic number	9	10
electronic config.	$1s^22s^22p^5$	$1s^22s^22p^6$
box config. for 2p	↑↓ ↑↓ ↑	↑↓ ↑↓ ↑↓

In the general trend across the period, we might expect the ionisation energy of oxygen to be higher than that of nitrogen; in fact, the ionisation energy of nitrogen is the higher of the two. Nitrogen has three electrons in the p orbitals, each of them unpaired; oxygen has four electrons, with two of them paired. The repulsion between the electrons in the pair makes it easier to remove one of them. It takes less energy to ionise an atom of oxygen, even though the nuclear charge is larger than in an atom of nitrogen. (Note, you will not be asked to explain this in the AS examination.)

The general trend, of increasing ionisation energies across a period, is re-established in atoms of fluorine and neon, because of the effect of larger nuclear charge.

SAQ

7　In terms of their electronic configurations, explain the relative first ionisation energies of:
　a　sodium, magnesium and aluminium
　b　silicon, phosphorus and sulfur.

Answer

First ionisation energies in groups

Elements are placed in groups in the Periodic Table, as they show many similar physical and chemical properties. Note how elements in groups occupy similar positions on the plot of first ionisation energy against atomic number (Figure 7.17). This is evidence that the elements in groups have similar electronic configurations in their outer orbitals. For example:

Group 1 (alkali metals, Li–Cs)	all have one (s^1) electron in their outer shells
Group 2 (alkaline earth metals, Be–Ba)	all have two (s^2) electrons in their outer shells
Group 7 (halogens, F–At)	all have seven ($s^2 p^5$) electrons in their outer shells

The first ionisation energies generally decrease down a vertical group, with increasing atomic number. This shows the combined result of several factors. With increasing proton number, in any group:
- the positive nuclear charge increases
- the atomic radius increases so the distance of the outer electrons from the nucleus also increases with each new shell
- the shielding effect of the filled inner electron shells increases as the number of inner shells grows.

The combined distance and shielding effects overcome the effect of the increasing nuclear charges from element to element down any group. Therefore the first ionisation energies decrease down a group.

SAQ

8 Helium has the highest first ionisation energy in the Periodic Table. Suggest which element is likely to have the lowest first ionisation energy and why.

[Hint]

[Answer]

First ionisation energies and reactivity of elements

Ionisation energies give a measure of the energy required to remove electrons from atoms and form positive ions. The lower the first ionisation energy of an element, the more easily the element forms positive ions during reactions:

$$M(g) \longrightarrow M^+(g) + e^-$$

This is the main reason for the metallic nature of the elements on the 'left' of the Periodic Table (Groups 1, 2 and 3) and the increasingly metallic nature of elements down all groups.

- In elements with low first ionisation energies, one or more electrons are relatively free to move from atom to atom in the metallic bonding of the structure.
- The characteristic chemical properties of metallic elements include the formation of positive ions. In the reactions of metals with oxygen, chlorine or water, for example, formation of positive ions is one of several stages involving enthalpy changes. The elements with low first ionisation energies usually do react more quickly and vigorously. In any period, the Group 1 elements (alkali metals) have the lowest first ionisation energies and are the most reactive metals. They also have lower first ionisation energies going down the group, with increasing atomic number, and become much more reactive.

The factors that affect the values of ionisation energies also influence the reactivities of many elements. We shall examine the effect of ionisation energies on reactivity for Group 2 metals in Chapter 8.

Successive ionisation energies and the Periodic Table

Successive ionisation may be interpreted in terms of the position of an element in the Periodic Table. In Chapter 5, the pattern of successive ionisation energies for an element was used to:

- provide evidence for the general pattern of electron shells
- predict the simple electronic configuration of an element
- confirm the position of an element in the Periodic Table.

For any one element, successive ionisation energies steadily increase as electrons are removed. A large increase occurs between two successive ionisation energies when the next electron is removed from a lower electron shell.

For example, carbon has the following successive ionisation energies:

$1090, 2350, 4610, 6220, 37\,800, 47\,300\,kJ\,mol^{-1}$

The first four values show a steady increase followed by a very large increase at the fifth value. Hence, a total of four electrons are removed from the outer shell of carbon. Carbon has the simple electronic configuration 2,4, which also confirms the position of carbon in Group 4.

SAQ

9 a Phosphorus is in Group 5. Between which of the first eight ionisation energies will there be a large rise in ionisation energy?

b In which group of the Periodic Table would you place the element with the following successive ionisation energies?

$1680, 3370, 6040, 8410, 11\,000,$
$15\,200, 17\,900, 92\,000,$
$106\,000\,kJ\,mol^{-1}$

Hint

Answer

Summary

- Early attempts to explain periodic patterns (regularly repeating variations) in the properties of the known elements were based on their relative atomic masses. The modern Periodic Table is based on the elements in order of their atomic numbers.

- A group in the Periodic Table contains elements with the same outer-shell electronic configuration but very different atomic numbers; the elements and their compounds have many similar chemical properties.

- The elements in a block have their outermost electrons in the same type of subshell. For example, s-block elements have their outermost electrons in an s subshell.

- Periods in the Periodic Table are sequences of elements whose outermost electrons are in the same shell. Neighbouring members differ by one proton and one electron.

- Periodic variations may be observed across periods in physical properties such as ionisation energies, electron configurations, atomic radii, boiling points and melting points.

- The main influences on ionisation energies and atomic radii are: the size of the positive nuclear charge; the distance of the electron from the nucleus; the shielding effect on outer electrons by electrons in filled inner shells.

- Ionisation energies decrease down a group and tend to increase across a period; atomic radii increase down a group due to the effect of increased shielding, and decrease across a period due to increasing nuclear charge.

- The drop in ionisation energy between Groups 2 and 3 is due to a change from s to p subshells.

- The drop in ionisation energy between Groups 5 and 6 is due to commencement of electron pairing in a Group 6 p orbital causing electron-pair repulsion.

- Across a period (left to right, from Group 1 to Group 7), the structures of the elements change from giant metallic, through giant molecular to simple molecular. Group 0 elements consist of individual atoms.

- Successive ionisation energy data of an element may be interpreted in terms of the position of the element within the Periodic Table.

- Data on electronic configurations, atomic radii, melting points and boiling points may be interpreted to demonstrate periodicity.

- Chemically, the elements change from reactive metals, through less reactive metals and less reactive non-metals to reactive non-metals. Group 0 contains the extremely unreactive noble gases.

Glossary

Questions

1   The diagram below shows the variation in the boiling points of elements across Period 3 of the Periodic Table.

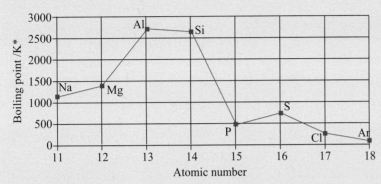

   a   Copy the table below for the elements Mg, Si and S.
   i   Complete the structure column using the word giant or simple.
   ii   Complete the bonding column using the word metallic, ionic or covalent.

element	structure	bonding
Mg		
Si		
S		

[3]

   b   Explain why silicon has a much higher boiling point than phosphorus.   [2]   Hint
   c   Explain why the boiling point increases from sodium to aluminium.   [2]   Hint

   * Note that to convert kelvin (K) to Celsius you subtract 273 from the kelvin value.

OCR Chemistry AS (2811) Jan 2007   [Total 7]

Answer

2   This question refers to the elements in the first three periods of the Periodic Table.

			H														He
Li	Be									B	C	N	O	F	Ne
Na	Mg									Al	Si	P	S	Cl	Ar

   a   Identify an element from the first three periods that fits each of the following descriptions.
   i   The element that forms a 2– ion with the same electronic configuration as Ne.   [1]
   ii   The element that forms a 3+ ion with the same electronic configuration as Ne.   [1]
   iii   The element that has the electronic configuration $1s^2\ 2s^2\ 2p^6\ 3s^2\ 3p^3$.   [1]

continued

    iv An element that forms a compound with hydrogen with tetrahedral molecules. [1]

    v An element that forms a compound with hydrogen with pyramidal molecules. [1]

    vi The element that forms a chloride XCl_2 with a molar mass of $95.3\,g\,mol^{-1}$. [1]

    vii The element with the largest atomic radius. [1]

    viii The element in Period 3 with the highest boiling point. [1]

b The diagram below shows the variation in the first ionisation energies of elements across Period 2 of the Periodic Table.

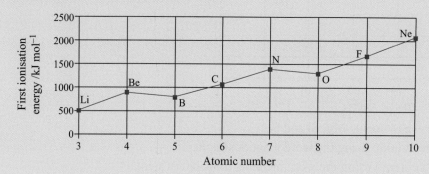

    i Define the term first ionisation energy. [3]   Hint

    ii Explain why the first ionisation energies show a general increase across Period 2. [2]

    iii Explain why the first ionisation energy of B is less than that of Be. [2]   Hint

    iv Estimate a value for the first ionisation energy of the element with atomic number 11. Explain how you made your choice. [2]

OCR Chemistry AS (2811) Jan 2002   [Total 17]

Answer

3 The first ionisation energies of the elements Na to K are represented below.

  a Explain why

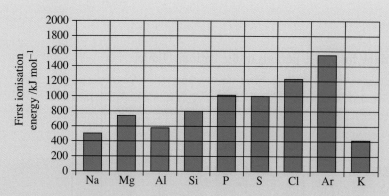

    i the first ionisation energies show an overall increase from Na to Ar [3]

    ii the first ionisation energy of Al is less than that of Mg. [2]

  b Explain the difference between the first ionisation energies of Ar and K. [3]   Hint

OCR Chemistry AS (2811) Jan 2005   [Total 8]

Answer

Chapter 8

Group 2

e-Learning

Objectives

The elements of Group 2 are often called the **alkaline earth metals**. They are:

beryllium	Be	[He]2s^2
magnesium	Mg	[Ne]3s^2
calcium	Ca	[Ar]4s^2
strontium	Sr	[Kr]5s^2
barium	Ba	[Xe]6s^2
radium	Ra	[Rn]7s^2

Beryllium is markedly different from the other members of the group, and so we shall not consider it here. Nor shall we consider radium (the element discovered and isolated by Marie Curie), as all its isotopes are radioactive.

The alkaline earth metals from magnesium to barium are silvery-white metals, with low melting and boiling points compared with transition metals like iron. They are good conductors of heat and electricity. The white colour that is often seen is an oxide film – the metals themselves are shiny but react quickly with air, forming an oxide film which prevents further reaction. These metals burn in air; magnesium with a very bright white light, and the others with characteristic flame colours – calcium is brick red, strontium is red and barium is green. Their physical properties are listed in Table 8.1.

Marie Curie

Marie Curie, *née* Marja Skłodowska, was born in Poland in 1867. Her first job was as a governess, which she took to pay for her sister's medical training in France. After her sister qualified, Marie also went to Paris to study at the Sorbonne, where she obtained the highest marks in physics. She also met and married Pierre Curie, who was a research scientist.

Marie Curie suspected that in the uranium ore called pitchblende there was another radioactive substance as well as uranium. She treated tonnes of pitchblende and eventually isolated a small quantity of radium chloride – radium was unknown before this. Pierre Curie helped her in this work, which was remarkable in its detective work, as 10 tonnes of pitchblende contain about 1 g of radium. The Curies were awarded the Nobel Prize in 1903 (along with Becquerel, who discovered radioactivity in 1896) for this work.

In 1906 Pierre, by then Professor of Physics at the Sorbonne, was killed by a horse-drawn carriage and Marie took over his post. She was the first woman to hold this position. In 1911 she was awarded a second Nobel Prize for her discovery of radium and polonium (which she named after Poland). The Curies' daughter Irene, with her husband Frederic Joliot, was also awarded a Nobel Prize for chemistry in 1935.

Figure 8.1 Marie Curie

	Mg	Ca	Sr	Ba
Atomic number	12	20	38	56
Metallic radius /nm	0.160	0.197	0.215	0.224
Ionic radius /nm	0.072	0.100	0.113	0.136
First ionisation energy /kJ mol^{-1}	738	590	550	503
Second ionisation energy /kJ mol^{-1}	1451	1145	1064	965
Third ionisation energy /kJ mol^{-1}	7733	4912	4210	3390
Melting point /°C	649	839	769	725
Boiling point /°C	1107	1484	1384	1640

Table 8.1 Physical properties of Group 2 elements.

The metals of Group 2 all have two electrons in their outer s subshell, and these are lost when the metal reacts. This means that they always form an ion of oxidation number +2 in their compounds, such as Mg^{2+} and Ca^{2+}. It also means that they are less reactive than Group 1 metals, because they have to lose two outer shell electrons, whereas Group 1 metals lose only one.

SAQ

1   a   Using Table 8.1, describe, for magnesium to barium, the trend in: **i** their metallic radii and **ii** their first ionisation energies.

   b   Explain the trends that you have described in **a**.

   c   Predict and explain the trend in electronegativity of the Group 2 elements.

General properties of the Group 2 elements

The general properties of the Group 2 elements magnesium to barium are as follows:

● they are all metals
● they are good conductors of heat and electricity
● their compounds are all white or colourless (unless the anion is coloured)
● in all their compounds they have an oxidation number +2
● their compounds are ionic
● they are called alkaline earth metals because their oxides and hydroxides are basic
● they react with acids to give a salt plus hydrogen.

Compared with the metals of Group 1:

● they are harder and denser
● they have higher melting points
● they exhibit stronger metallic bonding (because they have two outer shell electrons instead of one).

SAQ

2   Explain why magnesium oxide has an extremely high melting point, with reference to its structure and bonding.

Reactions of Group 2 elements

These elements are powerful reducing agents.

The reactions of the Group 2 metals with water

All the Group 2 metals, from magnesium to barium, react with water, producing hydrogen. In effect, they reduce water to hydrogen and in the process are oxidised from the element (oxidation state = 0) to their ions (oxidation state = +2) as we will see.

Magnesium reacts very slowly with cold water. A little hydrogen and magnesium hydroxide are formed over a few days.

$$Mg(s) + 2H_2O(l) \longrightarrow Mg(OH)_2(s) + H_2(g)$$

The aqueous solution is weakly alkaline as magnesium hydroxide is very sparingly soluble. When water, as steam, is passed over heated magnesium, there is a rapid reaction (Figure 8.2).

Figure 8.2 Apparatus to show the reaction of magnesium with steam. The steam is generated by heating mineral wool soaked in water at the left-hand end of the test-tube.

Hydrogen is released in the reaction and magnesium oxide remains:

$$Mg(s) + H_2O(g) \longrightarrow MgO(s) + H_2(g)$$

The reactivity of the elements with water increases down the group from magnesium to barium (Figure 8.3). Unlike magnesium, the metals calcium to barium all react readily with cold water to form a cloudy white precipitate of the hydroxide (which is sparingly soluble).

Figure 8.3 Barium reacts readily with cold water, producing a rapid stream of bubbles of hydrogen.

During the reaction with water, the metal atoms lose electrons which are transferred to hydrogen atoms in water molecules. For example:

$$Ca(s) + 2H_2O(l) \longrightarrow Ca(OH)_2(s) + H_2(g)$$

Ox. states Ca: 0   H: +1   Ca: +2   H: 0
                     O: −2   O: −2
                                H: +1

In this example, each calcium atom loses two electrons and changes oxidation state from 0 in Ca(s) to +2 in Ca(OH)$_2$(s). This is oxidation.

Meanwhile, two hydrogen atoms gain one electron each and change oxidation state from +1 in H$_2$O(l) to 0 in H$_2$(g). This is reduction. Notice that two water molecules are required for each calcium atom. Only one of the two hydrogen atoms in a water molecule is reduced, the oxidation state of the second hydrogen atom is unchanged at +1 in Ca(OH)$_2$(s).

Two electrons are lost from each metal atom in the reaction. Down the group, the first two ionisation energies decrease from magnesium to barium. Consequently, the reactivity of the metals increases down the group as less energy is required to remove the two electrons. A similar trend in reactivity is found when the Group 2 metals react with oxygen.

The reactions of the Group 2 metals with oxygen

The reactions of these metals with oxygen are vigorous once they get started.

Magnesium metal is normally covered with a layer of its oxide. It burns rapidly in air or oxygen with a brilliant white flame (Figure 8.4). This reaction is much used in fireworks and warning flares. White magnesium oxide solid is formed.

$$2Mg(s) + O_2(g) \longrightarrow 2MgO(s)$$

Figure 8.4 Magnesium ribbon burning in air. This reaction was used in the first photographic flash, and is still used in fireworks and flares.

Uses of two Group 2 compounds

Magnesium hydroxide is a weak alkali and is used in indigestion remedies to neutralise excess acid in the stomach. It is also used in toothpastes, where it helps to neutralise acids in the mouth which encourage tooth decay. Representing acid as $H^+(aq)$, the following reaction occurs:

$$Mg(OH)_2(s) + 2H^+(aq) \longrightarrow Mg^{2+}(aq) + 2H_2O(l)$$

Solid calcium hydroxide is used on acidic soil, to reduce the acidity of the soil. This can help to increase crop yields. Representing acid as $H^+(aq)$, the following reaction occurs:

$$Ca(OH)_2(s) + 2H^+(aq) \longrightarrow Ca^{2+}(aq) + 2H_2O(l)$$

SAQ

5   Compare the calcium hydroxide reaction with that of magnesium hydroxide in indigestion remedies above.
    Suggest why calcium hydroxide is not used in indigestion remedies. *Answer*

6   a   Predict what you would observe *Hint* when calcium carbonate is added to dilute hydrochloric acid. Write a balanced equation, including state symbols, for any reaction that you have predicted.
    b   Explain why calcium hydroxide might be added to acidic soil.
    c   Magnesium hydroxide is used in antacid tablets. They relieve excessive acidity in the stomach by neutralising some of the hydrochloric acid present. Write a balanced equation, including state symbols, for the reaction of magnesium hydroxide with hydrochloric acid. *Answer*

SAQ

3   Some magnesium ribbon (0.20 g) *Hint* was heated in a crucible until it began to burn. When the burning finished, a white powder remained in the crucible. What was this white powder? Calculate the mass of powder that was formed. *Answer*

4   a   Write an equation, including state symbols, for the burning of strontium in oxygen.
    b   Describe what you might observe during this reaction.
    c   Identify the element which is *Hint* oxidised and the element which is reduced. Explain your answer in terms of electron transfer and oxidation states.
    d   Explain the increasing reactivity of the Group 2 metals going down the group. *Answer*

Names of some calcium compounds

Several calcium compounds have common names based on the word 'lime', derived from limestone. Limestone is one of the most widespread types of rock in the UK.

Names of some calcium compounds		
limestone	contains mainly calcium carbonate	$CaCO_3$
quicklime	calcium oxide	CaO
slaked lime	solid calcium hydroxide	$Ca(OH)_2(s)$
lime water	a solution of calcium hydroxide (only sparingly soluble)	$Ca(OH)_2(aq)$

Chalk and lime chemistry

The white cliffs of Dover are composed of chalk. The cliffs are the sedimentary remains of tiny marine invertebrates. Chemically, they are composed of calcium carbonate, $CaCO_3$. Limestone is a similar sedimentary deposit which also contains calcium carbonate (Figure 8.5). Limestone and chalk are used in large quantities to manufacture quicklime (calcium oxide) and cement.

Figure 8.5 These limestone cliffs are sedimentary deposits that contain calcium carbonate.

Strong heating of calcium carbonate produces calcium oxide, CaO, and carbon dioxide.

$$CaCO_3(s) \longrightarrow CaO(s) + CO_2(g)$$

This type of reaction, where a compound is broken down by heat, is known as **thermal decomposition** (see also Chapter 16).

Traditionally, chalk or limestone was heated in a lime kiln using fuels such as wood or coal (Figure 8.6).

Figure 8.6 Old lime kilns near Agrigento in Sicily.

Other Group 2 carbonates

The carbonates of other Group 2 elements undergo thermal decomposition in a similar way. For example, magnesium carbonate breaks down when heated to give magnesium oxide and carbon dioxide.

$$MgCO_3(s) \longrightarrow MgO(s) + CO_2(g)$$

Magnesium carbonate decomposes at a lower temperature than calcium carbonate. Strontium carbonate decomposes at a higher temperature, and barium carbonate decomposes at an even higher temperature. The carbonates of Group 2 elements therefore become more stable to thermal decomposition as we go down Group 2.

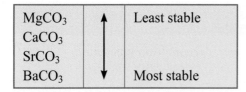

MgCO₃ CaCO₃ SrCO₃ BaCO₃	↑ ↓	Least stable Most stable

Trend in stability to thermal decomposition of Group 2 carbonates.

SAQ

7 Strontium carbonate decomposes when heated.
   a What are reactions of this type called?
   b Write a balanced chemical equation for this reaction.
   c Compare the ease of decomposition of strontium carbonate with that of calcium carbonate and barium carbonate.

Answer

Calcium oxide reacts vigorously with water (see page 93) to produce calcium hydroxide, $Ca(OH)_2$ (slaked lime).

$$CaO(s) + H_2O(l) \longrightarrow Ca(OH)_2(s)$$

The name 'quicklime' for CaO derives from the vigour of this reaction. Here, 'quick' is used in the sense of 'alive'. During the Black Death, corpses were buried under a layer of quicklime. The quicklime reacted with moisture from the corpses, helping to control the spread of disease. In more recent times also, quicklime has been used in this way, for example following the tragic earthquake in Turkey in 1999.

Further addition of water to calcium hydroxide produces the saturated aqueous solution that we call 'lime water'. Calcium hydroxide is only slightly soluble in water; the saturated solution has a concentration of approximately $1.5 \times 10^{-2}\,mol\,dm^{-3}$. Chemists often describe compounds which are only very slightly soluble as being 'sparingly soluble'. As the concentration of calcium hydroxide in lime water is low, the solution has a low concentration of hydroxide ion, OH⁻. The pH of lime water is about 9–10.

SAQ

8 Barium oxide reacts with water. The product of the reaction dissolves in water to produce a strongly alkaline solution.
   a Write a balanced chemical equation for the reaction of barium oxide with water.
   b Name the product of this reaction.
   c Explain why the solution produced is 'strongly alkaline'.

Answer

You will probably be familiar with lime water as the reagent used to identify carbon dioxide gas. When a sample of carbon dioxide is bubbled through lime water, a cloudy white precipitate forms. The precipitate is solid calcium carbonate.

$$Ca(OH)_2(aq) + CO_2(g) \longrightarrow CaCO_3(s) + H_2O(l)$$

Figure 8.7 shows a straightforward method for performing the lime water test for carbon dioxide.

air squeezed from teat pipette bulb before placing in test tube

teat pipette flushed with lime water by squeezing bulb several times

CO_2 gas collected from just above liquid

sample bubbled through lime water

reaction mixture containing marble chips and dilute HCl

collecting gas sample

testing gas sample

cloudy white precipitate

Figure 8.7 The lime water test for carbon dioxide.

Other uses of Group 2 elements and their compounds

The elements of Group 2 and their compounds are widely used in commerce and industry.

Magnesium burns with a bright white light, and is used in flares, incendiary bombs and tracer bullets. It was once used in photographic flash bulbs.

Magnesium has such a strong reducing power that it is widely used to protect steel objects such as ships, outboard motors and bridges from corrosion. Its strong reducing power also means that it can be used in the extraction of less reactive metals such as titanium in the Kroll process, which takes place at 1000 °C under an argon atmosphere:

$$2Mg(s) + TiCl_4(g) \longrightarrow Ti(s) + 2MgCl_2(l)$$

Magnesium oxide is a refractory material, which means that it is resistant to heat (its melting point is 2800 °C). Its main use is for the lining of furnaces.

Magnesium fluoride is used to coat the surface of camera lenses, to reduce the amount of reflected light. This thin coating is responsible for the violet colour on the surface of the lens.

Calcium carbonate is an important compound which occurs in vast quantities in sedimentary rocks, such as limestone, chalk and dolomite. Marble is also a form of calcium carbonate, in which the marbling effect is caused by the presence of iron oxides. The largest use of calcium carbonate is in the manufacture of cement.

Cement is made by heating a finely ground mixture of limestone with clay (aluminosilicate) at 1500 K in a rotary kiln. This process results in clinker, which is reground and mixed with about 3% gypsum (calcium sulfate). This is a complex process, with the final cement product being a mixture of calcium silicates and calcium aluminates.

The setting of cement is also a complex process – the cement reacts with water and carbon dioxide from the air.

The annual production of cement worldwide is about 1 billion tonnes.

Lime or quicklime, which is *calcium oxide*, was used in the past in cement, mortar (Figure 8.8) and plaster manufacture. It still has a very important role in the extraction of iron, as it reacts with impurities in the ore to form a molten slag:

$$CaO(s) + SiO_2(s) \longrightarrow CaSiO_3(l)$$
$$\text{basic} \quad \text{acidic} \quad \text{'slag'}$$
$$\text{oxide} \quad \text{oxide}$$

Calcium oxide is the origin of the theatrical term 'limelight', because it glows with a bright white light when strongly heated and was originally used in stage lighting.

Figure 8.8 a Calcium oxide was used in lime mortar before the introduction of cement mortar. **b** Addition of water to dry calcium oxide causes the solid to crumble, in an exothermic reaction producing calcium hydroxide.

continued

Lime mortar, which is prepared by mixing sand and calcium oxide with water, was widely used in bricklaying until the beginning of the 20th century.

Plaster of Paris, used to set broken bones and for modelling, is an insoluble form of hydrated *calcium sulfate*, $CaSO_4.H_2O$. When it is mixed with water, it further hydrates to $CaSO_4.2H_2O$ and sets hard.

The *hydrogencarbonates of calcium and magnesium* are responsible for temporary hardness of water. These compounds form when rain-water containing dissolved carbon dioxide trickles over limestone and other, similar rocks.

Barium sulfate is insoluble and, in suspension, is given to patients as a 'barium meal'. The barium ions coat the walls of the stomach and digestive tract and, as they are opaque to X-rays, they make any imperfections visible by X-ray photography. Soluble barium compounds are toxic, but barium sulfate is safe to use because its solubility is so low – for this reason, the presence of a barium sulfate precipitate is also used as a laboratory test for the sulfate ion.

Summary

Glossary

- The Group 2 elements magnesium to barium are typical metals with high melting points and good conductivity of heat and electricity.

- Progressing down Group 2 from magnesium to barium, the atomic radius increases. This is due to the addition of an extra shell of electrons for each element as the group is descended.

- Many of the compounds of Group 2 elements are commercially important. For example, slaked lime (calcium hydroxide) is used to neutralise acid soil. Magnesium hydroxide is used in antacid tablets to neutralise excess acid in the stomach.

- Reactivity of the elements with oxygen or water increases down Group 2 as the first and second ionisation energies decrease.

- The reactions of Group 2 elements with oxygen or water are redox reactions. The elements are powerful reducing agents. Redox reactions may be explained in terms of electron transfer (oxidation is loss of electrons, reduction is gain of electrons) or change of oxidation states (oxidation increases oxidation state, reduction decreases oxidation state).

- The Group 2 elements magnesium to barium react with water to produce hydrogen gas and the sparingly soluble metal hydroxide. As the hydroxide solutions have a pH of 8 or higher, they are called the alkaline earth elements.

- The Group 2 elements magnesium to barium burn in air with characteristic flame colours to form the oxide as a white solid. Magnesium burns with a bright white light, and the other flame colours are: calcium – brick red; strontium – red; barium – green.

- The carbonates of the Group 2 elements decompose when heated. The carbonates of the elements at the top of Group 2 decompose more easily than the carbonates of the elements at the bottom of Group 2.

- Calcium carbonate (as chalk or limestone) may be decomposed by heat to form calcium oxide (quicklime) which reacts violently with water to form calcium hydroxide (slaked lime). Calcium hydroxide is sparingly soluble in water, forming lime water.

Questions

1 This question concerns elements and compounds from Group 2 of the Periodic Table.

  a State the trend in reactivity of the Group 2 elements with oxygen. Explain your answer. [4]

  b The reactions of strontium are typical of a Group 2 element. Write the formulae for substances A–D in the flow chart below. [4]

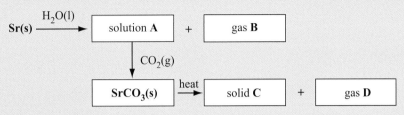

OCR Chemistry AS (2811) May 2002 [Total 8]

2 The elements calcium and strontium in Group 2 of the Periodic Table both react with water.

  a A student reacted 0.20 g of calcium and 0.20 g of strontium separately with 250 cm³ of water. The student measured the volume of gas produced from each reaction. The student's results are shown below.

metal	calcium	strontium
volume of gas / cm³	120	55

  i Name the gas produced. [1]

  ii Write a balanced equation for the reaction of strontium with water. [1]

  iii Explain why the student obtained different volumes of gas by reacting water with the same mass of calcium and strontium. [2]

  iv Predict the pH of the solutions formed in each of these reactions. [1]

  b In their reactions, calcium and strontium each lose electrons to form ions with a 2+ charge. The first and second ionisation energies of calcium and strontium are shown below.

	1st ionisation energy /kJ mol⁻¹	2nd ionisation energy /kJ mol⁻¹
calcium	590	1145
strontium	550	1064

  i Write an equation, with state symbols, to represent the second ionisation energy of calcium. [2]

  ii Why are the second ionisation energies of calcium and strontium greater than their first ionisation energies? [1]

  iii Explain why the first and second ionisation energies of strontium are less than those of calcium. [3]

OCR Chemistry AS (2811) June 2006 [Total 11]

continued

3 Solid calcium hydroxide, $Ca(OH)_2$ is commonly known as 'slaked lime'. Over one million tonnes of slaked lime are produced annually in the UK.

    a Lime water is an aqueous solution of calcium hydroxide, $Ca(OH)_2$. Lime water is commonly used in the laboratory as a test for carbon dioxide. Suggest a value for the pH of lime water.   [1]

    b The ions present in $Ca(OH)_2$ are Ca^{2+} and OH^-.

        i Give the electronic configuration of a Ca^{2+} ion.   [1]

        ii How many moles of ions are in one mole of $Ca(OH)_2$?   [1]

        iii How many moles of electrons are in one mole of OH^- ions?   [1]

        iv Draw a dot-and-cross diagram of $Ca(OH)_2$. Show outer electron shells only.   [2]

    c Slaked lime (solid calcium hydroxide) can be prepared from calcium carbonate, $CaCO_3$, in two stages.

      Outline how this could be done in the laboratory. Include an equation for each stage.   [4]

    d The production of lime water is a small-scale use of calcium hydroxide.

      State a large-scale use of calcium hydroxide outside of the laboratory.   [1]

OCR Chemistry AS (2811) Jan 2007   [Total 11]

Hint

Answer

4 The Group 2 element radium, Ra, is used in medicine for the treatment of cancer. Radium was discovered in 1898 by Pierre and Marie Curie by extracting radium chloride from its main ore pitchblende.

    a Predict the formula of radium chloride.   [1]

    b Pierre and Marie Curie extracted radium from radium chloride by reduction. Explain what is meant by reduction, using this reaction as an example.   [2]

    c Radium reacts vigorously when added to water.

$$Ra(s) + 2H_2O(l) \longrightarrow Ra(OH)_2(aq) + H_2(g)$$

        i Use the equation to predict two observations that you would see during this reaction.   [2]

        ii Predict a pH value for this solution.   [1]

    d Reactions of the Group 2 metals involve removal of electrons. The electrons are removed more easily as the group is descended and this helps to explain the increasing trend in reactivity.

        i What is the removal of one electron from each atom in 1 mole of gaseous radium atoms called?   [2]

        ii Write down the equation for this process.   [2]

        iii Atoms of radium have a greater nuclear charge than atoms of calcium. Explain why, despite this, *less* energy is needed to remove an electron from a radium atom than from a calcium atom.   [3]

OCR Chemistry AS (2811) June 2005   [Total 13]

Hint

Answer

continued

5 Calcium oxide, CaO, is used for making cement which is widely used in the construction industry. Calcium oxide can be prepared as 'quicklime' by heating limestone in a lime kiln to about 550 °C. The calcium carbonate in the limestone decomposes into calcium oxide and carbon dioxide.

$$CaCO_3 \longrightarrow CaO + CO_2$$

   a Draw a dot-and-cross diagram of calcium oxide, showing outer electrons only. [2]

   b In $CaCO_3$ what is the oxidation state of

      i Ca [1]

      ii C? [1]

   c Calculate the mass of CaO that could be made from limestone containing 20 tonnes of $CaCO_3$. [2]

      (molar masses: $CaCO_3$ 100 g mol^{-1}; CaO, 56 g mol^{-1}, 1 tonne = 10^6 g)

   d When water is added to quicklime, a vigorous reaction takes place forming slaked lime, $Ca(OH)_2$.

      Write an equation for the formation of slaked lime in this reaction. [1]

   e Farmers often add 'lime' to acid soils. The lime is mostly present as slaked lime.

      A chemist neutralised 25.0 cm^3 0.200 mol dm^{-3} HCl with slaked lime.

Hint

$$Ca(OH)_2(s) + 2HCl(aq) \longrightarrow CaCl_2(aq) + 2H_2O(l)$$

      i What is the molar mass of $Ca(OH)_2$? [1]

      ii How many moles of HCl were neutralised? [1]

      iii Calculate the mass of $Ca(OH)_2$ that neutralises this HCl. [2]

      iv The chemist neutralised the same amount of HCl with NaOH. Explain why the chemist would

      need to use more moles of NaOH than $Ca(OH)_2$. [2]

OCR Chemistry AS (2811) June 2003   [Total 13]

Answer

Chapter 9

Group 7

Objectives

The elements of Group 7 are called the halogens:

fluorine	F	$[He]2s^2\,2p^5$
chlorine	Cl	$[Ne]3s^2\,3p^5$
bromine	Br	$[Ar]3d^{10}\,4s^2\,4p^5$
iodine	I	$[Kr]4d^{10}\,5s^2\,5p^5$
astatine	At	$[Xe]4f^{14}\,5d^{10}\,6s^2\,6p^5$

All the isotopes of astatine are radioactive and so this element will not be considered here. Also, we shall not include fluorine in *all* the discussions on Group 7, because its small size and high electronegativity give it some anomalous properties.

The name 'halogen' is derived from the Greek and means 'salt producing'. It was first used at the beginning of the nineteenth century because chlorine, bromine and iodine are all found in the sea as salts. Nowadays we still use the term, because the halogens are very reactive and react readily with metals to form salts.

The halogens are a family of non-metallic elements with some very similar chemical properties, although there are also clear differences between each element. Their reactivity decreases going down the group. Their chemical characteristics are caused by the outermost seven electrons – two electrons in the s subshell and five electrons in the p subshell. Therefore only one more electron is needed to complete the outer shell of electrons. As a result the most common oxidation state for the halogens is –1, although other oxidation states do exist, especially for chlorine, which exhibits a range of oxidation states from –1 to +7.

In compounds a halogen atom increases its share of electrons from seven to eight (a full outer shell) by either gaining an electron to form a halide (Cl^-, Br^-, I^-) in ionic compounds, or sharing an electron from another atom in a covalent compound.

The halogen elements form covalent diatomic molecules. The atoms are joined by a single covalent bond (Figure 9.1).

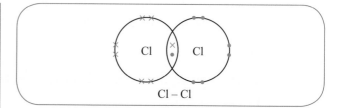

Figure 9.1 A dot-and-cross diagram of the covalent bonding in chlorine gas, $Cl_2(g)$.

Fluorine, chlorine and bromine are poisonous. Their melting and boiling points increase with increasing atomic number: fluorine and chlorine are gases at room temperature; bromine is a liquid; and iodine is a solid. This decrease in volatility is the result of increasingly strong **van der Waals' forces** as the number of electrons present in a molecule increases. Halogens form diatomic non-polar molecules, so only van der Waals' forces (see Chapter 6) are present between molecules.

The colour of the elements deepens with increasing atomic number: fluorine is a pale yellow gas; chlorine is a greenish yellow gas; bromine is a dark red liquid giving off a dense red-orange vapour; iodine is a shiny, grey-black crystalline solid which **sublimes** (changes directly from a solid to a gas) to a purple vapour.

The halogens are all oxidising agents, with fluorine the strongest. The oxidising ability is reflected by the reactivity – fluorine is the most reactive halogen. It is also reflected by the electronegativities – fluorine is the most electronegative element in the Periodic Table, chlorine the third. In many reactions, the halogen will gain an electron in its outer shell to form a negative ion. This is easier the smaller the atom, as the outer shell is nearer to the attractive force of the nucleus and there are fewer inner shells of electrons providing a shielding effect against the nuclear attraction. Look at Table 9.1 for a summary of their physical properties.

	F	Cl	Br	I
Atomic radius /nm	0.071	0.099	0.114	0.133
Ionic radius /nm	0.133	0.180	0.195	0.215
Electronegativity	4.0	3.0	2.8	2.5
Electron affinity /kJ mol^{-1}	−328	−349	−325	−295
Melting point /°C	−220	−101	−7	114
Boiling point /°C	−188	−35	59	184

Table 9.1 Physical properties of Group 7 elements. (The electron affinity is the enthalpy change for the process $X(g) + e^- \longrightarrow X^-(g)$, where X is the halogen.)

SAQ

1   a   What would you expect to be the physical state of astatine at room temperature? Explain your answer.

   b   Fluorine and chlorine both react with sodium metal to form a salt.
      i   Write a balanced chemical equation for each reaction.
      ii Which of the two reactions will be most vigorous. Why?
      iii What structure and bonding will the products of these reactions have?

   c   What is the oxidation number of chlorine in Cl_2, $CaCl_2$, Cl_2O_7 and ClO_2?

Hint

Answer

General properties of the Group 7 elements

The general properties of the Group 7 elements chlorine, bromine and iodine are as follows:

- they behave chemically in a similar way
- they are non-metals
- they all exist as diatomic molecules at room temperature
- their melting and boiling points increase with increasing atomic number
- the colour of the elements deepens with increasing atomic number
- they are very reactive and readily form salts
- in compounds a halogen atom increases its number of electrons in the outer shell from seven to eight by ionic or covalent bonding

- the reactivity of the elements decreases on descending the group
- they exhibit a range of oxidation states
- the electronegativity of the elements decreases on descending the group
- their oxidising ability decreases on descending the group.

SAQ

2   Draw dot-and-cross diagrams of NaCl, showing both ions; and of HCl, showing the covalent bond.

Answer

Uses

The halogens and their compounds have many commercial and industrial uses.

Chlorine is used in vast quantities for many different processes – twenty-nine million tonnes of chlorine are used worldwide annually. It is produced by the electrolysis of brine (see Chapter 6). Its main uses are in water purification, as a bleach and in the manufacture of various chemicals such as chloroethene, which is used to make PVC.

Chlorine is used to make a class of organic chemicals called CFCs (*chlorofluorocarbons*). CFCs have been used as aerosol propellants, refrigerants and as foaming agents in polymers. They are currently being withdrawn from many applications because they are pollutants which contribute to the destruction of the ozone layer. However, they are still used in fire extinguishers because they are inert and non-flammable. They are also vital constituents of artificial blood.

Clean water

Figure 9.2 Chlorine has killed bacteria in this swimming pool. It has also produced organochlorine compounds.

Chlorine is a good germicide – it is well known as the agent used to kill bacteria in swimming pools (Figure 9.2). Chlorine is also used by the water industry to kill bacteria in drinking water. Some environmental campaigners have tried to have this use of chlorine banned, because they are concerned about two issues.

- Firstly, chlorine itself is highly toxic.
- Secondly, chlorine reacts with traces of organic compounds in water, producing organochlorine compounds. Some of these organochlorine compounds are known to cause cancer and to act as mutagens.

Fears of this type led to a ban on the use of chlorine in drinking water in Peru in 1993. Unfortunately, as a consequence there was an outbreak of cholera, with 10 000 deaths.

The companies responsible for providing drinking water claim that it is scientifically proven that the benefits of chlorinating our drinking water far outweigh the risks. The problem can be seen as an ethical one. Should water companies continue to chlorinate drinking water, thus taking the decision away from consumers? Alternatively, should drinking water be supplied non-chlorinated, allowing consumers to choose to add (or not to add) their own chlorine or other germicide?

SAQ

3 a Describe the potential dangers of chlorinated drinking water. Use internet research to assess the level of risk involved.
   b Describe the potential dangers of unchlorinated drinking water.
   c Make and justify a decision on whether or not drinking water should be chlorinated by water companies.

Answer

Solvents containing chlorine, such as *dichloromethane*, CH_2Cl_2, are widely used to dissolve fats and oils.

Chlorine and some of its compounds are used as domestic and commercial bleaches. In the First World War chlorine and mustard gas ($ClCH_2CH_2SCH_2CH_2Cl$) were used with devastating effect as poison gases.

Fluorine is used, like chlorine, in CFCs. It is also used to make PTFE (polytetrafluoroethene), which is used as a lubricant, as a coating for non-stick cooking pans, as electrical insulation and in waterproof clothing.

Other uses of halogens

- *Hydrofluoric acid* (HF) is used to etch glass.
- *Bromochlorodifluoromethane* ($CClBrF_2$) is used in fire extinguishers.
- *Silver bromide* is used in photographic film. This use has decreased recently as digital cameras have become popular.
- *Iodine* is an essential part of our diet, and an imbalance can cause thyroid problems.
- A solution of *iodine in alcohol* is sometimes used as an antiseptic.

The fluoride controversy

Fluoride ions help to prevent tooth decay. Some children are given fluoride tablets; many toothpastes contain tin fluoride (SnF_2); and some water supplies are fluoridated with sodium fluoride.

Automatic fluoridation of the water we drink has caused some controversy. Many people feel that it is a good thing, as it is one of the factors linked to a reduction in the number of fillings in children's teeth. However, too much fluoride discolours teeth permanently, and can cause liver damage. Some people feel that water supplies should not be fluoridated, so that the freedom to choose to take fluoride supplements or not should be left to the individual (Figure 9.3).

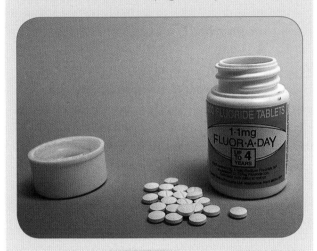

Figure 9.3 Fluoride tablets may be taken if drinking water is not fluoridated, to improve dental health.

The reactivity of the halogens: displacement reactions

The electron affinity of the halogens is shown in Table 9.1. The more negative the electron affinity, the greater the ease with which a halogen can capture an electron and the greater the oxidising power of the halogen. Electron affinity becomes less negative from chlorine to iodine so the oxidising power of the elements decreases down Group 7.

In most of their oxidising reactions the halogens react as X_2 molecules (X represents a halogen) and form hydrated halide ions, $X^-(aq)$. As the oxidising ability decreases from chlorine to iodine, any halogen can displace another lower in the group. This means that, if each halogen is reacted with an aqueous solution containing halide ions, a series of displacement reactions occurs:

- Chlorine displaces bromine and iodine:
$$Cl_2(aq) + 2Br^-(aq) \longrightarrow 2Cl^-(aq) + Br_2(aq)$$
$$Cl_2(aq) + 2I^-(aq) \longrightarrow 2Cl^-(aq) + I_2(aq)$$

- Bromine displaces iodine:
$$Br_2(aq) + 2I^-(aq) \longrightarrow 2Br^-(aq) + I_2(aq)$$

- Iodine does not displace either chlorine or bromine.

One of the problems with doing these displacement reactions is being able to see if a reaction has taken place – the halide ion solutions are all colourless and very dilute solutions of the halogens can also appear colourless. To avoid this problem, an organic solvent such as cyclohexane is added to the mixture, which forms a separate layer. The halogens are more soluble in organic solvents than in aqueous solution, so they are taken up by the cyclohexane and the colour is much more apparent (Figure 9.4).

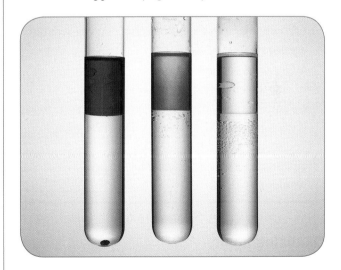

Figure 9.4 Colours of halogens in cyclohexane (upper layer) and water (lower layer).

Iodine is purple in cyclohexane, bromine is a strong orange-yellow colour and chlorine is pale yellow. If aqueous bromine is mixed with cyclohexane, the bromine dissolves in the cyclohexane, which turns orange. Then if aqueous potassium iodide is added, the cyclohexane turns purple. This shows us that bromine molecules have become bromide ions and displaced iodine from solution. The iodide ions have

Halogen	Halide ion		
	Chloride, Cl^-	Bromide, Br^-	Iodide, I^-
Chlorine, Cl_2	×	orange-yellow bromine released	purple iodine released
Bromine, Br_2	no reaction	×	purple iodine released
Iodine, I_2	no reaction	no reaction	×

Table 9.2 Displacement reactions of halogens in aqueous solution (the colours refer to the colours of the halogens in cyclohexane, see Figure 9.4).

become iodine molecules. Table 9.2 summarises the displacement reactions between the halogens and solutions of halide ions.

SAQ

4 From your knowledge of the structure and bonding of the halogens, explain why they are more soluble in organic solvents than in aqueous solution.

Hint

Answer

5 Bromine water (aqueous bromine, Br_2) is shaken with a small volume of cyclohexane, and then the following aqueous solutions are added to separate portions:
  a aqueous sodium iodide
  b aqueous chlorine
  c aqueous sodium astatide, NaAt.
  Each mixture is shaken again. Describe what you would expect to see. Write equations for any reactions that would occur.

Answer

Oxidising ability

The oxidising ability of the halogens means that they are useful in many ways. Chlorine and its aqueous solution, known as chlorine water, are often used as oxidising agents (chlorine water contains chlorine and chloric(I) acid, HClO). Chlorine is also used in industry as a bleach; it oxidises large coloured organic molecules, changing them to colourless compounds. In recent years controversy has arisen over the use of chlorine for bleaching paper – although very white paper pulp can be produced, the process results in the formation of dioxins, which are

poisonous and can accumulate in living organisms, as dioxins do not break down easily. Nowadays ozone is often used to bleach paper that does not have to be pure white, such as tissues, nappies and toilet paper.

The strong oxidising ability of chlorine is also used to treat drinking water, and to keep water in swimming pools free from contamination (see box on page 100).

Fluorine is rarely used as an oxidising reagent as it is difficult to handle.

Which halide?

Halides (ions of Group 7 elements) are extremely common, so a test to identify which halide is present is very useful. This test is based on the colour of the silver halides and the different solubilities of the silver halides in ammonia solution.

1 Acidify the unknown halide solution with dilute nitric acid.

2 Add aqueous silver nitrate (caution: silver nitrate is poisonous)
  ● a white precipitate of silver chloride forms if $Cl^-(aq)$ is present
  ● a cream precipitate of silver bromide forms if $Br^-(aq)$ is present
  ● a yellow precipitate of silver iodide forms if $I^-(aq)$ is present

3 Identification by colour is not completely reliable, so aqueous ammonia is added:
  ● white silver chloride dissolves in dilute aqueous ammonia forming a colourless solution
  ● cream silver bromide dissolves in concentrated aqueous ammonia forming a colourless solution
  ● yellow silver iodide does not dissolve in concentrated aqueous ammonia.

The colours of the silver halide precipitates are shown in Figure 9.5.

Figure 9.5 Colours of silver halide precipitates.

The equation for the precipitation of silver chloride is:

$$Ag^+(aq)+Cl^-(aq)\longrightarrow AgCl(s)$$

You will not be asked to write equations for the reactions of silver halide precipitates with ammonia for AS level chemistry.

Disproportionation reactions of chlorine

The reaction of chlorine with sodium hydroxide

The way in which chlorine reacts with aqueous sodium hydroxide depends on the temperature.

With *cold* (15 °C) dilute aqueous sodium hydroxide a mixture of chloride (Cl^-) and chlorate(I) (ClO^-) ions is formed:

$$Cl_2(g)+2NaOH(aq)\longrightarrow$$
$$NaCl(aq)+NaClO(aq)+H_2O(l)$$

This is an interesting reaction because it demonstrates **disproportionation** – a particular type of redox reaction in which the same species is oxidised and reduced at the same time. This happens to the chlorine – the ionic equation shows that the oxidation state of chlorine in the products of the reaction are both lower and higher than chlorine itself:

This reaction is used commercially to produce bleach, which is known as chloric(I) acid or hypochlorous acid, HClO. You can see this name on some bleach products. Household bleach is an aqueous solution of sodium chloride and sodium chlorate(I), NaClO, in a one-to-one mole ratio.

SAQ

6 Bromine reacts with cold dilute alkali to give a colourless solution containing Br^- and BrO^- ions.
   a Write a balanced chemical equation for this reaction.
   b What are the oxidation states of bromine before and after this reaction?
   c Name this type of reaction. Answer

The purification of drinking water

When chlorine is used to treat drinking water, disproportionation again occurs to form hydrochloric acid and chloric(I) acid:

$$Cl_2+H_2O\longrightarrow HCl+HClO$$

oxidation state of Cl   0   –1   +1

Bacteria in the water are killed by reactive oxygen atoms which are produced by a slow decomposition of the chloric(I) acid:

$$HClO\longrightarrow HCl+O$$

Summary

Glossary

- The halogens chlorine, bromine and iodine are covalent diatomic molecules at room temperature. They become increasingly less volatile and more deeply coloured on descending Group 7. Their boiling points increase as van der Waals' forces increase going down the group.

- The halogens have many characteristics in common. They are all reactive, but this reactivity decreases on descending the group.

- The halogens all have important industrial uses, especially chlorine, which is used in the manufacture of many other useful products. Possibly the most important use of chlorine is in the prevention of disease by chlorination of water supplies.

- All the halogens are good oxidising agents. Chlorine is a stronger oxidising agent than bromine or iodine.

- The order of reactivity can be determined by displacement reactions. A more reactive halogen can displace a less reactive halogen from a solution of one of its salts.

- The identification of a halide ion in solution is made after adding silver nitrate solution and then aqueous ammonia.

- Chlorine reacts with cold hydroxide ions in a disproportionation reaction. This reaction produces commercial bleach.

Questions

1 The halogens chlorine, bromine and iodine each exist as diatomic molecules at room temperature and pressure.
   a Draw a dot-and-cross diagram of a bromine molecule, showing outer electrons only.   [1]
   b The boiling points of the halogens chlorine to iodine are shown below.

Hint

halogen	boiling point /°C
chlorine	−35
bromine	59
iodine	184

   Explain why the halogens show this trend in boiling points.   [3]
   c When chlorine, Cl_2, is added to aqueous sodium bromide, NaBr, a reaction takes place.
      i   State what you would see in this reaction.   [1]
      ii   Write an equation for this reaction.   [1]
      iii What happens to electrons during this reaction?   [2]
      iv   Why does no reaction take place when bromine is added to aqueous sodium chloride?   [1]
      v   Describe a simple test to confirm the presence of iodide ions in aqueous sodium iodide.   [2]

OCR Chemistry AS (2811) June 2003   [Total 11]

Answer

continued

2   Chlorine is used in the preparation of many commercially important materials such as bleach and iodine.
   a   Bleach is a solution of sodium chlorate(I), NaOCl, made by dissolving chlorine in aqueous sodium hydroxide.

   $$Cl_2(g) + 2NaOH(aq) \longrightarrow NaOCl(aq) + NaCl(aq) + H_2O(l)$$

   Determine the changes in oxidation number of chlorine during the preparation of bleach and comment on your results.   [3]
   b   Iodine is extracted commercially from seawater with chlorine gas. Seawater contains very small quantities of dissolved iodide ions, which are oxidised to iodine by the chlorine gas.

        Hint

   i   Write an ionic equation for the reaction that has taken place.   [2]
   ii Use your understanding of electronic structure to explain why chlorine is a stronger oxidising agent than iodine.   [2]

OCR Chemistry AS (2811) June 2005   [Total 7]

     Answer

3   Aqueous silver nitrate can be used as a test for halide ions. A student decided to carry out this test on a solution of magnesium chloride. The bottle of magnesium chloride that the student used showed the formula $MgCl_2.6H_2O$.
   a   The student dissolved a small amount of $MgCl_2.6H_2O$ in water and added aqueous silver nitrate to the aqueous solution.

        Hint

   i   What is the molar mass of $MgCl_2.6H_2O$?   [1]
   ii   What would the student see after adding the aqueous silver nitrate, $AgNO_3(aq)$?   [1]
   iii Write an ionic equation for this reaction. Include state symbols.   [2]
   iv   Using aqueous silver nitrate, it is sometimes difficult to distinguish between chloride, bromide and iodide ions. How can aqueous ammonia be used to distinguish between these three ions?   [3]
   b   Domestic tap water has been chlorinated.
       Chlorine reacts with water as shown below.

   $$Cl_2(g) + H_2O(l) \longrightarrow HOCl(aq) + HCl(aq)$$

   When carrying out halide tests with aqueous silver nitrate, it is important that distilled or deionised water is used for all solutions, rather than tap water. Suggest why.   [1]

OCR Chemistry AS (2811) June 2006   [Total 8]

     Answer

continued

4 Chlorine can be prepared by reacting concentrated hydrochloric acid with manganese(IV) oxide.

$$4HCl(aq) + MnO_2(s) \longrightarrow Cl_2(g) + MnCl_2(aq) + 2H_2O(l)$$

a A student reacted 50.0 cm³ of 12.0 mol dm⁻³ hydrochloric acid with an excess of manganese(IV) oxide.

> Hint

   i Calculate how many moles of HCl were reacted. [1]

   ii Calculate the volume of $Cl_2(g)$ produced, in dm³.

      Under the experimental conditions, one mole of $Cl_2(g)$ occupies 24.0 dm³. [2]

b In this reaction, chlorine is oxidised.

   Use oxidation numbers to determine what is reduced. [2]

c Sodium reacts with chlorine forming the ionic compound sodium chloride, NaCl.

   i Write an equation, including state symbols, for this reaction. [2]

   ii Describe the structure of sodium chloride in the solid state. You may find it useful to draw a diagram. [2]

d *In this question one mark is available for the quality of spelling, punctuation and grammar.*

   Chlorine gas was bubbled through an aqueous solution of bromide ions and also through an aqueous solution of iodide ions. An organic solvent was then added and each mixture was shaken.

   ● State what you would see in each case.

   ● Write equations for any chemical reactions that take place.

   ● State and explain the trend in reactivity shown by these observations. [6]

   Quality of written communication [1]

OCR Chemistry AS (2811) Jan 2007   [Total 16]

> Answer

Chapter 10

Basic concepts in organic chemistry

Objectives

'Chains, Energy and Resources' seeks to provide you with a framework for the study of *organic* chemistry. Organic chemistry includes the study of compounds containing carbon and hydrogen *only* (that is, **hydrocarbons**, see Figure 10.1) and of compounds containing other elements *in addition* to carbon and hydrogen. This branch of chemistry is called 'organic chemistry' because it includes many of the complicated compounds essential to living organisms.

A study of organic chemistry can lead to the study of the chemistry of life itself – biochemistry.

Organic chemistry is a large subject, mainly because one of its 'essential ingredients', carbon, forms a much greater number and variety of compounds than any other element. Over 90% of known compounds contain carbon, despite the existence of several elements with much greater natural abundance.

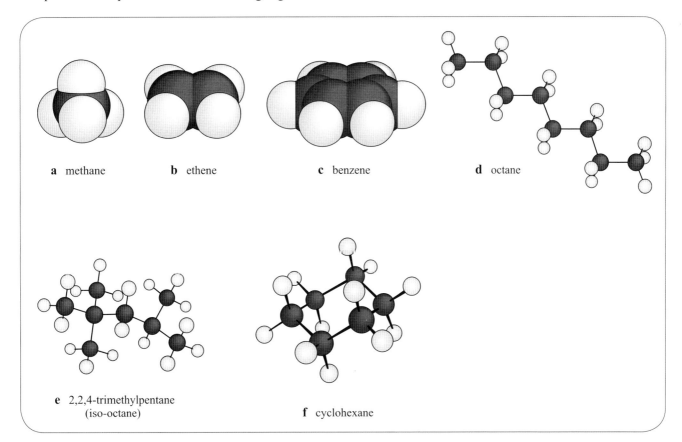

a methane   **b** ethene   **c** benzene   **d** octane

e 2,2,4-trimethylpentane
(iso-octane)

f cyclohexane

Figure 10.1 Examples of the variety of hydrocarbons. Chemists use various types of model for different purposes. The colours used in the modelling of molecules are shown in Table 10.1.
a–c These hydrocarbons are shown as space-filling models. Such models show the region of space occupied by the atoms and the surrounding electrons.
d–f These hydrocarbons are shown as ball-and-stick models, which enable bonds between atoms to be clearly seen.

Reasons for the greater number and variety of carbon compounds include the following:

- carbon readily bonds to itself and to most other elements, including metals
- carbon can bond in a variety of ways giving rise to chains, rings and even cages of carbon atoms.

Single carbon–carbon and carbon–hydrogen bonds are relatively unreactive, therefore the chemical behaviour of many organic compounds is caused by the other groups in their molecules. These groups are called **functional group**s (Table 10.2).

Colour	Atom/electron cloud
white	hydrogen
dark grey	carbon
red	oxygen
blue	nitrogen
yellow-green	fluorine
green	chlorine
orange-brown	bromine
brown	phosphorus
violet	iodine
pale yellow	sulfur
yellow ochre	boron
pink	lone pair electron clouds
green	π-bond electron clouds

Table 10.1 Colours used in molecular modelling in this text.

Class of functional group	Structure of functional group	Name of example	Structural formula of example
alkenes	$\begin{array}{c} \diagdown \quad \diagup \\ C=C \\ \diagup \quad \diagdown \end{array}$	ethene	$CH_2=CH_2$
arenes	⬡	benzene	⬡
halogenoalkanes	–X, where X = F, Cl, Br, I	chloromethane	CH_3Cl
alcohols	–OH	methanol	CH_3OH
aldehydes	$\begin{array}{c} O \\ \parallel \\ -C \\ \diagdown \\ H \end{array}$	ethanal	CH_3CHO
ketones	$\begin{array}{c} O \\ \parallel \\ -C-C \\ \end{array}$	propanone	CH_3COCH_3
carboxylic acids	$\begin{array}{c} O \\ \parallel \\ -C \\ \diagdown \\ OH \end{array}$	ethanoic acid	CH_3COOH
esters	$\begin{array}{c} O \\ \parallel \\ -C \\ \diagdown \\ O-C- \end{array}$	ethyl ethanoate	$CH_3COOC_2H_5$
amines	$-NH_2$	methylamine	CH_3NH_2

Table 10.2 Functional groups you will meet in the following chapters.

Types of formulae

As well as using different types of model to help visualise molecules, chemists also use different formulae. These include the following.

- *Molecular formula*

  This simply shows the number of atoms of each element present in the molecule, e.g. the molecular formula of hexane is C_6H_{14}.

- *General formula*

  A general formula may be written for each series of compounds. For example, the general formula for the alkanes is C_nH_{2n+2} (where *n* is the number of carbon atoms present).

- *Structural formula*

  This shows how the atoms are joined together in a molecule. The structural formula of hexane is $CH_3CH_2CH_2CH_2CH_2CH_3$. More information is conveyed. Hexane is seen to consist of a chain of six carbon atoms; the carbon atoms at each end are joined to one carbon and three hydrogen atoms; the carbon atoms between the two ends are joined to two hydrogen and two carbon atoms.

- *Displayed formula*

  This shows all the bonds and all the atoms. The displayed formula of hexane is:

  Displayed formulae are also called full structural formulae. One of their disadvantages is that they are a two-dimensional representation of molecules which are three-dimensional. Compare the displayed formula for hexane with the model in Figure 10.5a.

- *Skeletal formula*

  This shows the carbon skeleton only. Hydrogen atoms on the carbon atoms are omitted. Carbon atoms are not labelled. Other types of atom are shown as in a structural formula. Skeletal formulae are frequently used to show the structures of cyclic hydrocarbons. The skeletal formula of hexane is:

- *Three-dimensional formula*

  This formula gives the best representation of the shape of a molecule.

Examples of the different types of formula for the amino acid phenylalanine are shown in Figure 10.2. Phenylalanine is a common, naturally occurring amino acid.

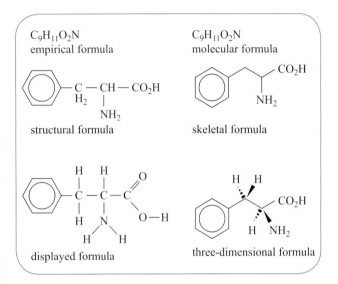

Figure 10.2 Different types of formula for phenylalanine. Note that the skeletal form of the phenyl ring ($-C_6H_5$) is acceptable in all of these formulae.

Functional groups

Organic chemistry can be studied in a particularly structured and systematic manner because each different group of atoms that becomes attached to carbon has its own characteristic set of reactions. Chemists refer to these different groups of atoms as *functional groups*. The functional groups that you will meet in the following chapters are shown in Table 10.2.

Table 10.2 provides you with the classes and structures of these functional groups. An example is also provided of a simple molecule containing each functional group. Each functional group gives rise to a **homologous series** (molecules with the same functional group but different length carbon chains). For example, the alcohol functional group gives rise to the homologous series of alcohols. The first four of these are methanol (CH_3OH), ethanol (CH_3CH_2OH),

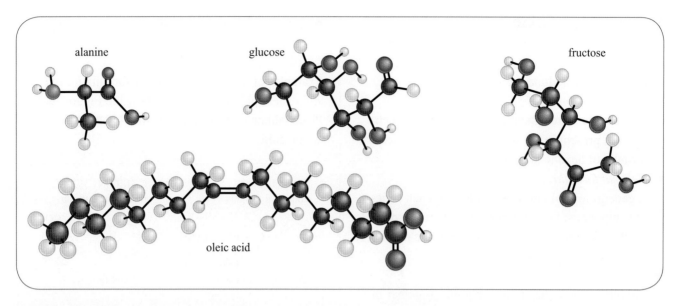

Figure 10.3 Ball-and-stick models of some naturally occurring molecules.

propan-1-ol ($CH_3CH_2CH_2OH$) and butan-1-ol ($CH_3CH_2CH_2CH_2OH$). The members of a homologous series all have similar chemical properties.

A **general formula** may be written for each homologous series. For example, the general formula of the aliphatic alcohols is $C_nH_{2n+1}OH$ (where n is the number of carbon atoms present).

Molecular modelling

Figure 10.3 shows a range of naturally occurring molecules. They are computer-produced images of ball-and-stick molecular models. In such models, atoms are shown as spheres. A single bond is represented by a line and a double bond by two lines. Different elements are distinguished by colour, as shown in Table 10.1.

SAQ

1 Draw the structural formulae for the molecules shown in Figure 10.3. Identify and label the functional groups present.

Answer

Various computer-produced images of molecular models will be used where appropriate throughout this book. Another type that will be used is a space-filling model. In space-filling models, atoms are shown including the space occupied by their electron

orbitals. As their orbitals overlap significantly, these images are very different image from the ball-and-stick images. Figure 10.4 shows these two types of model for alanine.

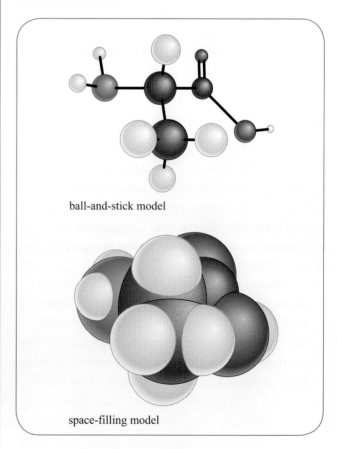

ball-and-stick model

space-filling model

Figure 10.4 Different model types for alanine.

Naming organic compounds

The names given in Figure 10.5 are the systematic names for the structural isomers of hexane (see page 113 for more about isomerism).

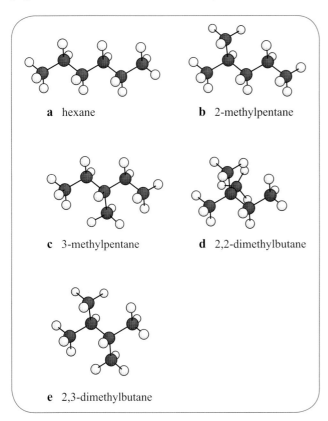

a hexane **b** 2-methylpentane

c 3-methylpentane **d** 2,2-dimethylbutane

e 2,3-dimethylbutane

Figure 10.5 Models and systematic names of C_6H_{14}.

Such names precisely describe the structure of a molecule and enable chemists to communicate clearly. International rules have been agreed for the systematic naming of most compounds. The basic rules for naming hydrocarbons are as follows.

1 The number of carbon atoms in the longest chain provides the stem of the name. Simple alkanes consist entirely of unbranched chains of carbon atoms. They are named by adding -ane to this stem as shown in Table 10.3.

Molecular formula	Number of carbon atoms in longest chain	Stem	Name
CH_4	1	meth-	methane
C_2H_6	2	eth-	ethane
C_3H_8	3	prop-	propane
C_4H_{10}	4	but-	butane
C_5H_{12}	5	pent-	pentane
C_6H_{14}	6	hex-	hexane
C_7H_{16}	7	hept-	heptane
C_8H_{18}	8	oct-	octane
C_9H_{20}	9	non-	nonane
$C_{10}H_{22}$	10	dec-	decane
$C_{20}H_{42}$	20	eicos-	eicosane

Table 10.3 Naming simple alkanes.

2 Branched-chain alkanes are named in the same way. The name given to the longest continuous carbon chain is then prefixed by the names of the shorter side-chains. The same stems are used with the suffix -yl. Hence $CH_3–$ is methyl (often called a methyl group). In general, such groups are called alkyl groups. The position of an alkyl group is indicated by a number. The carbon atoms in the longest carbon chain are numbered from one end of the chain. Numbering starts from the end that produces the lowest possible numbers for the side-chains. For example:

$$CH_3CHCH_2CH_2CH_3$$
$$|$$
$$CH_3$$

is 2-methylpentane, not 4-methylpentane.

3 Each side-chain must be included in the name. If there are several identical side-chains, the name is prefixed by di-, tri-, etc. For example 2,2,3-trimethyl- indicates that there are three methyl groups, two on the second and one on the third carbon atom of the longest chain. Note that numbers are separated by commas, whilst a number and a letter are separated by a hyphen.

4 Where different alkyl groups are present, they are placed in alphabetical order as in 3-ethyl-2-methylpentane.

5 Hydrocarbons may contain alkene or arene groups. These are represented as follows:

alkene: ethene

arene: benzene

displayed formulae   skeletal formulae

Hydrocarbons containing one double bond are called alkenes. The same stems are used but are followed by -ene. The position of an alkene double bond is indicated by the lower number of the two carbon atoms involved. This number is placed between the stem and -ene. Hence $CH_3CH=CHCH_3$ is but-2-ene.

6 The simplest arene is benzene. When one alkyl group is attached to a benzene ring, a number is not needed because all the carbon atoms are equivalent. Two or more groups will require numbers. For example:

methylbenzene   1,2-dimethylbenzene   1,4-dimethylbenzene

7 Halogeno compounds are named in the same way as alkyl-substituted alkanes or arenes:

$$CH_3CH_2CHBrCH_3$$
2-bromobutane

8 Aliphatic alcohols and ketones are named in a similar way to alkenes:

$CH_3CH_2CH_2OH$ $CH_3CH_2COCH_2CH_3$
propan-1-ol   pentan-3-one

9 Aliphatic aldehyde and carboxylic acid groups are at the end of a carbon chain, so they do not need a number. There is only one possible butanoic acid, $CH_3CH_2CH_2COOH$, or butanal, $CH_3CH_2CH_2CHO$. The names of ketones, aldehydes and carboxylic acids include the carbon atom in the functional group in the stem. Hence CH_3COOH is ethanoic acid.

10 Amines are named using the alkyl- or aryl- prefix followed by -amine. Hence $CH_3CH_2NH_2$ is ethylamine.

SAQ

2 Represent the compound 2-chloro-2-methylpropane by means of the following types of formulae:
a displayed   b structural
c skeletal   d molecular
e three-dimensional.

Answer

3 a Name the following compounds.

A $CH_3CH_2CH_2CH_2CH_2CH_2CH_3$

B

C

D

E

F

b Draw structural formulae for the following compounds:
    i propanal
    ii propan-2-ol
    iii 2-methylpentan-3-one
    iv propylamine.

                (Answer)

4 Draw displayed formulae for the following compounds:
    a 2,2,3-trimethylbutane
    b 3-ethylpent-2-ene
    c ethylbenzene.

                (Answer)

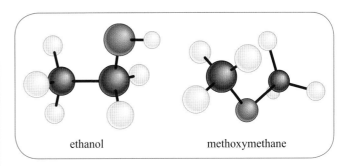

butane

methylpropane

SAQ

5 Copy the displayed formulae of the structural isomers of butane. Label each isomer with its appropriate boiling point.

                (Answer)

Isomerism

Most organic compounds have a molecular formula that is the same as one or more other compounds. These 'other compounds' consist of molecules made of the same atoms, but arranged in different ways. This property is called isomerism. **Isomers** have the same molecular formula but different structural formulae. Isomerism arises for a number of reasons, including the ability of carbon to bond to itself and to many other elements in the Periodic Table. The atoms present in a given molecular formula may be treated rather like Lego®, in that a given number of different pieces may be put together in a variety of ways.

Structural isomerism

Structural isomerism describes the situation where chemicals with the same formula behave differently because the structures of their molecules are different.

For example, the atoms of butane, C_4H_{10}, can be put together in two different ways. Try building models of these two isomers (or use a molecular modelling program on a computer). The two isomers behave in a very similar way chemically. The most noticeable difference in their properties is their boiling points. One isomer is more compact. This reduces the intermolecular forces as the molecules cannot approach each other so closely. This isomer has a boiling point of −11.6 °C. The isomer which is less compact has a boiling point of −0.4 °C. The displayed formulae and the names of these two structural isomers are:

The molecular formula C_2H_6O provides a very different example of structural isomerism. It has two isomers: ethanol, C_2H_5OH, and methoxymethane, CH_3OCH_3. Molecular models of these two isomers are shown in Figure 10.6.

ethanol   methoxymethane

Figure 10.6 Structural isomers of C_2H_6O.

Ethanol is an alcohol whilst methoxymethane is the simplest member of the homologous series of ethers, which are characterised by a –COC– group. As they contain different functional groups, they have very different chemical and physical properties. Ethanol is able to form intermolecular hydrogen bonds; methoxymethane has weaker dipole–dipole intermolecular forces. Consequently the boiling point of ethanol (78.5 °C) is considerably higher than that of methoxymethane (−25 °C). Alcohols take part in many different reactions. Ethers, apart from being highly flammable, are relatively inert.

It is quite easy to mistake the flexibility of molecular structures for isomerism. For example, if you build a model of pentane, C_5H_{12}, you will find it is very flexible (Figure 10.7 shows three of the possibilities).

The flexibility of a carbon chain arises because atoms can rotate freely about a carbon–carbon single bond. You should be careful when drawing displayed formulae of isomers. The following structures are not isomers, they are actually the same molecule.

$$H-\overset{\overset{\displaystyle H}{|}}{\underset{\underset{\displaystyle H}{|}}{C}}-\overset{\overset{\displaystyle H}{|}}{\underset{\underset{\displaystyle H}{|}}{C}}-\overset{\overset{\displaystyle H}{|}}{\underset{\underset{\displaystyle H}{|}}{C}}-\overset{\overset{\displaystyle H}{|}}{\underset{\underset{\displaystyle H}{|}}{C}}-\overset{\overset{\displaystyle H}{|}}{\underset{\underset{\displaystyle H}{|}}{C}}-H$$

Compare the displayed formulae with the models in Figure 10.7. Displayed formulae give a false impression of these structures. Remember that there is a tetrahedral arrangement of atoms round each carbon atom (with bond angles of 109.5°, not 90° as it appears in displayed formulae).

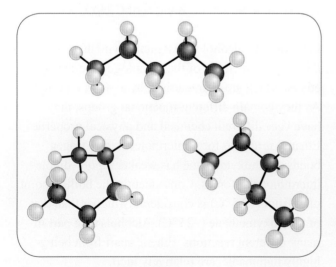

Figure 10.7 Models showing the flexibility of pentane. These forms are the same molecule, they are not isomers. The flexibility is due to the free rotation about the C–C single bond.

6 Draw the displayed formulae for the five structural isomers of hexane. Name each one.

Answer

Stereoisomerism

In stereoisomerism, the same atoms are joined to each other in different spatial arrangements. There are two types of stereoisomerism – Z/E and optical isomerism. Optical isomerism is studied in *Chemistry 2*.

Z/E isomerism

Alkenes are hydrocarbons containing one or more double bonds between carbon atoms.

Whilst atoms on either side of a carbon–carbon single bond can rotate freely, those either side of a carbon–carbon double bond cannot. This means that stereoisomerism is possible if there are two different groups (one of which might be an H atom) attached to each carbon atom in the C=C bond. Look at the example of 1,2-dibromoethene below (Figure 10.8).

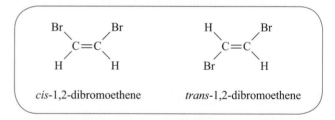

cis-1,2-dibromoethene   *trans*-1,2-dibromoethene

Figure 10.8 The two stereoisomers cannot be superimposed on each other. Try this for yourself with molecular models.

If the stereoisomer has two H atoms, one attached to each carbon of the double bond, then in this simple case the isomers can be distinguished by the prefix *cis*- or *trans*- in front of their name. In Figure 10.8 both Br atoms are on the same side of the double bond in *cis*-1,2-dibromoethene. They are on opposite sides in *trans*-1,2-dibromoethene. (Remember that the *trans*- isomer has the same group diagonally 'across' the double bond.)

In more complex stereoisomers a system has had to be devised to distinguish the different isomers by their names. This is based on a set of group priority rules (you don't need to know them!), which assigns a *Z* (from the German word 'zusammen' meaning 'together') or *E* (German for 'entgegen' meaning 'opposite') to the stereoisomers. In Figure 10.8, where *cis/trans* notation was adequate, *Z* is equivalent to *cis*, and *E* is equivalent to *trans*. Look at the *Z* and *E* stereoisomers of bromo-1,2-dichloroethene in Figure 10.9.

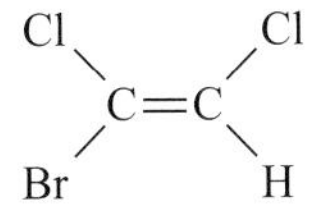

Z-bromo-1,2-dichloroethene

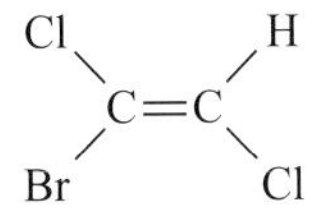

E-bromo-1,2-dichloroethene

Figure 10.9 The *Z/E* isomers of bromo-1, 2-dichloroethene

Try making models of but-2-ene, $CH_3CH=CHCH_3$. Two **Z/E isomers** are possible. You should obtain models similar to those shown in Figure 10.10. (In *cis*-but-2-ene or (*Z*)-but-2-ene, the methyl groups are on the *same* side of the double bond. In *trans*-but-2-ene or (*E*)-but-2-ene, they lie *across* the double bond.)

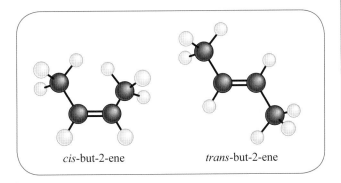

cis-but-2-ene   trans-but-2-ene

Figure 10.10 *Z/E* or *cis/trans* isomers of but-2-ene.

SAQ

7 **a** Draw the *cis/trans* isomers of 1,2-dichloroethene, CHCl=CHCl, and label them as *cis*- or *trans*-.

   b Copy the following structures and indicate which can exhibit *Z/E* isomerism by drawing the second isomer and labelling the two isomers as *Z* or *E* (or use *cis*- or *trans*- where appropriate).

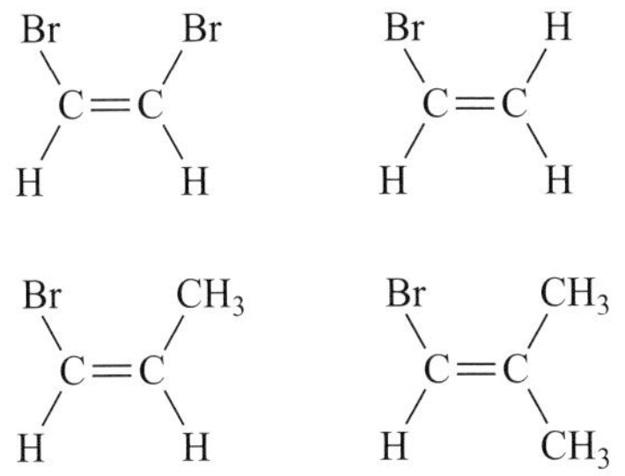

Answer

Organising organic reactions

There are several ways of organising the study of organic reactions. In this book the information is organised by functional group, so that subsequent chapters provide you with details of the typical reactions of the functional groups. Before you study these reactions, you need to know a little about the general types of reaction that occur.

You should be familiar with acid–base and reduction–oxidation (**redox**) reactions. Organic compounds frequently exhibit both these types of reaction. For example, ethanoic acid behaves as a typical acid, forming salts when reacted with alkalis such as aqueous sodium hydroxide:

$$CH_3COOH(aq) + NaOH(aq) \longrightarrow CH_3COONa(aq) + H_2O(l)$$

As with other acid–base reactions, a salt (sodium ethanoate) and water are formed.

Ethanol is readily oxidised in air to ethanoic acid (wine or beer soon become oxidised to vinegar if left exposed to the air):

$$CH_3CH_2OH(aq) + O_2(g) \longrightarrow CH_3COOH(aq) + H_2O(l)$$

In this redox reaction, oxygen is reduced to water.

There are several other types of reaction. These are substitution, addition, elimination and hydrolysis.

- **Substitution** involves replacing an atom (or a group of atoms) by another atom (or group of atoms). For example, the bromine atom in bromoethane is substituted by the –OH group to form ethanol on warming with aqueous sodium hydroxide:

$$CH_3CH_2Br(l) + OH^-(aq) \longrightarrow CH_3CH_2OH(aq) + Br^-(aq)$$

- **Addition** reactions involve two molecules joining together to form a single new molecule. If ethene and steam are passed over a hot phosphoric acid **catalyst**, ethanol is produced:

$$CH_2=CH2(g) + H_2O(g) \longrightarrow CH_3CH_2OH(g)$$

- **Elimination** involves the removal of a molecule from a larger one. The addition of ethene to steam may be reversed by passing ethanol vapour over a hot catalyst such as aluminium oxide. A water molecule is eliminated:

$$CH_3CH_2OH(g) \longrightarrow CH_2=CH_2(g) + H_2O(g)$$

- **Hydrolysis** reactions involve breaking covalent bonds by reaction with water. The substitution of the bromine atom in bromoethane (above) by hydroxide is also a hydrolysis. The reaction proceeds much more slowly in water:

$$CH_3CH_2Br(l) + H_2O(l) \longrightarrow CH_3CH_2OH(aq) + HBr(aq)$$

What is a reaction mechanism?

A balanced chemical equation shows the reactants and the products of a chemical change. It provides no information about the *reaction pathway*. The reaction pathway will include details of intermediate chemical species (molecules, radicals or ions) which have a short-lived existence between reactants and products. The **activation energy** for a reaction is the energy required to form these short-lived species (Figure 10.11a). You will learn more about activation

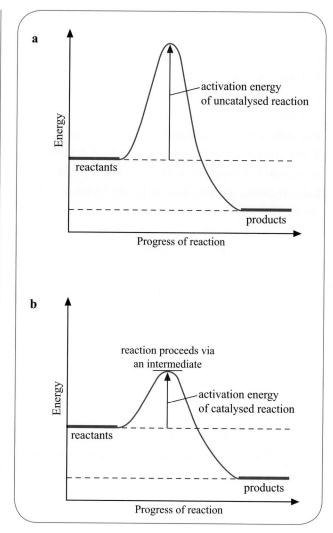

Figure 10.11 Activation energy diagrams for a reaction:
a without a catalyst **b** with a catalyst.

energy in Chapter 17. Catalysts are frequently used in reactions to increase the rate of reaction. They do this by providing an alternative reaction pathway with a lower activation energy (Figure 10.11b).

If a reaction pathway with a lower activation energy is found, more molecules will have sufficient kinetic energy to react. Catalysts take part in the reaction mechanism, but they are recovered unchanged at the end of the reaction. Hence the catalyst does not appear in the balanced chemical equation for the reaction.

A reaction mechanism is described using equations for the steps involved. You will meet the following organic mechanisms in this book:

● free-radical substitution (Chapter 11)
● electrophilic addition (Chapter 12)
● nucleophilic substitution (Chapter 14).

The following mechanisms are studied in *Chemistry 2*:

● electrophilic substitution
● nucleophilic addition.

The terms free radical, electrophilic and nucleophilic refer to the nature of the attacking species in the reaction. The attacking species is the reactant that starts a reaction by 'attacking' a bond on another reactant. These terms will be explained in the next section.

Breaking bonds in different ways

A covalent bond consists of a pair of electrons lying between the nuclei of two atoms. The negatively charged electrons attract both nuclei, binding them together. Such a bond may be broken in two different ways. We will consider these possibilities for hydrogen chloride. The dot-and-cross diagram for hydrogen chloride is:

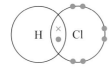

The bond may be broken so that each element takes one of the covalent bond electrons:

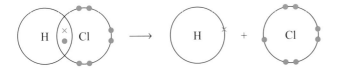

Each element now has a single unpaired electron. (In this example they have also become atoms.) Atoms (or groups of atoms) with unpaired electrons are known as **free radicals** (or radicals). When a covalent bond is broken to form two free radicals, the process is called **homolytic fission**. If a bond breaks homolytically, the energy is usually provided by ultraviolet light or high temperature.

An unpaired electron is represented by a dot. Using dots for the unpaired electrons, the homolytic fission of bromomethane to form a methyl radical and a bromine radical may be represented as follows:

$$H_3C\text{–}Br \longrightarrow CH_3\bullet + Br\bullet$$

Alternatively, a covalent bond may be broken so that one element takes both covalent bond electrons. Hydrogen chloride would form hydrogen ions and chloride ions:

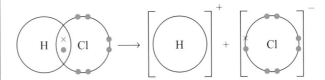

Notice that the more electronegative element takes both electrons. When a covalent bond is broken to form two oppositely charged ions, the process is called **heterolytic fission**. The bond in hydrogen chloride breaks heterolytically when the gas dissolves in water to form hydrochloric acid.

The movement of *two* electrons from the bond to the same atom is shown in diagrams of mechanisms by a curly arrow. Using curly arrows, the heterolytic fission of bromomethane to form a positive methyl ion and a bromide ion may be represented as follows:

$$H_3C\text{–}Br \longrightarrow CH_3^+ + Br^-$$

Positively charged ions that contain carbon, such as CH_3^+, are known as **carbocations**. (A negatively charged ion such as CH_3^- is known as a *carbanion*.)

Free radicals, carbocations and carbanions are all highly reactive species. They react with molecules, causing covalent bonds to break and new covalent bonds to form.

Carbocations and carbanions are examples of reagents known as electrophiles and nucleophiles respectively. An **electrophile** (electron-lover) is an electron pair acceptor which is attracted to an electron-rich molecule, leading to the formation of a new covalent bond between the electrophile and the molecule under attack. Electrophiles must be capable of accepting a pair of electrons. A **nucleophile** (nucleus-lover) is an electron pair donor which is attracted to an atom with a partial or full positive charge, leading to the formation of a new covalent bond between the nucleophile and the atom under attack. Nucleophiles must possess a lone pair of electrons for this new bond.

SAQ

8 Draw dot-and-cross diagrams for the following species: Br•, Cl$^-$, CH$_3^+$, CH$_3^-$, CH$_3$•, NH$_3$, BF$_3$. Classify them as free radicals, electrophiles or nucleophiles. What do you notice about the outer electron shells of free radicals, electrophiles and nucleophiles?

Hint

Answer

Practical techniques

Many organic reactions proceed slowly. They often require heating for a period of time. As the reaction mixtures often contain volatile reactants or solvent, the heating must be carried out under *reflux*. This means a condenser is placed in the neck of the reaction flask so that the volatile components are condensed and returned to the flask. Figure 10.12 illustrates the arrangement together with a cross-section diagram (of the type you might reproduce in an examination answer). Note that the water flows into the condenser at the lower (hotter) end. This provides the most rapid cooling of the vapour back to liquid. The liquid which is returned to the flask is still close to its boiling point.

The time required for the reflux period will depend on the rate of the reaction. Many reactions require a short period of reflux (perhaps 10 to 30 minutes); some reactions may require as long as 24 hours. The use of thermostatically controlled heating mantles (shown in the cross-section diagram in Figure 10.12) allows long refluxes to be carried out safely overnight.

After reflux, the reaction mixture is likely to consist of an equilibrium mixture containing both reactant and product molecules. These can often be separated by a simple *distillation*. A photograph of distillation apparatus appears in Figure 10.13.

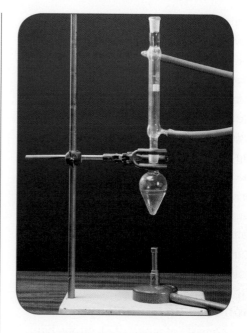

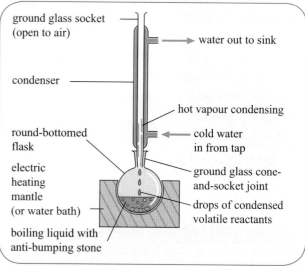

ground glass socket (open to air)

water out to sink

condenser

hot vapour condensing

round-bottomed flask

cold water in from tap

electric heating mantle (or water bath)

ground glass cone-and-socket joint

drops of condensed volatile reactants

boiling liquid with anti-bumping stone

Figure 10.12 The apparatus for carrying out a reaction under reflux.

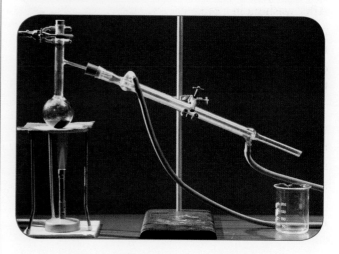

Figure 10.13 The apparatus for a distillation.

Compare the water flow with the flow for the reflux apparatus. For distillation, although the cold water enters at the lower end of the condenser (as with reflux), this entry point is further from the flask. The water not only condenses the vapour, but also cools the liquid to bring it close to room temperature.

After distillation, further purification may require washing the impure product with water in a separating funnel (Figure 10.14). This enables the separation of immiscible liquids.

Figure 10.14 A separating funnel enables the pink organic layer to be separated from the aqueous layer.

After washing in a separating funnel, the liquid may require drying. It is placed in a stoppered flask together with an anhydrous salt such as calcium chloride. This absorbs excess water. After drying, the liquid will need to be filtered and redistilled.

Where the product is a solid, distillation is inappropriate. Solid products may crystallise in the reaction flask or may be precipitated by pouring the reaction mixture into water. Rapid separation of the solid is achieved by vacuum filtration (Figure 10.15).

After separation, the solid is purified by recrystallisation from a suitable solvent. The aim of

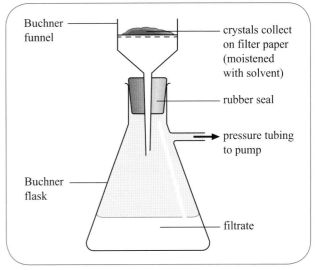

Figure 10.15 The apparatus required for a vacuum filtration.

recrystallisation is to use just enough hot solvent to completely dissolve all the solid. On cooling, the product crystallises, leaving impurities in solution. The purity of a compound may be checked by finding its melting point. A pure compound will usually have a sharp melting point. This means that the point from where it begins to soften to where it is completely liquid is a narrow range of temperature (1 or 2 °C). An impure compound will melt over a larger range of temperature. The melting-point apparatus shown in Figure 10.16 enables the melting of individual crystals to be seen.

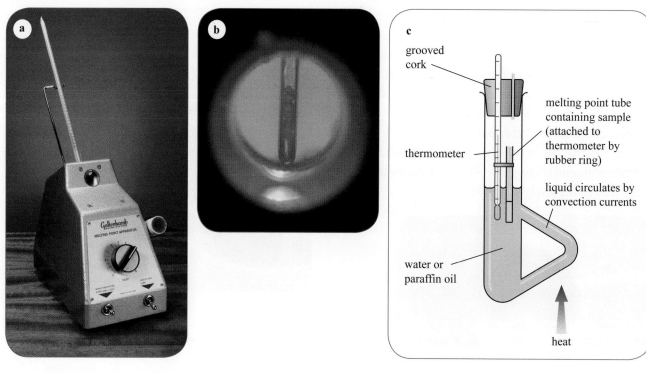

Figure 10.16 a A simple apparatus used to determine melting point.
b Close-up of crystals melting in the apparatus.
c Diagram of a Thiele tube apparatus, also used to determine melting point.

Calculation of percentage yields

We can calculate the mass of a product that can be made from given masses of starting materials using the balanced chemical equation and moles (see Chapter 2). This will be the theoretical yield, assuming maximum possible conversion of reactants to products. However, organic reactions often give yields much less than 100%. This is hardly surprising when the product is subjected to recrystallisation or distillation. Material is lost each time the product is transferred from one piece of equipment to another. In addition, many reactions produce equilibrium mixtures.

In order to find the actual yield of the product, you first calculate the maximum mass of product that you could obtain from the starting material. This may involve a preliminary calculation to decide if one or more of the reagents is in excess. If a reagent is in excess, the other reagent will limit the maximum yield of product. We will use the synthesis of aspirin as an example.

2.0 g of 2-hydroxybenzoic acid is refluxed with 5.5 g of ethanoic anhydride. The products are aspirin and ethanoic acid. The aspirin is easily separated as a solid. The equation for the reaction, together with the relative molecular masses of the compounds, is:

2-hydroxybenzoic acid $M_r = 138$
ethanoic anhydride $M_r = 102$
aspirin $M_r = 180$

$$\text{amount of 2-hydroxybenzoic acid used} = \frac{2.0}{138}$$

$$= 0.0145\,\text{mol}$$

$$\text{amount of ethanoic anhydride used} = \frac{5.5}{102}$$

$$= 0.054\,\text{mol}$$

As 0.0145 mol of 2-hydroxybenzoic acid requires only 0.0145 mol of ethanoic anhydride, a large excess of ethanoic anhydride has been used.

The reaction equation shows us that one mole of 2-hydroxybenzoic acid produces one mole of aspirin.

Hence maximum yield of aspirin

$$= \frac{2.0}{138} \times 180 = 2.6\,g$$

A student making aspirin whilst studying this module prepared 1.2 g of recrystallised aspirin. So his percentage yield was

$$\frac{1.2}{2.6} \times 100 = 46\%$$

SAQ

9 A student prepared a sample of 1-bromobutane, C_4H_9Br, from 10.0 g of butan-1-ol, C_4H_9OH (the other reagents being in excess). After purification she found she had made 12.0 g of 1-bromobutane. What was the percentage yield?

> Answer

Atom economy

Chemists working in industry are becoming increasingly aware of the need to conserve resources and reduce the waste produced in the reactions involved in a process. We can use the '**atom economy**' of a reaction to measure their success. It is defined as:

atom economy =

$$\frac{\text{molecular mass of the desired products}}{\text{sum of molecular masses of all products}} \times 100\%$$

A reaction with a high atom eceonomy is preferable because it produces a low percentage of waste products.

Look at Worked example 1.

This substitution reaction has a relatively low atom economy. The waste produced in this case is an acidic gas (HCl) that cannot be allowed to escape into the atmosphere from the processing plant. It has to be removed by neutralising it with sodium hydroxide solution. This reduces the efficiency of the process.

However, despite its low atom economy, the original reaction could still produce a high percentage yield of the desired product if it produces close to the maximum mass of $C_2H_2Cl_2$ possible. This would be calculated from the starting masses of reactants, the mass of $C_2H_2Cl_2$ produced and the balanced equation (see page 120).

Worked example 1

Poly(1,1-dichloroethene) is used as a plastic wrapping material. It is made from the monomer 1,1-dichloroethene. Its formula is $C_2H_2Cl_2$. The monomer can be produced by heating ethane (C_2H_6) with chlorine gas:

$$C_2H_6 + 3Cl_2 \longrightarrow C_2H_2Cl_2 + 4HCl$$

To calculate the atom economy we need to work out the relative molecular masses of the products formed in the reaction:

$$C_2H_2Cl_2 = (2 \times 12.0) + (2 \times 1.0) + (2 \times 35.5) = 97.0$$

$$4HCl = 4 \times (1.0 + 35.5) = 146.0$$

atom economy =

$$\frac{\text{molecular mass of the desired products}}{\text{sum of molecular masses of all products}} \times 100\%$$

$$\frac{\text{molecular mass of } C_2H_2Cl_2}{\text{sum of molecular masses of } C_2H_2Cl_2 + 4HCl} \times 100\%$$

$$= \frac{97.0}{97.0 + 146} \times 100\%$$

$$= 39.9\%$$

An example of a reaction in industry with a high atom economy is the manufacture of ethanol (C_2H_5OH) by heating ethene (C_2H_4) and steam in the presence of a catalyst:

$$C_2H_4 + H_2O \longrightarrow C_2H_5OH$$

The atom economy of this reaction is 100%, as the desired product (ethanol) is the only product. (In fact any addition reaction that gives only one product will have an atom economy of 100%.) This reaction is reversible but any unconverted ethene gas and steam can be recycled into the reactor continuously to reduce any wastage.

SAQ

10 a The monomer tetrafluoroethene (C_2F_4) can be used to make non-stick linings for pans. Calculate the atom economy of the following reaction used to make the monomer:

$$2CHF_3 \longrightarrow C_2F_4 + 2HF$$

b Why will any substitution reaction have an atom economy of less than 100% whereas an addition reaction can have an atom economy of 100%?

(Answer)

(Extension)

Summary

(Glossary)

- All organic compounds contain carbon and hydrogen. Most organic compounds also contain other elements, such as oxygen, nitrogen and chlorine.

- Chemists use a wide variety of formulae to represent organic molecules. These include general, empirical, molecular, structural, skeletal, displayed and three-dimensional formulae.

- Functional groups, which have their own characteristic reactions, are attached to the hydrocarbon framework of an organic molecule. Alkenes, arenes, halogen atoms, alcohols, aldehydes and ketones, carboxylic acids, esters and amines are examples of functional groups.

- Various types of molecular model (ball-and-stick, space-filling) are used to visualise organic molecules.

- Organic molecules are named in a systematic way, which relates to their structures.

- Organic molecules with the same molecular formula but with different structural formulae are called isomers. Two common types of isomerism are structural and Z/E isomerism. Structural isomers have different structural formulae and Z/E isomers have different displayed formulae.

- The study of organic reactions is traditionally organised by functional group. Each functional group has its own characteristic reactions.

- Reactions may also be studied by type or by mechanism. Organic compounds may show the following types of reaction: acid–base, redox, substitution, addition, elimination or hydrolysis.

- Reaction mechanisms may involve electrophiles, nucleophiles or free radicals. Each of these reagents is capable of forming a new covalent bond to the atom attacked. Electrophiles are electron-pair acceptors. Nucleophiles are electron-pair donors. Free radicals have an unpaired electron.

- Covalent bonds may be broken homolytically to form two free radicals, each with an unpaired electron.

- Covalent bonds may be broken heterolytically to form one cation and one anion.

- A curly arrow shows the movement of two electrons in a reaction mechanism.

- Practical techniques used in the preparation of organic compounds include reflux, distillation, vacuum filtration, separation of immiscible liquids in a separating funnel and recrystallisation.

continued

- Most organic preparations involve equilibrium reactions and/or lead to losses of product during separation and purification. The percentage yield indicates the proportion of the maximum yield that has been obtained.

- In order to reduce waste, industrial chemists aim to use reactions with a high atom economy as calculated by the equation:

$$\text{atom economy} = \frac{\text{molecular mass of the desired products}}{\text{sum of molecular masses of all products}} \times 100\%$$

Questions

1 The structures of some saturated hydrocarbon compounds are shown below. They are labelled **A**, **B**, **C**, **D** and **E**.

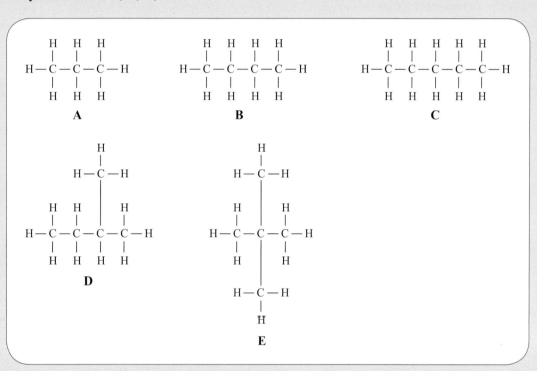

a The general formula of an alkane is C_nH_{2n+2}.

    i **A**, **B** and **C** are successive members of the alkane series.

       What is the *molecular* formula of the next member in the series? [1]

    ii What is the *empirical* formula of compound **A**? [1]

b Three of the compounds **A** to **E** are structural isomers of each other.

    i Identify, by letter, which of the compounds are the three structural isomers. [1]

    ii Explain what is meant by the term *structural isomer*. [1]

OCR Chemistry AS (2812) June 2001   [Total 4]

Answer

continued

2 a Explain, using examples, the following terms used in organic chemistry:
  i *general formula*   [2]
  ii *homologous series*.   [2]
  b The Cl–Cl bond can be broken either by homolytic fission or by heterolytic
    fission. Explain, with the aid of suitable equations, what you understand by
    the terms *homolytic fission* and *heterolytic fission*.   [4]

[Total 8]

Answer

3 a An organic compound of bromine, **X**, has a molecular mass of 136.9 and the
    following percentage composition by mass: C, 35.0%; H, 6.6%; Br, 58.4%.
  i Calculate the empirical formula of compound **X**.   [2]   Hint
  ii Show that the molecular formula of **X** is the same as the empirical formula.   [1]
  b i Draw displayed formulae for all the possible structural isomers of **X**.   [4]   Hint
    ii Name the structural isomers that you have drawn in **i**.   [4]

[Total 11]

Answer

Chapter 11

Alkanes

Background

e-Learning

Objectives

Physical properties of alkanes

As the name suggests, **hydrocarbons** are compounds containing only carbon and hydrogen. The **homologous series** (see Chapter 10) of alkanes are hydrocarbons with the general formula C_nH_{2n+2}. (NB This general formula works for straight and branched-chain alkanes but not for cycloalkanes.)

Alkanes are non-polar molecules containing only C–H and C–C covalent bonds. As all the C–C bonds are single bonds, alkanes are described as saturated hydrocarbons. Unsaturated hydrocarbons, such as the *alkenes* (Chapter 12), contain one or more double bonds between carbon atoms.

The geometry of alkane molecules is based on the tetrahedral arrangement of four single bonds round each carbon atom. All bond angles are 109.5°. The molecules can rotate freely about each carbon–carbon single bond. This freedom to rotate allows a great degree of flexibility to alkane chains. The shape of an ethane molecule is shown in Figure 11.1.

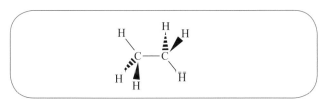

Figure 11.1 The shape of an ethane molecule.

The physical states of alkanes at room temperature and pressure change, from gases to liquids to solids, as the number of carbon atoms in the alkane molecule increases. We say that the **volatility** of the alkanes decreases with increasing number of carbon atoms in the alkane molecule. Volatility is the ease with which a liquid turns to vapour. We can examine this trend by plotting graphs of the melting or boiling points of alkanes against the number of carbon atoms present. Figure 11.2 shows such a graph of the melting points for the *straight-chain* alkanes, butane to dodecane and

eicosane. The term 'straight-chain' indicates that no branching is present in the molecule's carbon chain.

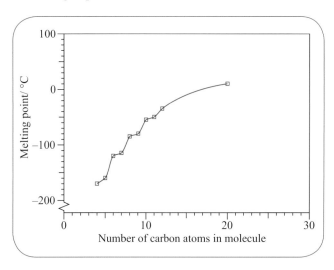

Figure 11.2 Melting points of straight-chain alkanes.

SAQ

1   Table 11.1 contains the boiling points of the first twelve alkanes and eicosane, $C_{20}H_{42}$. Plot these boiling points on the vertical axis of a graph against the number of carbon atoms present. Compare the shape of your graph with the melting point graph (Figure 11.2).

Answer

2   Volatility is determined by the strength of the intermolecular forces between alkane molecules.

   a   Name the type of intermolecular force found in alkanes, and explain why this type of intermolecular force is present.

Hint

   b   Explain the trends in the melting and boiling points of the alkanes in terms of the intermolecular forces.

Answer

Alkane	Molecular formula	Boiling point /K
methane	CH_4	109
ethane	C_2H_6	185
propane	C_3H_8	231
butane	C_4H_{10}	273
pentane	C_5H_{12}	309
hexane	C_6H_{14}	342
heptane	C_7H_{16}	372
octane	C_8H_{18}	399
nonane	C_9H_{20}	424
decane	$C_{10}H_{22}$	447
undecane	$C_{11}H_{24}$	469
dodecane	$C_{12}H_{26}$	489
eicosane	$C_{20}H_{42}$	617

Table 11.1 The boiling points of straight-chain alkanes.

Table 11.2 shows the boiling points of pentane and its two isomers. 2-methylbutane has one methyl group as a branch, 2,2-dimethylpropane has two methyl group branches. Again, a trend is apparent. As the number of branches in the chain increases, the boiling point decreases. The isomers all have the same number of carbon and hydrogen atoms, so we cannot explain the trend in terms of an increasing number of electrons. Look at the space-filling models for these isomers, also shown in the table.

As the isomers become more branched, the overall shape of the molecule changes from a long sausage shape (pentane) to a more spherical shape (2,2-dimethylpropane). The long sausage shape of pentane molecules allows them to pack more closely together than the more spherical shape of 2,2-dimethylpropane. The intermolecular forces increase when the molecules approach more closely so the boiling point of pentane is higher than that of 2,2-dimethylpropane. The boiling point of 2-methylbutane, with only one branching point in the carbon chain, lies between the other two isomers.

Crude oil

Crude oil is a complex mixture of hydrocarbons. The composition of oil from different places varies considerably (Figure 11.4 and Figure 11.5).

Three main series of hydrocarbons are present: arenes, cycloalkanes and alkanes. Arenes are hydrocarbons containing one or more benzene rings (see Chapter 10). At a given boiling point, the densities of these decrease in, the order arenes > cycloalkanes > alkanes. This provides a method for comparing the compositions of different oils.

Alkane	Structural formula	Boiling point /K	Space-filling model
pentane	$CH_3CH_2CH_2CH_2CH_3$	309	
2-methylbutane	CH_3 | $CH_3CHCH_2CH_3$	301	
2,2-dimethylpropane	CH_3 | CH_3CCH_3 | CH_3	283	

Table 11.2 The boiling points of pentane and its isomers. Note: To convert kelvin (K) to Celsius, subtract 273 from the kelvin value.

Exploiting fossil fuels

The fossil deposits of crude oil and natural gas have been the primary sources of alkanes since early in the 20th century. Much of the wealth of the modern industrialised world can be ascribed to the exploitation of this natural resource.

The vast majority of these deposits have been used to provide fuel for heating, electricity generation and transport. Smaller, but significant, proportions have been used to produce lubricants and to provide a source of hydrocarbons for the chemical process industry. In the UK, the chemical and petrochemical industries are by far the biggest contributors towards a positive balance in the value of manufacturing trade with the rest of the world. The UK chemical industry employs approximately 250 000 people and in 1996 produced about 11% of total industrial output. Figure 11.3 shows just how dependent the UK is on its chemical industry for its contribution to manufacturing exports.

However, the burning of fossil fuels has caused severe environmental problems on our planet which must be addressed urgently (see pages 133 to 136).

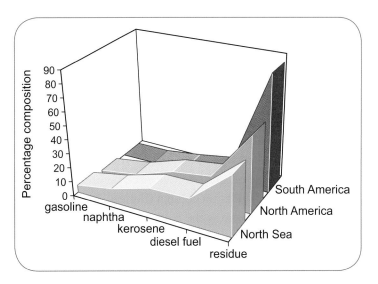

Figure 11.4 Breakdown of compositions of oils by oil fractions. North Sea oil contains a higher proportion of the gasoline and naphtha fractions than oils from North or South America.

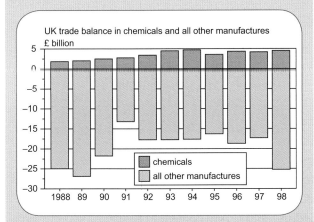

Figure 11.3 UK trade balance in chemicals and all other manufactures. The chemical industry is UK manufacturing's number one exporter. With exports of £22.5 bn and imports of £18.1 bn, it earned a trade surplus of £4.4 bn in 1998.

Figure 11.5 Californian crude is rich in cycloalkanes!

Separating the hydrocarbons in crude oil

Crude oil must be refined so that use can be made of the wide variety of hydrocarbons present. As crude oil is such a complex mixture, it is first broken down into *fractions*. This separation into fractions is known as *fractional distillation*. This technique relies on differences in boiling points of the different molecules in the liquid. Each fraction consists of a mixture of hydrocarbons with a much narrower range of boiling points than the full range of hydrocarbons in crude oil. Fractional distillation can be successful even where differences in boiling point are small. Further distillation processes in an oil refinery can be used to separate the hydrocarbons in a fraction. We shall consider fractional distillation in more detail by looking at one type of column for such a distillation.

The crude oil must be vaporised before it enters a fractional distillation column (Figure 11.6). This is achieved by passing the crude oil through pipes in a furnace where the oil is heated to about 350 °C.

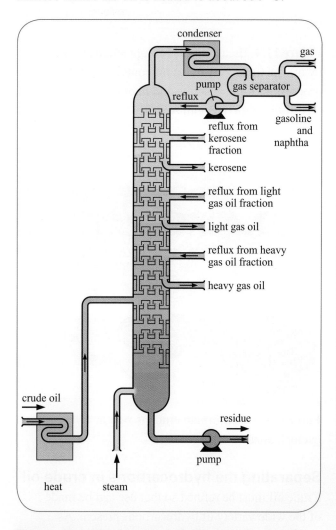

Figure 11.6 The fractional distillation of crude oil.

The resulting mixture of liquid and vapour is fed into the distillation column near the bottom. The column (which may be up to 100 m in height) is divided by a number of steel 'trays' (40 to 50 in a 100 m column). Vapour passes up the column through the trays *while*

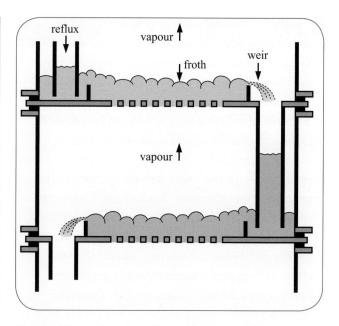

Figure 11.7 Trays in a fractionating column.

liquid flows down the column from tray to tray over a 'weir' (Figure 11.7).

There is a temperature gradient between the bottom of the column, which is hot, and the top of the column, which is cool. The most volatile hydrocarbons are collected from the top of the column; these hydrocarbons have the lowest boiling points. The least volatile hydrocarbons are collected from the bottom of the column; these hydrocarbons have the highest boiling points. From the bottom to the top of the column, increasingly volatile hydrocarbons will be found.

Once operating, a column may be kept in a *steady state* by maintaining the input of crude oil at a flow rate which balances the total of the flow rates at which the fractions are removed. When a steady state exists, the compositions of the liquid and vapour at any one tray do not vary. This enables the various fractions to be drawn from the column at appropriate points. An individual tray will contain a mixture of hydrocarbons with quite a narrow range of boiling points.

A fractional distillation column is designed to separate crude oil into several fractions. These include refinery gases, gasoline, naphtha, kerosene, gas oil (diesel oil), and residue.

● The refinery gases consist of simple alkanes containing up to four carbon atoms. They are used as fuels or as a source (or feedstock) for building other molecules.

- Gasoline contains alkanes with five to ten carbon atoms and is used as petrol.
- Naphtha is the fraction of crude oil which is the most important source of chemicals (or feedstock) for the chemical process industry.
  The other fractions are of lesser importance
- Kerosene is used for jet fuels and for domestic heating.
- Gas oil is used as diesel fuel and as a feedstock for catalytic cracking.
- The residue is used as a source of lubricating oils and waxes and bitumen. Bitumen mixed with crushed stone is the tarmac used for road surfaces.

Figure 11.8 shows fractional distillation columns in a modern oil refinery.

Figure 11.8 The skyline of an oil refinery is dominated by fractional distillation columns.

Further treatment

After distillation, the different hydrocarbon fractions are treated in a variety of ways. These include processes such as vacuum distillation (to separate out less volatile components such as lubricating oils and waxes from the residue), desulfurisation (to remove sulfur) and cracking (to produce more gasoline and alkenes). There are insufficient gasoline and naphtha fractions from the primary distillation to satisfy the demand for petrol, so higher boiling fractions are cracked to produce more gasoline and naphtha. Modern petrol engines require higher proportions of branched-chain alkanes, cycloalkanes and arenes to promote efficient combustion. These are produced by re-forming and isomerisation.

Cracking involves heating the oil fraction with a catalyst. Under these conditions, high-molecular-mass alkanes are broken down into low-molecular-mass alkanes as well as alkenes. Both C–C and C–H bonds are broken in the process. As the bond-breaking is a random process, a variety of products, including hydrogen, is possible and some of the intermediates can react to produce branched-chain alkane isomers. For example, a possible reaction equation for decane is:

$$CH_3CH_2CH_2CH_2CH_2CH_2CH_2CH_2CH_2CH_3 \longrightarrow$$
$$\text{decane}$$

$$CH_3CH_2CH=CH_2 \quad + \quad H_3C-\overset{\overset{\displaystyle H}{|}}{\underset{\underset{\displaystyle CH_3}{|}}{C}}-CH_2CH_2CH_3$$
$$\text{but-1-ene} \qquad\qquad\qquad \text{2-methylpentane}$$

The chemical industry uses alkenes to make a variety of products that include polymers. Branched alkanes are important components of petrol.

SAQ

3 Write balanced equations showing the structural formulae for four sets of possible products formed on cracking pentane.

(Answer)

In the catalytic cracker (Figure 11.9) the hot, vaporised oil fraction and the catalyst behave as a fluid. This is the result of the hot oil vapour providing lift and support for the finely divided solid catalyst particles. The seething mixture is called a *fluidised bed*. Unfortunately some of the hydrocarbon mixture is broken down to carbon, which blocks the pores of the catalyst. The fluidised bed of the catalyst is pumped into a regeneration chamber, where the carbon coke is burnt off in air at a high temperature, allowing the catalyst to be recycled.

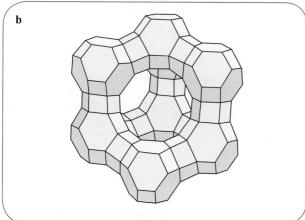

Figure 11.9

a A catalytic cracker occupies the bulk of the central part of this photograph.

b A computer graphic showing the framework of zeolite Y, a modern catalyst used to crack hydrocarbons.

Re-forming involves the conversion of alkanes to cycloalkanes, or of cycloalkanes to arenes. Re-forming reactions are catalysed by bimetallic catalysts. For example, a cluster of platinum and rhenium atoms is very effective at removing hydrogen from methylcyclohexane to form methylbenzene:

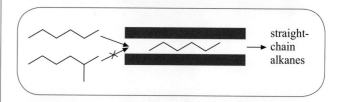

A catalyst containing clusters of platinum and iridium atoms enables conversion of straight-chain alkanes to arenes:

$$CH_3CH_2CH_2CH_2CH_2CH_3 \longrightarrow \bigcirc + 4H_2$$

These metal clusters are between 1 and 5 nm in diameter and are deposited on an inert support such as aluminium oxide. The rhenium and iridium help prevent the build-up of carbon deposits, which reduces the activity of the catalysts.

Isomerisation involves heating the straight-chain isomers in the presence of a platinum catalyst:

$$CH_3CH_2CH_2CH_2CH_2CH_3 \longrightarrow H_3C - \underset{\underset{CH_3}{|}}{\overset{\overset{CH_3}{|}}{C}} - CH_2CH_3$$

The resulting mixture of straight- and branched-chain isomers then has to be separated. This is done by using a molecular sieve, which is made of a type of zeolite that has pores through which the straight-chain isomers can pass (Figure 11.10).

Figure 11.10 Shape selectivity by a zeolite catalyst – separation of isomers by a molecular sieve.

The branched-chain isomers are too bulky and so are separated off; the straight-chain molecules are recycled to the reactor to be converted to branched-chain isomers.

Chemical properties of alkanes

Combustion in air

Alkanes make excellent fuels. When they burn, a large amount of heat energy is released per gram of fuel burnt. Complete combustion in an excess of air produces carbon dioxide and water. For example, butane is used as a fuel in camping-gas stoves. The equation for the complete combustion of butane is:

$$C_4H_{10}(g) + 6\tfrac{1}{2} O_2(g) \longrightarrow 4CO_2(g) + 5H_2O(l)$$

Natural gas, used for cooking and heating in many homes (Figure 11.11), is predominantly methane. Whether using natural gas or butane camping gas, it is important to ensure a good supply of air. If there is insufficient oxygen, there will be incomplete combustion, with the formation of carbon monoxide instead of carbon dioxide. For example, with methane the equation for complete combustion is:

$$CH_4(g) + 2O_2(g) \longrightarrow CO_2(g) + 2H_2O(l)$$

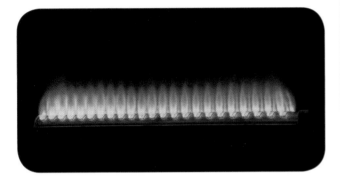

Figure 11.11 In a gas boiler the complete combustion of natural gas produces carbon dioxide and water.

The equation for incomplete combustion and formation of carbon monoxide is:

$$CH_4(g) + 1\frac{1}{2} O_2(g) \longrightarrow CO(g) + 2H_2O(l)$$

Carbon monoxide is a poisonous gas which bonds to the iron of haemoglobin in the blood in preference to oxygen. Carbon monoxide is colourless and odourless, so its presence is not obvious. Early symptoms of poisoning by carbon monoxide include the skin flushing red, headache and nausea. Many deaths result from the use of faulty gas fires in poorly ventilated rooms. It is important to have all gas equipment serviced annually and checked for carbon monoxide emissions. In the UK, legislation now requires adequate ventilation wherever there are gas installations. Property landlords are required by law to have their gas equipment checked annually.

SAQ

4 A principal component of petrol is an isomer of octane (C_8H_{18}).
 a Write balanced equations for the combustion of octane:
  i in a limited supply of air with the formation of carbon monoxide and water and
  ii in a supply of air which ensures complete combustion with the formation of carbon dioxide and water.
 b Using your equations from part **a**, calculate:
  i the additional number of moles of oxygen required   `Hint` to prevent the formation of carbon monoxide on combustion of one mole of octane and
  ii the additional volume of air required (assume one mole of a gas occupies 24.0 dm^3 and that air contains 20% oxygen).   `Answer`

Fuels

Many substances burn in reactions with oxygen, with transfer of energy to the surroundings (Figure 11.12). Only those used on a large scale, however, are properly described as fuels.

Figure 11.12 Gases from oilfields are often disposed of by being burnt as controllable 'flares'. This is a waste of gas but the costs of collection, storage and transportation are higher than the income available from selling the gas for other uses.

Fuel	Formula	Relative molecular mass	Energy released per mole /kJ mol^{-1}	Energy released per kilogram /kJ kg^{-1}
carbon (coal)	C(s)	12.0	−393	−32 750
methane	CH$_4$(g)	16.0	−890	−55 625
octane	C$_8$H$_{18}$(l)	114.0	−5512	−48 350
methanol	CH$_3$OH(l)	32.0	−715	−22 343
hydrogen	H$_2$(g)	2.0	−286	−143 000

Table 11.3 Comparison of fuels in terms of energy released.

Oxidation of the chemicals in fuels such as coal, petroleum and gas provides over 90% of the energy used in most industrialised countries; hydroelectricity and nuclear power together supply about 9%.

What makes a good fuel?

The essential reaction for any chemical fuel is:

fuel + oxygen (or other oxidiser) ⟶
           oxidation products + energy transfer

SAQ

5 Write balanced equations for the complete combustion of the fuels shown in Table 11.3.

Hint
Answer

Though different fuels are needed for different purposes, the ideal characteristics include the following.

● *A fuel should react with an oxidiser to release large amounts of energy.*
  It is interesting to compare fuels on the basis of energy per unit amount of material (mole) and energy per unit mass (kilogram) (Table 11.3). Remember that fuels are usually purchased in kilograms or tonnes, not in moles.

SAQ

6 From the data in Table 11.3, compare hydrogen with methane. Why are the values for the energy released per kilogram so different when compared with the energy released per mole?

Hint
Answer

● *A fuel must be easy to oxidise, ignite quickly and sustain burning without further intervention.*
  Gaseous or easily vaporised fuels usually perform well, as they mix easily and continuously with air/oxygen, which helps the reaction. Solid fuels like coal are sometimes powdered, which helps them to ignite and sustain burning.
● *A fuel should be readily available in large quantities and at a reasonable price.*
  The availability and price of oil, for example, affect national economies so much that governments can fall and countries go to war when they change.

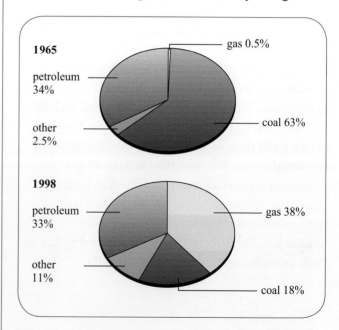

Figure 11.13 Changing use of fuel in Britain, 1965–98. Use of coal has greatly declined as more methane gas is used for domestic and industrial energy supply, and many more gas-fuelled power stations are built for generating electricity.

The price of any fuel includes many factors, including the costs of finding it, extraction, refining and transportation, and of course taxes. During the 1960s Britain changed the main gas supply from 'coal gas' (mainly hydrogen and carbon monoxide) to 'natural gas' (methane). This required a large-scale and expensive programme of adapting gas burners in industries and homes to suit the slower burning rate of methane. The advantages were that large supplies of methane were becoming available from gasfields near the British coast and that methane was thought to be a much 'cleaner' and safer fuel than the coal gas produced in dirty gas-works in most towns (Figure 11.13).

- *A burning fuel should not release products that are difficult to dispose of, or are unpleasant or harmful.* This is a considerable problem for most fuels (see below). Hydrogen is the only fuel with a safe, non-polluting product from its oxidation reaction.
- *A fuel should be convenient to store and should transport safely and without loss.*
Over the ages, people have tackled many problems of fuel storage, from how to keep wood dry to how to keep liquid oxygen extremely cold and safe for space flight (Figure 11.14).

Figure 11.14 A space shuttle is propelled by fuels including liquid oxygen.

If gases such as methane and hydrogen are to be used as alternatives to petrol in vehicles, the problems of storage of large amounts of gas must be solved. People are worried about storing these gases under high pressure in cylinders. Scientists are developing some interesting ways, however, of storing hydrogen as its solid compounds, such as the hydrides of metals: $FeTiH_2$ or $LaNi_5H_7$ or $MgNiH_4$. These hydrides release hydrogen when warmed gently and they may enable safe hydrogen-fuelled motor transport.

SAQ

7 **a** Why is oxygen transported into space in liquid form instead of as gas?
   b Why are large quantities of liquid or gaseous fuels often stored in spherical tanks?

Hint

Answer

Problems with chemical fuels

Reliance upon the main chemical fuels (coal, oil and gas) is increasingly a matter of worldwide concern.

These fuels are 'fossil fuels', formed over millions of years. They are, in effect, non-renewable resources, yet we are consuming them extremely quickly. It is predicted that most of the Earth's oil reserves will be consumed over the next hundred years. Britain's oil- and gasfields will be empty long before that, if used at the present rate.

Fossil fuels are also the raw materials that supply the feedstock for most of our chemical industry. They are processed by distillation, cracking and re-forming to yield carbon-based compounds which can then be made into polymers, medicines, solvents, adhesives, and many other things. How long can we afford to carry on burning such a useful source of chemicals?

Oxidation of the carbon-based compounds in fuels produces vast amounts of carbon dioxide (CO_2). At one time carbon dioxide was considered to be a relatively harmless gas. Now it is known to be a major contributor to the 'greenhouse effect', which causes an increase in atmospheric temperatures (see Chapter 18). Some governments are so concerned about this effect, which could bring about disastrous climatic change, that many means of reducing carbon dioxide levels in the atmosphere are being considered. Britain set itself a target of reducing CO_2 emissions by 35% of the 1992 levels by the year 2000, and lower targets have been set since then. The simplest solution would be an outright ban on the use of coal, oil and methane, but this would be impractical. The established industrial economies in the western world depend heavily on these fuels. New economies in countries such as India and China also rely heavily on fossil fuels to power their development. However, we may see increasing 'carbon taxes' (extra taxes levied upon the use of carbon-based fuels) and other means of restricting their use.

Spillage of fuel often causes great damage to local environments (Figure 11.15). This damage ranges from streams and ponds polluted by leaky fuel tanks to major disasters when oil tankers break open. There can be immense loss of animal and plant life and enormous costs of cleaning up.

Figure 11.15 The oil tanker *Braer* broke open and spilled large amounts of oil around the Shetland Islands in 1993.

Inefficient burning of carbon-based fuels in defective furnaces and domestic gas fires and in poorly tuned engines produces the very poisonous gas, carbon monoxide. Instead of:

$$C(s) + O_2(g) \longrightarrow CO_2(g)$$

partial oxidation gives:

$$2C(s) + O_2(g) \longrightarrow 2CO(g)$$

Inhalation of carbon monoxide may cause death because it interferes with the transport of oxygen in the bloodstream. Other dangerous gases produced by the burning of fuels include nitrogen oxides and sulfur oxides, which form strongly acidic solutions

in water (hence 'acid rain') (Figure 11.16). Power stations can remove acidic gases from their chimneys by reacting them with calcium oxide, which is basic.

A large variety of compounds, including carcinogens, appear in the smoke from burning coal and wood (Figure 11.17).

Figure 11.16 These trees in the Czech Republic were killed by the effects of acid rain, caused mainly by the sulfur oxides produced from burning coal.

Figure 11.17 In this power station the 'smoke' from the cooling towers is actually water vapour. The polluting gases are coming from the single tall chimney.

SAQ

8　What are the reactions of nitrogen(IV) oxide (NO_2) and the sulfur oxides (SO_2 and SO_3) with water? Write equations for these reactions.

Hint

Answer

Alternatives to fossil fuels

Biofuels

Plants can be grown to be used *directly* as fuels, e.g. wood. Plants can also be grown for *conversion* into fuels, e.g. sugar from sugar cane is easily fermented into ethanol. This can be used directly as an alternative to petrol or mixed with petrol. There is increasing use of renewable natural oils, such as rapeseed (see Figure 11.18) or sunflower oil, as part of diesel fuels. The **biodiesel** can be added to normal diesel to make it last longer. In France, about 5% of their diesel is biodiesel. Engines don't even have to be changed to run on biodiesel.

Figure 11.18 This field of rapeseed can be harvested and used to make biodiesel. Even recycled oil from chip shops can be used as the starting material for biodiesel!

Plants remove carbon dioxide from the air and use it for photosynthesis. The carbon dioxide is used to make glucose which is then converted to cellulose and other plant material. If crops are used either directly as, or for conversion into, a fuel, the carbon dioxide released to the atmosphere simply replaces the carbon dioxide removed during plant growth. However, the planting, growing, harvesting, processing and transportation involved in producing biofuels means they are not totally 'carbon-neutral'. Nevertheless, they can help tackle the problem of global warming by not adding as much carbon dioxide to the atmosphere as fossil fuels do when burnt.

There are also some other positive effects for the environment. Biodiesel does not produce sulfur dioxide as it burns. The fuel does not contain impurities of sulfur compounds like those from crude oil. Therefore, it is better for people with breathing problems and reduces the acidity of rain. Burning biodiesel also produces fewer particulates than burning ordinary diesel. Particulates contain volatile unburnt hydrocarbons that can cause cancer, so using biodiesel also reduces this danger (see Figure 11.19).

Accidental oil spills on waterways are also less damaging with biodiesel as it is biodegradable and breaks down if spilt in water.

Figure 11.19 Biodiesel is a 'cleaner' fuel than diesel from crude oil.

However, on a global scale, the use of large areas of farmland to produce fuel instead of food could result in food shortages. People are also worried about the destruction of habitats of some endangered species. For example, large areas of tropical forest are being turned into plantations for the production of palm oil to make biodiesel.

Scientists working for Shell are exploring the potential for growing forests of fast-cropping trees and using the biomass (plant material) as a renewable energy source. The biomass is dried and chipped before being converted to gas and bio-oil by heating in the absence of air. The gas or bio-oil is then used to fuel a gas turbine to generate electricity. Greater overall efficiency results when the biomass is first converted to gas and bio-oil.

Large municipal landfill sites produce significant quantities of biogas by anaerobic decay of biological materials. In the past, this gas often seeped into the atmosphere where it could form an explosive mixture with air. Now, it is collected in pipes and often flared for safe disposal. Biogas is mainly composed of methane, which has a much greater greenhouse effect than its combustion product carbon dioxide. In some places, the collection and combustion of biogas from landfill sites is being used to generate electricity.

As a future energy source, biomass has several advantages. It is renewable; it helps to reduce waste; and in some forms it can be used with simple technology. It has one main disadvantage: there is not a large enough supply of biomass to replace fossil fuels at present rates of use.

The substitution reactions of alkanes

Alkanes are remarkably inert compounds. A reason for the inertness of alkanes arises from their lack of polarity. As carbon and hydrogen have very similar electronegativities, alkanes are non-polar molecules. Consequently, alkanes are not readily attacked by common chemical reagents. Most reagents that you have met are highly polar compounds. For example, water, acids, alkalis and many oxidising and reducing agents are polar, and they usually initiate reactions by their attraction to polar groups in other compounds. Such polar reagents do not react with alkanes.

Some non-polar reagents will react with alkanes. The most important of these are the halogens, which, in the presence of ultraviolet light, will *substitute* hydrogen atoms in the alkane with halogen atoms. For example, when chlorine is mixed with methane and exposed to sunlight, chloromethane is formed and hydrogen chloride gas is evolved:

$$CH_4(g) + Cl_2(g) \longrightarrow CH_3Cl(g) + HCl(g)$$

Because the reaction requires ultraviolet light, it is called a photochemical reaction.

Further substitution is possible, in turn producing dichloromethane, trichloromethane and tetrachloromethane. Other halogens, such as bromine, produce similar substitution products. With hexane, for example, bromine produces bromohexane (Figure 11.20):

$$C_6H_{14}(l) + Br_2(l) \longrightarrow C_6H_{13}Br(l) + HBr(g)$$

Figure 11.20 The reaction of bromine with hexane in ultraviolet light. The bromine is decolourised, showing that the reaction has occurred.

The substitution mechanism

The overall equation for a reaction gives no clue to the stages involved between reactants and products. The sequence of stages is known as the *mechanism* of a reaction. For example, the energy of ultraviolet light is sufficient to break the Cl–Cl bond. Absorption of light energy causing a bond to break is known as *photodissociation*. **Homolytic fission** occurs and two chlorine atoms are formed, each having seven electrons in their outer shell. The chlorine atoms each have one unpaired electron and are thus **free radicals** (Chapter 10). Free radicals react very rapidly with other molecules or chemical species. As homolytic fission of a chlorine molecule must occur before any chloromethane can be formed, it is known as the **initiation** step.

$$Cl–Cl(g) \xrightarrow{\text{UV light}} Cl\bullet(g) + Cl\bullet(g)$$

The reaction of a chlorine free radical with a methane molecule produces hydrogen chloride and a $CH_3\bullet$ (methyl) free radical. The dot indicates the unpaired electron. The carbon atom in this $CH_3\bullet$ fragment also has seven electrons in its outer shell. A methyl free radical can react with a chlorine molecule to produce chloromethane and a new chlorine free radical:

$$Cl\bullet(g) + H–CH_3(g) \longrightarrow Cl–H(g) + CH_3\bullet(g)$$
$$CH_3\bullet(g) + Cl–Cl(g) \longrightarrow CH_3Cl(g) + Cl\bullet(g)$$

These two steps enable the reaction to continue. In the first step, a chlorine free radical is used up. The second step releases a new chlorine free radical, which allows repetition of the first step. The reaction will continue for as long as there is a supply of methane molecules and undissociated chlorine molecules. The two steps constitute a *chain reaction* and are known as the **propagation** steps of the reaction. More propagation steps can occur, involving the products of earlier propagation steps:

$$Cl\bullet(g) + H–CH_2Cl(g) \longrightarrow Cl–H(g) + CH_2Cl\bullet(g)$$
$$CH_2Cl\bullet(g) + Cl–Cl(g) \longrightarrow CH_2Cl_2(g) + Cl\bullet(g)$$

The reaction to form various substituted chloromethanes and hydrogen chloride ceases when the supply of reagents is depleted.

There is a variety of possible **termination** steps, which take place when two free radicals meet:

$$Cl\bullet(g) + CH_2Cl\bullet(g) \longrightarrow CH_2Cl_2(g)$$
$$CH_2Cl\bullet(g) + CH_2Cl\bullet(g) \longrightarrow C_2H_4Cl_2(g)$$

These also include recombination of two chlorine free radicals to form chlorine molecules. Alternatively, two methyl free radicals can combine to form an ethane molecule:

$$Cl\bullet(g) + Cl\bullet(g) \longrightarrow Cl_2(g)$$
$$CH_3\bullet(g) + CH_3\bullet(g) \longrightarrow CH_3CH_3(g)$$

These, or any other, termination steps will remove free radicals and disrupt the propagation steps, thus stopping the chain reaction. As you can see from the reactions above, a mixture of products is obtained, making free radical reactions unsuitable for the preparation of a pure product.

The four steps (initiation, two propagation steps and one of two termination steps) involved in the formation of chloromethane and hydrogen chloride from methane and chlorine constitute the mechanism of this reaction. As the reaction is a **substitution** involving free radicals, it is known as a *free-radical substitution*.

SAQ

9   a   Which of the following reagents are likely to produce free radicals in ultraviolet light? HCl(aq), Br_2(l), NaOH(aq), Cl_2(g), $KMnO_4$(aq).

      b   Write balanced equations for the reactions of butane with those reagents that produce free radicals.

Hint

Answer

Summary

Glossary

- Alkanes are saturated hydrocarbons with the general formula C_nH_{2n+2}. At room temperature the alkanes from methane to butane are gases; pentane to $C_{16}H_{34}$ are liquids; $C_{17}H_{36}$ and above are waxy solids.

- The melting and boiling points of alkanes increase with the length of the hydrocarbon chains. The increase may be explained in terms of increasing attraction between the non-polar alkanes with increasing chain length. The intermolecular forces are instantaneous dipole–induced dipole (or van der Waals') forces.

- For a given straight-chain alkane, the boiling points of its branched-chain isomers are lower because the branched molecules cannot approach as closely.

- Alkanes are relatively unreactive as they are non-polar. Most reagents are polar and do not usually react with non-polar molecules.

- Currently, natural gas and crude oil are our major sources of hydrocarbons. The majority of hydrocarbons release large amounts of energy on combustion and are used as fuels for electricity generation, industry, homes or transport.

- Alkanes are widely used as fuels. When they burn completely they produce carbon dioxide and water; unfortunately they produce toxic carbon monoxide gas when they burn in a limited supply of oxygen, e.g. in a car engine.

- A significant proportion of hydrocarbons from fossil fuels are converted into a wide range of chemical products.

- Cracking of the less useful fractions from crude oil produces a range of more useful alkanes and alkenes. The branched-chain alkanes are suitable for petrol and the alkenes are used to make polymers and other chemical products.

- Isomerisation converts straight-chain alkanes into more of the branched-chain alkanes. Re-forming converts alkanes to cycloalkanes or arenes. Branched-chain alkanes, cycloalkanes and arenes all improve the efficiency of combustion in modern petrol engines.

- Our reserves of fossil fuels such as gas and oil are limited; once these reserves are exhausted, alternative sources of energy will be needed by society. In the search for alternatives, chemists and other scientists are now working to develop renewable fuel sources such as biofuels.

- Chlorine atoms or bromine atoms can substitute for hydrogen atoms in alkanes in the presence of ultraviolet light, producing halogenoalkanes.

- The Cl–Cl or Br–Br bond undergoes homolytic fission in ultraviolet light producing reactive Cl or Br free radicals. This initiation step of free-radical substitution is followed by propagation steps involving a chain reaction which regenerates the halogen free radicals. Termination of the reaction may occur, for example, when two free radicals combine.

Questions

1 The hydrocarbons in crude oil can be separated by fractional distillation.
   a Explain what is meant by the terms
      i *hydrocarbons*, [1]
      ii *fractional distillation*. [1]
   b Undecane, $C_{11}H_{24}$, can be isolated by fractional distillation.
      Calculate the percentage composition by mass of carbon in undecane. [3]

   Hint

   c Undecane can be cracked into nonane and compound **A**. One molecule of nonane
      contains nine carbon atoms.
      i Write a balanced equation for this reaction. [2]
      ii Name compound **A**. [1]
   d Hydrocarbons of formula C_5H_{12} can also be isolated from crude oil
      i Draw the three structural isomers of C_5H_{12} (Isomers **B**, **C** and **D**). [3]
      ii Isomers **B**, **C** and **D** can be separated by fractional distillation. State the order,
         lowest boiling point first, in which they would distil. [1]
      iii Justify the order stated in **d ii**. [1]
      iv Write a balanced equation for the <u>complete</u> combustion of pentane, C_5H_{12}. [2]
      v Why do oil companies isomerise alkanes such as pentane? [1]

   Hint

OCR Chemistry AS (2812) Jan 2001 [Total 16]

   Answer

2 Crude oil is first separated by fractional distillation. The fractions can then be refined
   further by cracking, re-forming and isomerisation.
   The reaction sequence below shows the production of heptane, C_7H_{16}, from fractional
   distillation of crude oil, followed by cracking, re-forming and isomerisation.

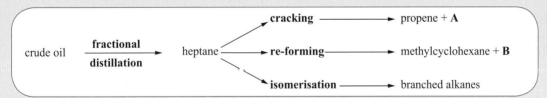

   a The cracking of heptane produces propene and **A**.
      Write a balanced equation for this cracking of heptane. [1]
   b The re-forming of heptane produces methylcyclohexane and **B**.
      i Show the structural formula of methylcyclohexane. [1]
      ii Write a balanced equation for this re-forming. [1]

   Hint

continued

c The isomerisation of heptane produces *seven* branched alkanes, five of which are shown below.

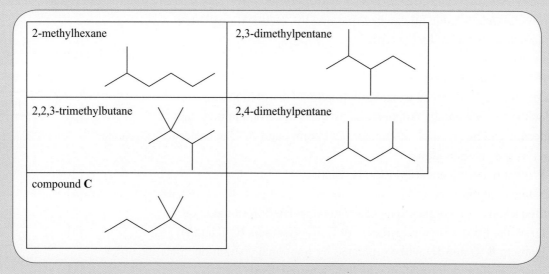

i Name compound **C**. [1]

ii Draw skeletal formulae for the other <u>two</u> branched
    alkanes formed by the isomerisation of heptane. [2]

iii Predict which of 2-methylhexane, 2,3-dimethylpentane and
    2,2,3-trimethylbutane has the lowest boiling point. [1]

iv Explain why 2-methylhexane, 2,3-dimethylpentane and
    2,2,3-trimethylbutane have different boiling points. [2]

OCR Chemistry AS (2812) Jan 2007 [Total 9]

Hint

Answer

3 The refined fractions of crude oil are used to make many organic compounds.
   In turn, these compounds are used to manufacture a great variety of products.
   The reaction sequence below shows the production of hexane from fractional
   distillation of crude oil followed by cracking, re-forming and isomerisation.

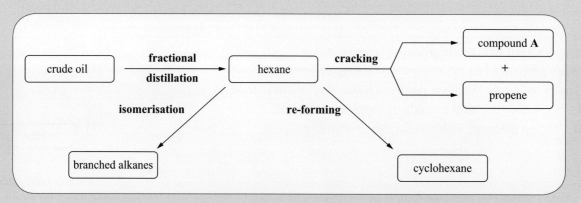

continued

a The cracking of hexane produces propene and compound **A**.
  i Complete a balanced equation for this cracking of hexane.
  $C_6H_{14} \longrightarrow$ ....+............................ [1]
  ii Name compound **A**. [1]
b The re-forming of hexane produces cyclohexane.
  Write a balanced equation for this re-forming. [1]
c The isomerisation of hexane produces <u>four</u> branched alkanes.
  Show the structural formulae and names of <u>two</u> of these branched isomers. [4]
d State why hydrocarbons such as hexane are both re-formed and isomerised by oil companies. [1]
e Crude oil and its fractions are described as non-renewable fossil fuels. To reduce the demand for fossil fuels, ethanol can be mixed with petrol. Ethanol is an example of a renewable biofuel.
  i Explain what is meant by a *biofuel*. [1]
  ii Why are fossil fuels *non-renewable* whereas ethanol is *renewable*? [2]

OCR Chemistry AS (2812) May 2002 [Total 11]

Answer

4 Crude oil is a complex mixture of hydrocarbons. Initial separation is achieved by fractional distillation. The separate fractions are then further refined to produce hydrocarbons such as decane.
  a **i** State what is meant by the term *hydrocarbon*. [1]
    ii A molecule of decane contains ten carbon atoms. State the molecular formula of decane. [1]
    iii Deduce the empirical formula of decane. [1]
  b Dodecane, $C_{12}H_{26}$, is a straight-chain alkane that reacts with chlorine to produce a compound with molecular formula $C_{12}H_{25}Cl$.
    $$C_{12}H_{26}+Cl_2 \longrightarrow C_{12}H_{25}Cl+HCl$$
    The reaction is initiated by the formation of chlorine free radicals from chlorine.
    i What is meant by the term *free radical*? [1]
    ii State the conditions necessary to bring about the formation of the chlorine free radicals from Cl_2. [1]
    iii State the type of bond fission involved in the formation of the chlorine free radicals. [1]
    iv The chlorine free radicals react with dodecane to produce $C_{12}H_{25}Cl$. Write equations for the <u>two</u> propagation steps involved. [2]
    v How many different structural isomers can be formed when chlorine reacts with dodecane to form $C_{12}H_{25}Cl$? [1]
  c Dodecane, $C_{12}H_{26}$, can be cracked into ethene and a straight-chain alkane such that the molar ratio ethene : straight-chain alkane is 2 : 1.
    i Write a balanced equation for this reaction. [2]
    ii Name the straight-chain alkane formed. [1]
  d Straight-chain alkanes such as heptane, C_7H_{16}, can be isomerised into branched-chain alkanes and re-formed into cyclic compounds.
    i Using <u>skeletal</u> formulae, write an equation to show the isomerisation of heptane into 2,2,3-trimethylbutane. [2]
    ii Write a balanced equation to show the re-forming of heptane into methylcyclohexane. [2]

OCR Chemistry AS (2812) Jan 2006 [Total 16]

Answer

Chapter 12

Alkenes

e-Learning

Objectives

The alkenes are a homologous series of **unsaturated hydrocarbons**. Each alkene molecule contains at least one C=C double bond. The family of alkenes includes a number of biologically important molecules. Many of these are based on the simple diene (i.e. an alkene with two C=C double bonds), isoprene (2-methylbuta-1,3-diene):

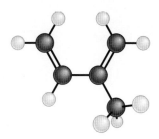

Some trees can be tapped for their latex or natural rubber (Figure 12.1). Latex is a polymer of isoprene.

Figure 12.1 Scraping the bark off a rubber tree in this way causes the liquid rubber to accumulate at one point, where it can be collected.

The natural oil, limonene, present in the rind of oranges and lemons is derived from two isoprene units:

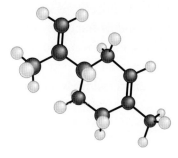

Alkenes are used to make many chemicals that feature prominently in modern life. Some examples of these chemicals are shown in Figure 12.2.

Figure 12.2 A range of products are produced from alkenes, including poly(chloroethene) window frames, ethane-1,2-diol (used in antifreeze) and industrial methylated spirits (mainly ethanol with methanol added to avoid alcohol tax – used as a solvent).

Physical properties of alkenes

Simple alkenes are hydrocarbons that contain one carbon–carbon double bond. The simplest alkene is ethene, $CH_2=CH_2$. The general formula of the homologous series of alkenes with one double bond is C_nH_{2n}.

SAQ

1   Draw a dot-and-cross diagram for ethene. Predict the shape of the molecule and give estimates of the bond angles.

> Hint
>
> Answer

Bonding in alkenes: σ and π bonds

A single covalent bond, such as C–C or C–H, consists of a σ bond. Double bonds such as C=C consist of one σ ('sigma') bond and one π ('pi') bond. Each carbon atom in ethene uses three of its four outer electrons to form three σ bonds, two to hydrogen atoms and one to the other carbon atom. The fourth electron on each carbon atom is found in a p orbital. Overlap of two p orbitals on neighbouring carbon atoms produces a π bond. To ensure maximum overlap, ethene must be a planar molecule.

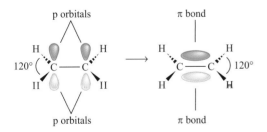

overlap of p orbitals produces
π molecular orbitals

Notice the bond angle of 120°. The carbon atoms involved in the double bond are in the centre of three areas of high electron density (bonding pairs of electrons). This means we get bond angles of 120° around the carbons at either end of the double bond.

Compounds which contain π bonds, such as ethene, are called unsaturated compounds. The term *unsaturated* indicates that the compound will combine by *addition* reactions with hydrogen or other chemicals, losing its multiple bonds.

Saturated compounds contain only *single* carbon–carbon bonds. The terms 'saturated' and 'unsaturated'

are often used in connection with oils and fats. The molecules in vegetable oils contain several double bonds – they are described as *polyunsaturated*. In hard margarine, hydrogen has been added to these double bonds so the margarine becomes saturated. However, several of the fatty acids which are essential to our diet are polyunsaturated and so, to ensure that these fatty acids are retained, much modern margarine is only partially saturated.

Z/E (cis/trans) isomerism

Many alkenes exhibit **Z/E isomerism** (also called *cis/trans* isomerism when there are two hydrogen atoms and two 'non-hydrogen' groups around a double bond – see Chapter 10); we shall consider an example here. Natural rubber is a polymer of 2-methylbuta-1,3-diene (or isoprene, page 142). The repeating unit contains a carbon–carbon double bond. All the links between the isoprene units are on the same side of this double bond. This arrangement is described as the *cis* isomer:

cis-poly(2-methylbuta-1,3-diene): natural rubber

Natural rubber is an elastic material used for balloons, rubber gloves and condoms.

Another possible arrangement has the links between each 2-methylbuta-1,3-diene unit, on alternate sides of the double bond. As they lie across the double bond, this is the *trans* isomer. It is found naturally as gutta-percha, which is a hard, grey, inelastic, horny material obtained from the percha tree in Malaysia. It is still used in the manufacture of some types of golf ball. The different properties of natural rubber and gutta-percha is a good example of the importance of the shape of molecules in organic chemistry.

trans-poly(2-methylbuta-1,3-diene): gutta-percha

Both *cis*- and *trans*-2-methylbuta-1,3-diene can be manufactured from 2-methylbuta-1,3-diene using appropriate Ziegler–Natta catalysts. Such catalysts were developed by the German chemist Karl Ziegler and the Italian chemist Giulio Natta, and are based on triethylaluminium and titanium(IV) chloride. Ziegler and Natta made a substantial contribution to the development of polymers and were jointly awarded the Nobel Prize for Chemistry in 1963.

Z/E (or *cis/trans*) isomerism is frequently encountered in alkenes, and arises because rotation about a double bond cannot occur unless the π bond is broken. In addition to a double bond, the molecule can have two identical groups, one on each of the two carbon atoms involved in the double bond. The other two groups must be different to this identical pair. But-2-ene is the simplest alkene to show *cis/trans* isomerism:

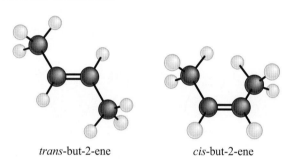

trans-but-2-ene   *cis*-but-2-ene

In the *cis* isomer, two methyl groups are on the same side of the double bond; in the *trans* isomer they are on opposite sides.

SAQ

2 Consider the following.

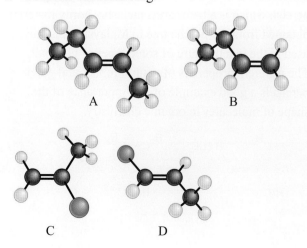

A   B

C   D

a Which, apart from A, can exist as *cis/trans* isomers?

b Draw and name the structural formulae for the pair of *cis/trans* isomers for A.

Answer

Characteristic reactions of alkenes

Alkenes undergo addition reactions. In these reactions the π bond is broken. The π bond is weaker than the C–C σ bond. This allows alkenes to react with a variety of reagents.

Addition reactions to the double bond

The characteristic reaction of an alkene involves a simple molecule (such as hydrogen, water or bromine) joining across the double bond to form a single product. These **addition reactions** are the second of the important types of organic reactions that you will meet.

Addition of hydrogen

Adding hydrogen converts the unsaturated alkene to a saturated alkane. Hydrogen gas and a gaseous alkene are passed over a finely divided nickel catalyst supported on an inert material. The equation for the addition of hydrogen to cyclohexene is:

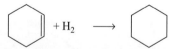

Another example is the manufacture of margarine from vegetable oil over a nickel catalyst at a temperature of about 200 °C and a hydrogen pressure of up to 1000 kPa (Figure 12.3).

Figure 12.3 Making this margarine involved the addition of hydrogen to C=C double bonds in molecules of vegetable oil.

The molecules of sight

Many organic chemicals perform very specialised jobs in living organisms. A compound that is essential to our eyesight is called retinal. This compound is present in the rod and cone cells of the eye. One of its isomers is responsible for the absorption of light. Each molecule has several double bonds, all locked into position, preventing rotation. When this isomer absorbs light, the π bond in one of its double bonds is broken. This allows the retinal molecule to change its shape by rotating around the single bond left behind. After bond rotation, the π bond is re-formed, fixing the molecule in its new shape and preventing further rotation. This is shown in the reaction sequence in Figure 12.4. The π bond that breaks and re-forms is labelled A.

The dramatic change of shape affects the shape of a protein that the retinal is attached to. This causes a signal to be sent via the optic nerve to the brain. The new *trans*-retinal isomer breaks away from the protein and is converted back to *cis*-retinal, ready for further light absorption.

In the *cis* isomer, the two carbon chains are on the same side of the double bond; in the *trans* isomer, they are on opposite sides. The breaking of one of the bonds in a double bond by light absorption is not a common reaction for alkenes, but if it didn't happen in retinal we wouldn't be able to see.

cis-retinal

+*hf*

only one of the double bonds (labelled A here) is broken by the absorption of light energy, *hf*

trans-retinal

Figure 12.4 The change of shape of a retinal molecule by rotation about the single bond that results from absorption of light.

Addition of halogens

When an alkene such as propene is bubbled through a solution of bromine at room temperature, the bromine solution is rapidly decolourised from its characteristic orange colour (Figure 12.5).

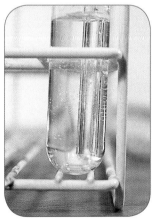

Figure 12.5 The reaction of propene with a solution of bromine. This characteristic reaction provides a test for an alkene (Table 12.1).

Test	Observation if an alkene is present
shake alkene with bromine water	orange bromine water is decolourised

Table 12.1 Simple test for an alkene.

Unlike free-radical substitution on an alkane, this reaction does not require ultraviolet light and will occur in total darkness. The bromine joins to the propene to form 1,2-dibromopropane:

$$CH_3CH=CH_2 + Br_2 \longrightarrow CH_3CHBrCH_2Br$$

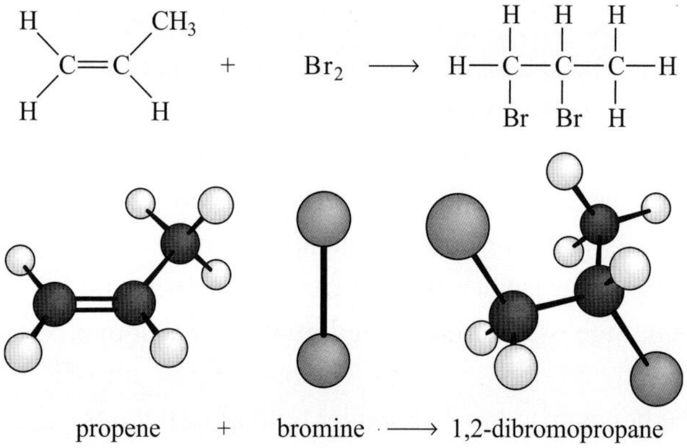

propene + bromine ⟶ 1,2-dibromopropane

Chlorine and iodine produce similar addition products with alkenes. Fluorine is too powerful an oxidant and tends to ignite hydrocarbons!

Addition of hydrogen halides

Hydrogen halides also add readily to alkenes. If ethene gas is bubbled through concentrated hydrochloric acid at room temperature, chloroethane is produced:

$$CH_2=CH_2(g) + HCl(aq) \longrightarrow CH_3CH_2Cl(l)$$

Alkenes such as propene can give rise to two different products:

$$CH_3CH=CH_2 + HBr \longrightarrow CH_3CHBrCH_3 \text{ or}$$
$$CH_3CH_2CH_2Br$$

The normal product is 2-bromopropane, $CH_3CHBrCH_3$.

Addition of steam

This is a route to making alcohols. Industrially, steam and a gaseous alkene are passed over a solid catalyst. A temperature of 330 °C and a pressure of 6 MPa are used in the presence of a phosphoric acid, H_3PO_4, catalyst. The addition of steam to ethene produces ethanol:

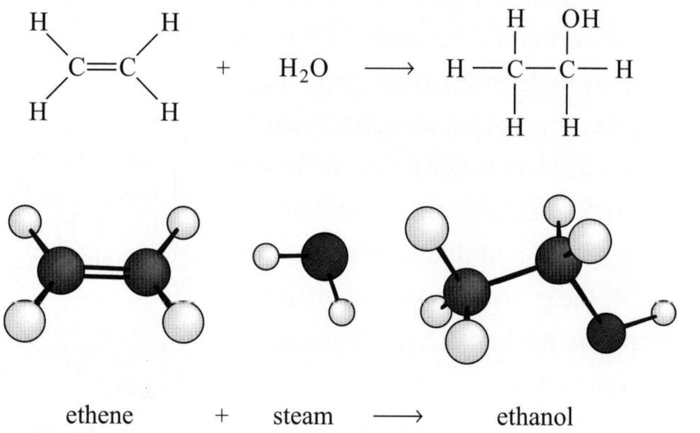

ethene + steam ⟶ ethanol

The mechanism of addition

Although bromine and ethene are non-polar reagents, they react together quickly at room temperature. The bromine molecule becomes polarised when close to a region of negative charge (high electron density) such as the ethene π bond. The π bond then breaks, with its electron pair forming a new covalent bond to the bromine atom, which carries a partial positive charge. At the same time, the bromine molecule undergoes heterolytic fission (Chapter 10). Heterolytic fission involves both electrons in the bond moving to the same atom. This produces a bromide ion and a positively charged carbon atom (a **carbocation**) in the ethene molecule (Figure 12.6).

Carbocations are highly reactive and the bromide ion rapidly forms a second carbon–bromine covalent bond to give 1,2-dibromoethane.

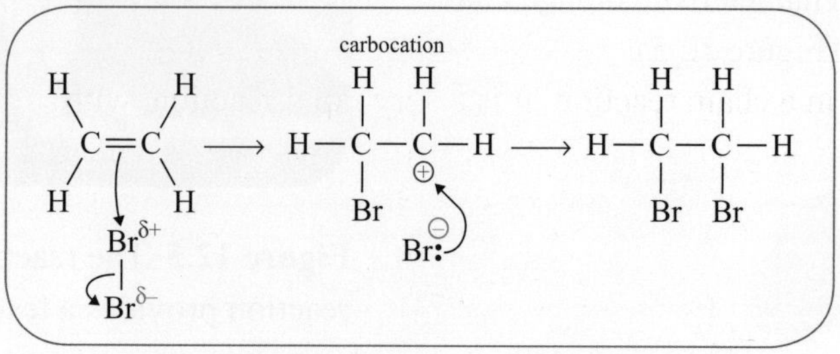

Figure 12.6 The formation of a carbocation in the bromination of ethene. This is the mechanism of this electrophilic addition reaction.

In this mechanism, the polarised bromine molecule has behaved as an electrophile. An **electrophile** is a reactant that accepts a pair of electrons to form a new covalent bond. The mechanism of this reaction is *electrophilic addition*.

SAQ

3 a Draw a dot-and-cross diagram of the carbocation formed in an electrophilic addition to ethene. How many electrons are there on the positively charged carbon atom? Explain how this atom completes its outer electron shell when it combines with a bromide ion.

Hint

b Suggest a mechanism for the addition of hydrogen chloride to ethene.

c If the reaction of bromine with ethene is carried out in ethanol containing some lithium chloride, a second chlorine-containing product is formed as well as 1,2-dibromoethane. Suggest a structure for this second product.

Hint

Answer

Polymerisation of alkenes

During polymerisation, an alkene undergoes an addition reaction to itself. When the addition reaction starts, one alkene molecule joins to a second one, and then to a third one, and so on. In this way a long molecular chain is built up. This type of reaction is called **addition polymerisation**. Many useful polymers are obtained via addition polymerisation of different alkenes.

Poly(ethene) was first produced accidentally by two scientists (Eric Fawcett and Reginald Gibson) in 1933. The reaction involves ethene adding to itself in a chain reaction. It is a very rapid reaction, with chains of up to 10 000 ethene units being formed in one second. The product is a high-molecular mass alkane. It is a polymer and is a member of a large group of materials generally known as plastics. The alkene from which a polymer is made is called the **monomer** and the section of polymer that the

Industrial importance of alkenes

Alkenes are used as starting material in the manufacture of many important materials (see Figure 12.2). The presence of the double bonds in alkenes makes them far more reactive than the corresponding alkanes.

One product made from ethene is ethane-1,2-diol, commonly used as antifreeze. A mixture of this and water freezes at $-37\,^{\circ}C$ so it will only freeze at exceptionally low temperatures.

The antifreeze is also used at airports to de-ice aeroplanes before they take off in cold climates (see Figure 12.7).

However, ethane-1,2-diol is toxic and can harm the wildlife living around an airfield. So chemists have developed propane-1,2-diol as a safe alternative. Propane-1,2-diol is made from propene, and is even approved for use in cosmetics and snack foods.

By far the most important use of the alkenes is in the manufacture of plastics. This is discussed in the section on polymerisation.

Figure 12.7 This plane is being de-iced with ethane-1,2-diol. The compound is recycled in modern airports to stop it entering water courses.

monomer forms is called the *repeat unit* (often shown within brackets in structural formulae):

n is very large, e.g. up to 10 000

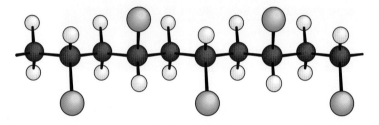

Other important poly(alkene)s include poly(chloroethene) and poly(phenylethene).

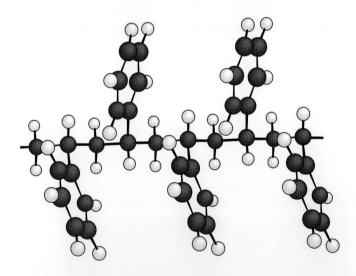

poly(chloroethene)

poly(phenylethene)

They are more commonly known as PVC and polystyrene respectively. Note how the systematic name is derived by putting the systematic name of the alkene in brackets, and prefixing this with 'poly'.

The skeletal formulae of the monomers, chloroethene (traditionally called vinyl chloride) and phenylethene (styrene), are as follows:

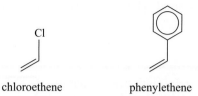

chloroethene   phenylethene

Note that when a benzene ring becomes a group attached to an alkene its name changes to phenyl (from 'phene', an old name for benzene). The phenyl group may also be written as C_6H_5-.

SAQ

4 a Write balanced equations for the formation of poly(chloroethene) and poly(phenylethene) using displayed formulae. Show the repeat unit in brackets.

b A polymer which is often used to make plastic boxes for food storage has the structure:

$$CH_3 \quad CH_3 \quad CH_3 \quad CH_3$$

Draw displayed formulae to show:
   i the repeat unit of this polymer
   ii the monomer from which it is made.
   Label your diagrams with the appropriate systematic names.

Answer

The chemical industry can now produce a wide variety of poly(alkene)s. It does this by using different monomers, and also by changing the conditions under which polymerisation occurs. The poly(alkene)s have many uses. Figure 12.8 shows some of these uses.

Figure 12.8 Some products made from poly(alkene)s.

Problems with poly(alkene)s

Our large-scale use of poly(alkene)s has created a major environmental problem. Figure 12.9 is a sight familiar to us all. These plastics are alkanes, so they break down very slowly in the environment. They are resistant to most chemicals and to bacteria (they are non-biodegradable). If plastics are dumped in landfill sites they cause the 'tip' to fill up very quickly. Local authorities cannot afford to allow this to happen.

Instead of dumping it in landfill sites we can collect waste poly(alkene)s, sort them and recycle them into new products (as in Figure 12.10). Unfortunately, the costs of recycling, in terms of the energy used to collect and reprocess waste plastic, is often greater than the energy used in making new material.

Figure 12.9 A beach littered with poly(alkene) waste products.

Figure 12.10 These poly(alkene)s will be recycled into new products. Different types of polymer must be separated if the plastics are to be recycled. A mixture of polymers will produce a very inferior plastic.

A third option is to burn the poly(alkene)s to provide energy. Some plastics are potentially good fuels as they are hydrocarbons and would reduce our need to burn oil or other fossil fuels. These plastics could be burnt with other combustible household waste. This would save considerable landfill costs and provide a substantial alternative energy source.

Unfortunately plastics tend to produce toxic gases when they burn, but this problem can be solved if a sufficiently technologically advanced incinerator is used. Modern technology is such that the waste could be burnt cleanly and with less pollution than that from traditional fossil-fuel power stations. The carbon dioxide produced would not add to the total emissions of this greenhouse gas as it would replace emissions from burning other fossil fuels. Other pollutant gases, such as hydrogen chloride from PVC, can be removed by the use of gas scrubbers. In a gas scrubber, acidic gases are dissolved and neutralised in a spray of alkali. European Union legislation requires household waste incinerators to use gas scrubbers. Figure 12.11 shows a modern waste incinerator in Vienna.

A fourth option is feedstock recycling. This means that the waste polymers are used as a source of hydrocarbons. In view of the limitations of mechanical recycling, BP (British Petroleum) developed a method for processing mixed and contaminated plastics. They have now built a pilot plant at their refinery site in Grangemouth, Scotland.

Figure 12.11 The incinerator in Vienna is not only clean, it is also a tourist attraction!

In the plant the polymers are cracked (see Chapter 11) to produce a mixture of hydrocarbons. The mixture contains alkanes, alkenes and arenes which provide additional feedstock for the main refinery. Alkenes, once separated, may once again be made into polymers such as poly(ethene) or poly(propene).

Chemists are also working to help solve the problems of plastic waste by producing plastics that breakdown once we discard them. These are called **degradable plastics**.

We are now making more plastics that do rot away in the soil when we dump them. Plastics that can be broken down by microbes are called *biodegradable* (see Figure 12.12). Chemists have found different ways to speed up the decomposition; one way uses granules of starch built into the plastic. The microbes in soil feed on the starch, breaking the plastic into small bits that will rot away more quickly.

Other types of plastic have been invented that are actually made by bacteria. ICI have made a plastic called poly(3-hydroxybutyrate) – known as PHB. The bacteria are fed on sugars or alcohol, or even carbon dioxide and hydrogen. The plastic is totally biodegradable. It also preserves our supplies of crude oil which traditional plastics use up. However, at present PHB costs about 15 times as much as traditional plastics.

Other plastics have been developed that are broken down by sunlight. Chemists have made long polymer molecules with groups of atoms along each chain that absorb the energy from sunlight. This splits the chain down into smaller molecules. These are called *photodegradable* plastics.

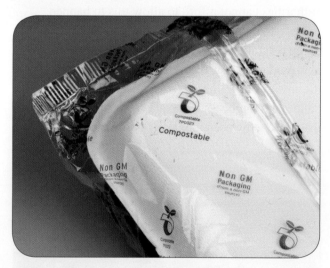

Figure 12.12 This plastic packaging is biodegradable.

Summary

- The homologous series of alkenes has the general formula C_nH_{2n}. Alkenes are unsaturated hydrocarbons containing one double bond. Their carbon–carbon double bond consists of a σ bond and a π bond.

- Ethene is a planar molecule, other alkenes are planar around the double bond and the four adjacent atoms. Many alkenes have Z/E isomers which arise because rotation about the double bond is prevented.

- Alkenes are more reactive than alkanes because they contain a π bond. The characteristic reaction of the alkene functional group is addition, which occurs across the π bond. For example ethene produces: ethane with hydrogen over a nickel catalyst; 1,2-dibromoethane with bromine at room temperature; chloroethane with hydrogen chloride at room temperature; ethanol with steam in the presence of H_3PO_4 catalyst.

- The mechanism of the reaction of bromine with ethene is electrophilic addition. Electrophiles accept a pair of electrons from an electron-rich atom or centre, in this case the π bond. A carbocation intermediate is formed after the addition of the first bromine atom. This rapidly reacts with a bromide ion to form 1,2-dibromoethane.

- Alkenes produce many useful polymers by addition polymerisation. For example, poly(ethene) from $CH_2=CH_2$, poly(propene) from $CH_3CH=CH_2$, poly(chloroethene) from $CH_2=CHCl$ and poly(tetrafluoroethene) from $CF_2=CF_2$ (see Chapter 14).

- The disposal of plastic waste is difficult as much of it is chemically inert and non-biodegradable. When burnt, the waste plastics may produce toxic products such as hydrogen chloride from PVC (poly(chloroethene)). However, acidic gases can be neutralised by a basic substance in 'scrubbers'. Whilst much waste plastic is recycled, the costs of collecting and sorting most domestic waste plastic are too high to make recycling worthwhile. Use of the energy released on combustion (for heating buildings) is a better option for domestic waste, but treatment of flue gases is again required to remove toxic pollutants. Another option is feedstock recycling where the polymers are cracked to form alkenes. The alkenes are separated and used as feedstock to make new polymers.

- Chemists are also developing many new degradable plastics, designed to decompose once discarded.

Questions

1 Dibromoethane, $C_2H_2Br_2$, has two structural isomers **C** and **D**, shown below.

    a i Name compound **C**. [1]

      ii What is the empirical formula of compound **D**? [1]

    b The reaction between either **C** or **D** and Br_2 can be used to show the presence of the C=C double bond.

      i State what you would see when isomer **C** reacts with Br_2. [1]

continued

ii State the type of mechanism for the reaction between isomer **C** and Br_2. [2]

c Isomers **C** and **D** can both behave as alkenes. For each of the following reactions, state the conditions, if any, and identify the organic products that can be formed.

i Isomer **C** reacts with H_2.

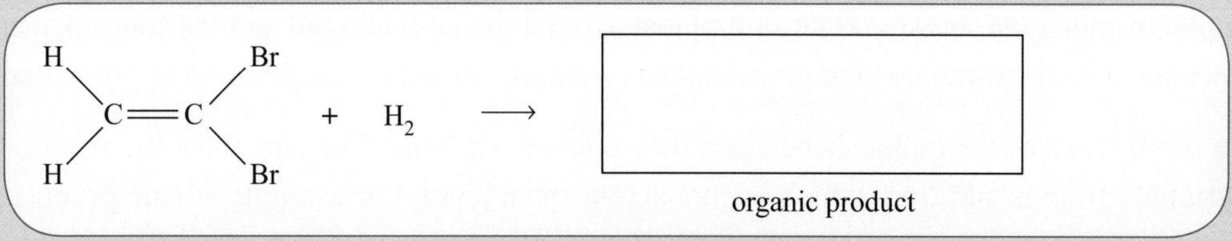

organic product

[2]

ii Isomer **C** reacts with HBr.

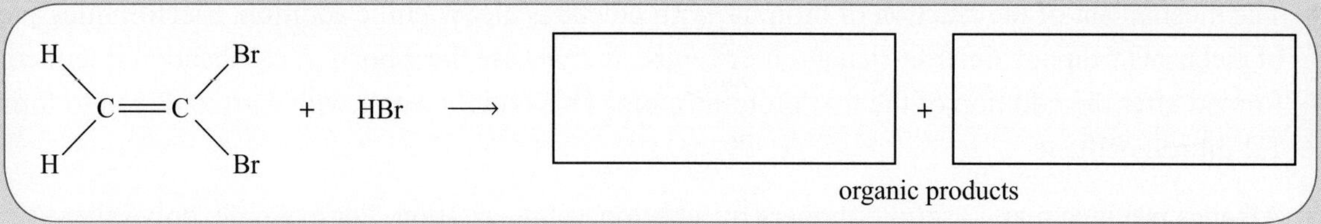

organic products

[2]

iii Isomer **D** reacts with H_2O.

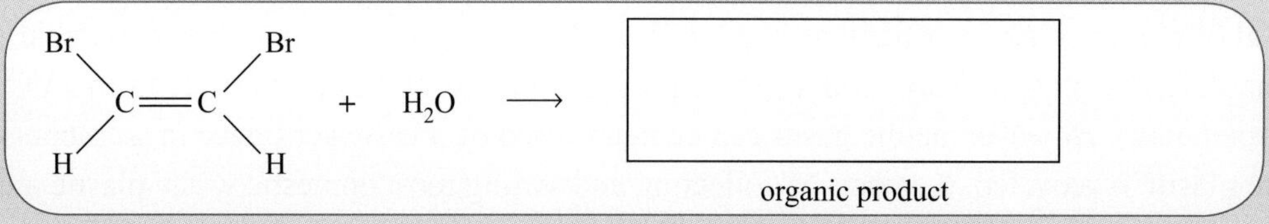

organic product

[3]

OCR Chemistry AS (2812) June 2003 [Total 12]

2 a Propan-2-ol can be formed by the hydration of an alkene in the presence of a catalyst.

i Suggest a suitable catalyst for this reaction. [1]

ii This is an electrophilic addition reaction. What is meant by the term *electrophile*? [1]

b A mechanism for the reaction in **a** is shown below.

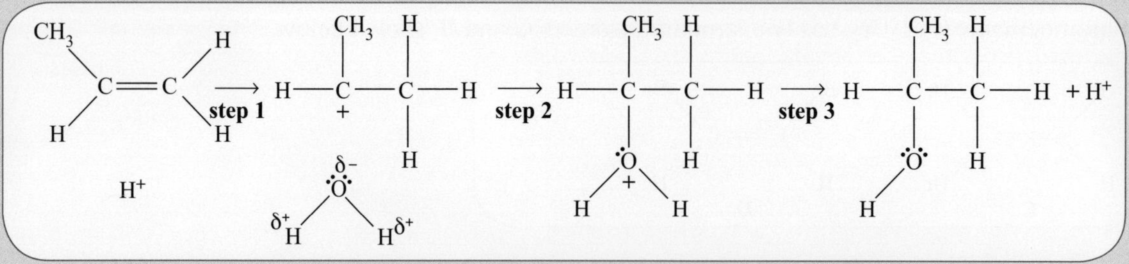

i Copy the mechanism and add 'curly arrows' to the mechanism to show the movement of electron pairs in steps **1**, **2** and **3**. [3]

ii Suggest, with a reason, the role of the H^+. [1]

OCR AS Chemistry 2812 Jan 2007 [Total 6]

continued

3 Ethene can be used to manufacture chloroethene $H_2C=CHCl$. This involves the following reactions.

    step 1 $H_2C=CH_2+Cl_2\longrightarrow ClH_2C–CH_2Cl$

    step 2 $ClH_2C–CH_2Cl\xrightarrow{500\,°C}H_2C=CHCl+HCl$

  a Copy and complete, with the aid of curly arrows, the mechanism involved in step **1**. Show any relevant dipoles and charges.

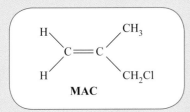

                                                                                [4]

  b The chloroethene (also known as vinyl chloride) produced can be polymerised to form poly(chloroethene) or PVC.

    i Draw a section of the polymer, PVC, to show <u>two</u> repeat units.   [1]

    ii Describe the difficulties in the disposal of polymers in general and identify a specific additional problem with the disposal of PVC.   [3]

    iii Outline the role of chemists in minimising damage to the environment during the disposal process.   [1]

OCR Chemistry AS (2812) Jan 2003   [Total 9]

                                                                             Answer

4 Methyl allyl chloride, MAC, is an important industrial chemical. It is used as an intermediate in the production of synthetic fibres, pharmaceuticals and epoxy resins. The structural formula of MAC is shown below.

$$\begin{array}{ccc} H & & CH_3 \\ & C=C & \\ H & & CH_2Cl \end{array}$$

MAC

  a Give the <u>systematic</u> chemical name of MAC.   [1]

  b MAC contains the alkene group and can undergo polymerisation. Draw a section of the polymer, poly(MAC), showing <u>two</u> repeat units.   [2]

OCR Chemistry AS (2812) Jan 2007   [Total 3]

continued

5 a Propene, C_3H_6, readily undergoes electrophilic addition reactions. Show, with the aid of curly arrows, the mechanism of the electrophilic addition reaction of propene with bromine.

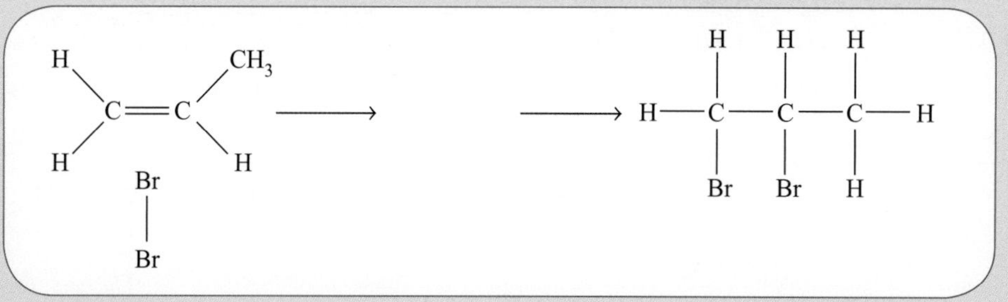

Answer

[4]

b Propene also reacts as shown below.

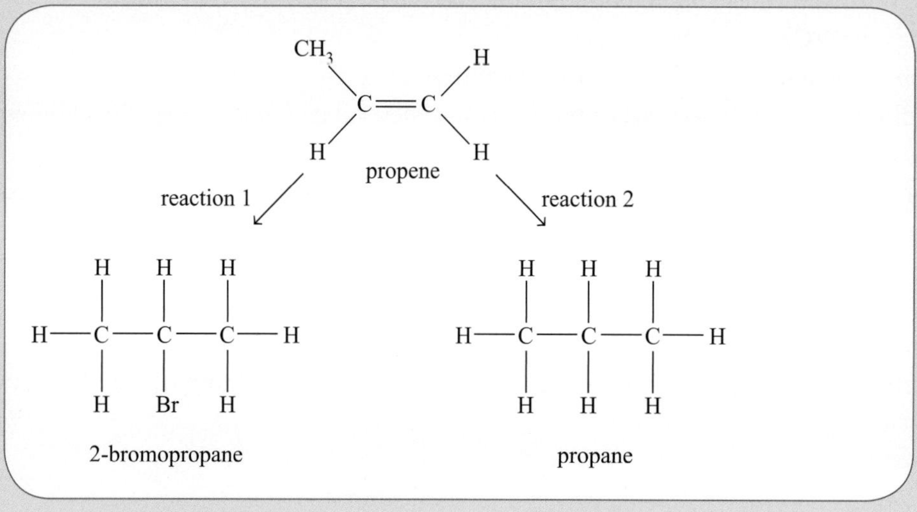

i State a suitable reagent for reaction 1. [1]

ii State a suitable reagent and conditions for reaction 2. [2]

iii In the presence of an acid catalyst, propene can react with steam to form a mixture of two alcohols. Draw the structures of the two alcohols. [2]

c The scientists Ziegler and Natta were awarded a Nobel Prize for Chemistry in 1963 for their work on polymerisation. Part of this work involved the polymerisation of propene into poly(propene).

i What type of polymerisation forms poly(propene)? [1]

ii Draw a section of poly(propene) to show <u>two</u> repeat units. [1]

iii State <u>two</u> difficulties in the disposal of poly(propene). [2]

OCR Chemistry AS (2812) Jan 2001 [Total 13]

Answer

Chapter 13

Alcohols

e-Learning

Objectives

The homologous series of aliphatic alcohols has the general formula $C_nH_{2n+1}OH$. They are named by replacing the final '-e' in the name of the alkane with '-ol'. For example C_2H_5OH is ethanol. If an alcohol has three or more carbon atoms the position of the alcohol group is indicated by a number. For example, $CH_3CH_2CH_2OH$ is propan-1-ol. $CH_3CH(OH)CH_3$ is propan-2-ol. (Note that the usual rules apply, i.e. use the lowest possible number to indicate the position of the alcohol group.)

Physical properties of alcohols

Miscibility with water

Miscibility is a measure of how easily a liquid mixes; it is the equivalent of solubility for solids. The miscibility of alcohols with water is due to their ability to form hydrogen bonds. Methanol and ethanol are freely miscible in water in all proportions. When water and ethanol mix, some of the hydrogen bonds between the molecules in the separate liquids are broken. These are replaced by hydrogen bonds between water molecules and ethanol molecules. There is no significant gain or loss in energy. Figure 13.1 shows a molecular model of hydrogen bonds between water and ethanol.

The miscibility of alcohols in water decreases with increasing length of the hydrocarbon chain. Although the hydroxyl group can still form hydrogen bonds to water, the long hydrocarbon chain disrupts hydrogen bonding between other water molecules. The hydrocarbon chains do not form strong intermolecular

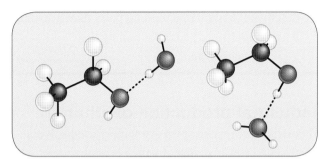

Figure 13.1 Hydrogen bonding between ethanol and water (the hydrogen bonds are represented by dotted lines). The bonds form between an oxygen lone pair and a hydrogen atom.

bonds with water molecules, because the hydrocarbon chains are essentially non-polar and can only form weak van der Waals' forces.

Volatility of alcohols

Hydrogen bonding between alcohol molecules reduces their **volatility** considerably. Volatility is the ease with which a liquid turns to vapour. The intermolecular forces are significantly stronger in alcohols than in alkanes of comparable size and shape. Table 13.1 shows the boiling points of some alcohols together with the boiling points of comparable alkanes. To allow for the atoms in the –OH group, each alcohol is compared to the alkane with one more carbon atom. The higher the boiling point, the less volatile the liquid.

All the alcohols in Table 13.1 have boiling points significantly above those of the comparable alkanes.

Alcohol	Boiling point /°C	Intermolecular forces	Alkane	Boiling point /°C	Intermolecular forces
ethanol	79	hydrogen bonds	propane	−42	van der Waals'
propan-1-ol	98	hydrogen bonds	butane	0	van der Waals'
butane-1-ol	117	hydrogen bonds	pentane	36	van der Waals'

Table 13.1 Boiling points of alcohols and alkanes.

SAQ

1 Explain why:
   a ethanol is highly miscible with water   Hint
   b pentan-1-ol is miscible with water, but less so than ethanol   Hint
   c ethanol has a higher boiling point than propane
   d propan-1-ol has a higher boiling point than ethanol.   Answer

Industrial production of ethanol

Industrial methylated spirit is a widely used solvent. Methylated spirit is ethanol which is adulterated with methanol. Methanol is poisonous and so it is dangerous to drink methylated spirit. The coloured methylated spirit sold in hardware shops contains a purple dye with a foul taste as a further deterrent. The adulteration means that methylated spirit does not carry the high taxes imposed on alcoholic drinks.

Most industrial ethanol is made by the addition reaction of steam with ethene in the presence of a phosphoric acid catalyst:

$$CH_2{=}CH_2(g) + H_2O(g) \longrightarrow CH_3CH_2OH(g)$$

Production of ethanol by fermentation and distillation

Ethanol has been known to humans in the form of alcoholic drinks for many thousands of years. If ripe fruit is harvested and left, *fermentation* of the sugar in the fruit will produce ethanol and other compounds. Fermentation involves yeasts, which occur naturally on the skins of many ripening fruits such as grapes (Figure 13.2). Fermentation is an exothermic reaction which provides yeast with energy for its metabolism. Glucose (a sugar) is converted to ethanol and carbon dioxide by enzymes in the yeast:

$$C_6H_{12}O_6(aq) \longrightarrow 2C_2H_5OH(aq) + 2CO_2(g)$$

The reaction does not require oxygen (it is an *anaerobic* process), so fermentation is carried out with air excluded to prevent the oxidation of the ethanol to undesirable compounds such as aldehydes, which affect the flavour of the product and may cause headaches.

Fermentation stops when the ethanol concentration reaches about 15% by volume. This is because ethanol

Figure 13.2 Stainless steel fermentation vessels at a modern winery.

kills the yeast at this concentration. The higher concentration of alcohol found in spirits is produced by distillation of the fermented liquor (see Chapter 10).

The use and abuse of alcoholic drinks

Alcoholic drinks provide us with valuable nutrients, such as minerals and vitamins, as well as being a source of energy. It has been estimated that almost 25% of the dietary intake of energy (for both children and adults!) in seventeenth-century Britain came from alcoholic drinks.

The ethanol in these drinks also affects our behaviour, and when drunk in excess may cause liver damage or even death. Whilst death may result from long-term alcohol abuse, it may also be caused from excessive short-term consumption, known as 'binge drinking'. Examples of binge drinking include the consumption of eight pints of beer or half a bottle of spirits in a few hours. One reason people consume alcoholic drinks is because it makes them feel more relaxed and able to cope with stress (see Figure 13.3). They generally feel more cheerful, less anxious and less tense. These

continued

effects are produced by the ethanol depressing the activity of the central nervous system.

Even small quantities of alcohol affect our ability to concentrate when driving motor vehicles or operating machinery. It has been shown that the intake of only one unit of alcohol may be sufficient to affect us. A unit of alcohol is a rough measure of the quantity of alcohol consumed. It is approximately:

- half a pint of beer
- a glass of wine
- a single measure of spirits.

The number of units required to raise the blood alcohol concentration (BAC) over the legal limit depends on a number of factors such as gender, body weight, age and how quickly the alcohol is consumed. Our ability to drive is affected long before the legal limit is reached, so it is clearly very unwise to consume anything containing ethanol before driving. The BAC can remain above the legal limit until the morning after drinking or even longer.

Figure 13.3 Alcoholic drinks form a large part of many people's social lives.

The reactions of alcohols

Alcohol reactions may be divided into groups, according to which bonds are broken. The bonds present in a typical alcohol such as ethanol are shown in Table 13.2, together with their average bond enthalpies (see Chapter 16).

Bond	Bond enthalpy /kJ mol^{-1}
C–C	347
C–H	410
C–O	336
O–H	465

Table 13.2 Bonds and bond enthalpies in ethanol.

SAQ

2 When a bond is broken, is the energy absorbed or released? Place the bonds in Table 13.2 in order of increasing strength. [Answer]

Although the O–H bond is the strongest, it is also the most polar. The atoms involved in polar bonds are more susceptible to attack by polar reagents, so the O–H bond is not necessarily the most difficult bond to break in ethanol.

SAQ

3 a Apart from the O–H bond, which other bond in an alcohol is very polar? [Hint]

  b Why are these two bonds so polar?

  c Polar reagents include electrophiles and nucleophiles. Explain what is meant by each of these terms. [Answer]

We shall now look at the reactions of alcohols in order, according to which bonds are broken.

A reaction in which the O–H bond is broken

The formation of esters

When ethanol is warmed with ethanoic acid in the presence of a strong acid catalyst, an ester, ethyl ethanoate, is formed. During this reaction the O–H bond in ethanol is broken. Many esters have characteristic fruity odours and are found naturally in fruits. The equation for the formation of ethyl ethanoate is:

$$CH_3CH_2OH + H_3C - \overset{O}{\overset{\|}{C}} - OH \rightleftharpoons H_3C - \overset{O}{\overset{\|}{C}} - OCH_2CH_3 + H_2O$$

Concentrated sulfuric acid is usually used as the acid catalyst and the mixture is refluxed (Chapter 10). The impure ester is obtained from the reaction mixture by distillation. The reaction mixture contains an equilibrium mixture of reactants and products.

Esters may also be prepared by reaction of an alcohol with an acyl chloride (*acylation*). Acyl chlorides react very vigorously and exothermically with alcohols, releasing hydrogen chloride gas. No catalyst is required and the reaction mixture may require cooling to slow down the reaction. The equation for the reaction of ethanol with ethanoyl chloride is:

$$CH_3CH_2OH + H_3C - \overset{O}{\overset{\|}{C}} - Cl \rightleftharpoons H_3C - \overset{O}{\overset{\|}{C}} - OCH_2CH_3 + HCl$$

Reactions that may also involve breaking C–C or C–H bonds

Mild oxidation

Like halogenoalkanes (Chapter 14), aliphatic alcohols may be classed as primary, secondary or tertiary.

- In a *primary alcohol*, the –OH group is on a carbon atom which is bonded to only *one* other carbon atom (or none in the case of methanol, CH_3OH). Ethanol is a primary alcohol.
- In a *secondary alcohol*, the –OH group is on a carbon atom which is bonded to *two* other carbon atoms.
- In a *tertiary alcohol*, the –OH group is on a carbon atom which is bonded to *three* other carbon atoms.

Examples of primary, secondary and tertiary alcohols are:

$$H_3C-CH_2-CH_2-OH$$

propan-1-ol
primary alcohol

$$H_3C-\overset{CH_3}{\underset{H}{\overset{|}{\underset{|}{C}}}}-OH$$

propan-2-ol
secondary alcohol

$$H_3C-\overset{CH_3}{\underset{CH_3}{\overset{|}{\underset{|}{C}}}}-OH$$

2-methylpropan-2-ol
tertiary alcohol

SAQ

4 Molecular models for six isomers of pentanol, $C_5H_{11}OH$, are shown below. The oxygen atoms are coloured red. Draw and label the structural formulae for these isomers. Name those without names and classify each of the alcohols as primary, secondary or tertiary.

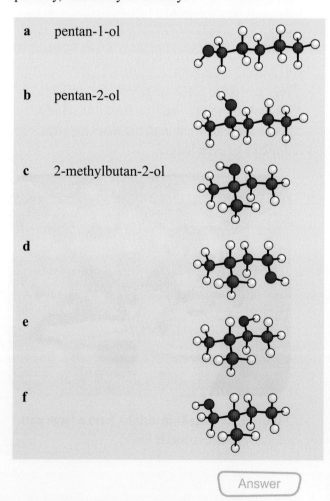

a	pentan-1-ol
b	pentan-2-ol
c	2-methylbutan-2-ol
d	
e	
f	

Answer

Primary and secondary aliphatic alcohols are oxidised on heating with acidified aqueous potassium dichromate(VI); tertiary alcohols remain unchanged with this reagent (Figure 13.4).

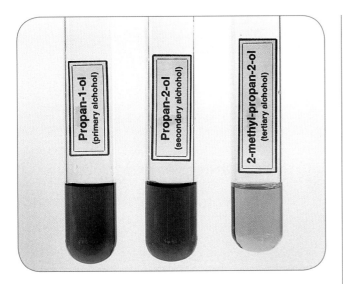

Figure 13.4 Before warming with the labelled alcohol, each of these tubes contained orange potassium dichromate(VI). After warming, the dichromate(VI) has been reduced to green chromium(III) by the primary and secondary alcohols only. This means only the primary and secondary alcohols have been oxidised.

As primary and secondary alcohols produce different, easily distinguished products, this reaction provides a useful means of identifying an unknown alcohol as primary, secondary or tertiary.

- Primary alcohols produce compounds called aldehydes on gentle heating with acidified dichromate(VI). As aldehydes are more volatile than their corresponding alcohols, they are usually separated by distillation as they are formed. On stronger heating under reflux with an excess of acidified dichromate(VI), the aldehydes are oxidised to carboxylic acids.
- Secondary alcohols produce compounds called ketones on heating with acidified dichromate(VI).
- Tertiary alcohols do not react with acidified dichromate(VI).

During the oxidation reactions that occur with primary and secondary alcohols, the orange colour of the dichromate(VI) ion, $Cr_2O_7^{2-}(aq)$, changes to the green colour of the chromium(III) ion, $Cr^{3+}(aq)$.

Ethanol, a primary alcohol, produces an aldehyde called ethanal on gentle heating with acidified dichromate(VI). You may prepare a sample of aqueous ethanal by distilling the aldehyde as it is formed when acidified dichromate(VI) is added

dropwise to hot ethanol. Simplified equations are frequently used for the oxidation of organic compounds, with the oxygen from the oxidising agent being shown as [O]:

$$CH_3CH_2OH + [O] \longrightarrow \underset{\text{ethanal}}{H_3C-\overset{\overset{\displaystyle O}{\|}}{C}-H} + H_2O$$

Ethanal has a smell reminiscent of rotting apples. Further oxidation, by refluxing ethanol with an excess of acidified dichromate(VI), produces ethanoic acid:

$$H_3C-\overset{\overset{\displaystyle O}{\|}}{C}-H + [O] \longrightarrow \underset{\text{ethanoic acid}}{H_3C-\overset{\overset{\displaystyle O}{\|}}{C}-OH}$$

You can separate aqueous ethanoic acid from the reaction mixture by distillation after it has been refluxing for 15 minutes. You can detect the ethanoic acid by its characteristic odour of vinegar and by its effect on litmus paper, which turns red.

SAQ

5 Propan-1-ol can be oxidised to propanal (CH_3CH_2CHO) and to propanoic acid (CH_3CH_2COOH).
   a Which reagents and conditions should be used to oxidise propan-1-ol to propanal?
   b Write a balanced chemical equation for this oxidation. Oxygen from the oxidising agent should be shown as [O].
   c What reagents and conditions should be used to oxidise propan-1-ol to propanoic acid?
   d Write a balanced chemical equation for this oxidation. Again, show oxygen from the oxidising agent as [O]. Answer

The secondary alcohol propan-2-ol, on heating with acidified dichromate(VI), produces a ketone called propanone. No other products can be obtained even with prolonged refluxing of an excess of the reactants.

$$\underset{\overset{\displaystyle |}{H}}{\overset{\overset{\displaystyle CH_3}{|}}{H_3C-C-OH}} + [O] \longrightarrow \underset{\text{propanone}}{H_3C-\overset{\overset{\displaystyle O}{\|}}{C}-CH_3} + H_2O$$

Typically, ketones have pleasant odours resembling wood and fruit. For example, heptan-2-one is present in oil of cloves as well as in some fruits.

Complete oxidation: combustion

Ethanol is used as a fuel in the form of methylated spirit; it burns with a pale blue flame, but it is rather volatile and the flame is hard to see in sunlight, so accidents can occur when refilling stoves. Many campers favour it for cooking as it may be carried in lighter containers than those needed for gas. The equation for the complete combustion of ethanol is:

$$C_2H_5OH(l) + 3O_2(g) \longrightarrow 2CO_2(g) + 3H_2O(l)$$
$$\Delta H = -1367.3 \, kJ \, mol^{-1}$$

C–C and C–H bonds are broken in this reaction, as ethanol is completely oxidised to carbon dioxide and water.

In some countries ethanol is blended with petrol to make a cheaper motor fuel. Methanol is used as a fuel for US ChampCar racing.

Dehydration to alkenes

The elimination of water from an alcohol will produce an alkene. For example, if ethanol vapour is passed over hot pumice, one water molecule is eliminated from each ethanol molecule. Both C–O and C–H bonds are broken, producing ethene and water:

$$C_2H_5OH(g) \longrightarrow CH_2=CH_2(g) + H_2O(g)$$

The pumice acts as a catalyst; the pores of the ceramic provide a large surface area. The high temperature, catalyst and large surface area all increase the rate of this reaction. The reaction is often referred to as **dehydration**, because a water molecule is removed. Figure 13.5 shows how you can prepare a small sample of ethene by this method. (Note the similarity to the cracking of an alkane – see Chapter 11.)

An alternative method of dehydrating an alcohol involves heating the alcohol with an excess of concentrated sulfuric acid at about 170°C. It is important to use an excess of acid, because an excess of ethanol leads to a different reaction producing an ether (ethoxyethane) and water:

$$2C_2H_5OH(l) \longrightarrow C_2H_5OC_2H_5(l) + H_2O(l)$$

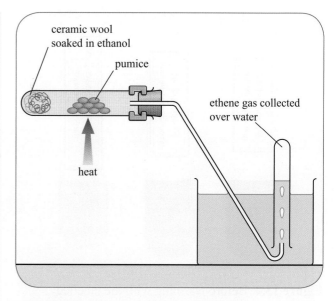

Figure 13.5 The dehydration of ethanol.

SAP

6 **a** Draw the displayed formula for the organic product produced when propan-2-ol vapour is passed over heated pumice.
   b Write a balanced equation for this reaction.

Answer

The uses of alcohols

Fuels

Alcohols have high enthalpies of combustion. The uses of ethanol and methanol as fuels have already been mentioned (see also Chapter 16). Unleaded petrol contains about 5% of methanol and 15% of an ether known as MTBE (which is made from methanol). The rapid increase in the number of vehicles which can use unleaded fuel caused MTBE production to grow faster than that of any other chemical.

Solvents

As alcohols contain the polar hydroxyl (–OH) group and a non-polar hydrocarbon chain, they make particularly useful solvents. They will mix with many other non-polar compounds and with polar compounds. Methanol and ethanol will also dissolve some ionic compounds.

Summary

Glossary

- Alcohols are miscible or partly miscible with water because hydrogen bonds can form between water molecules and the –OH group.

- Hydrogen bonding causes alcohols to be less volatile than hydrocarbons with molecules of similar size.

- Industrially, ethanol is produced by the addition reaction of steam to ethene using H_3PO_4 as a catalyst.

- Alcoholic drinks are produced by the fermentation of sugar. They contain ethanol. Excessive alcohol consumption has serious consequences.

- An alcohol will react with a carboxylic acid to form an ester and water.

- A primary alcohol can be oxidised to an aldehyde by heating the alcohol gently with acidified dichromate(VI) and distilling out the aldehyde immediately.

- A primary alcohol can be oxidised to a carboxylic acid by refluxing the alcohol with acidified dichromate(VI) and distilling out the carboxylic acid after at least fifteen minutes of refluxing.

- A secondary alcohol can be oxidised to a ketone by refluxing the alcohol with acidified dichromate(VI).

- Acidified dichromate(VI) changes colour from orange to green when a primary or secondary alcohol is oxidised by it.

- Tertiary alcohols cannot be oxidised by refluxing with acidified dichromate(VI).

- Complete oxidation of alcohols occurs on combustion to form carbon dioxide and water.

- Elimination of water from an alcohol produces an alkene; the reaction is a dehydration. Dehydration may be carried out by passing ethanol vapour over heated pumice.

- Both methanol and ethanol are useful fuels. Alcohols are also used as solvents.

Questions

1 The table below shows information about some alcohols which form part of an homologous series.

Name	Formula	Boiling point /°C	Relative molecular mass
methanol	CH_3OH	65	32
ethanol	C_2H_5OH	78	46
propan-1-ol	C_3H_7OH	97	60
butan-1-ol	C_4H_9OH		74
pentan-1-ol	$C_5H_{11}OH$	138	
hexan-1-ol	$C_6H_{13}OH$	158	102

continued

a i Identify the functional group common to all alcohols. [1]

    ii What is the general formula for these alcohols? [1]

    iii What is the formula of the next alcohol in the series? [1]

b Calculate the relative molecular mass of pentan-1-ol. [1]

c i Plot a graph of boiling point against number of carbon atoms in a molecule of the alcohol using axes as shown below.

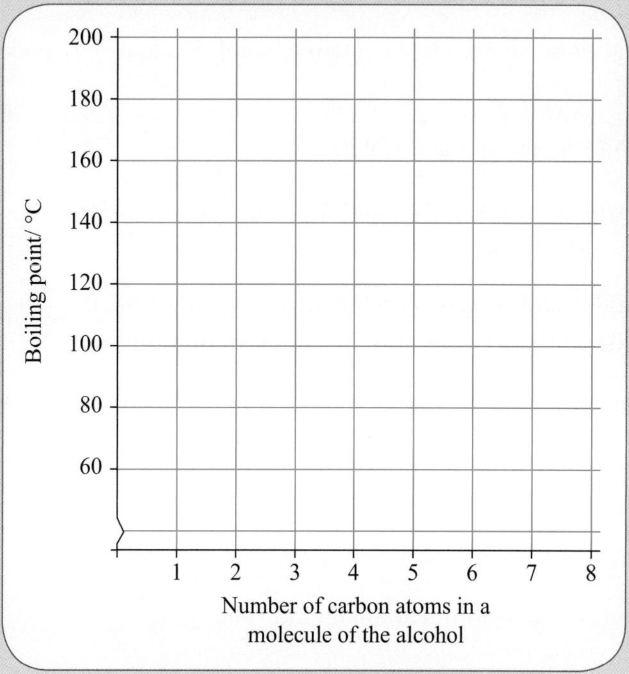

[2]

Use the graph to estimate the boiling points of

- butan-1-ol [1]
- $C_8H_{17}OH$. [1]

    ii State the connection between boiling point and the relative molecular mass of these alcohols. [1]

OCR Chemistry AS (2812) Jan 2001 [Total 9]

Answer

2 This question is about two alcohols, ethanol and propan-2-ol, $CH_3CH(OH)CH_3$.

    a Ethanol can be formed by fermentation of glucose, $C_6H_{12}O_6$.

       i Write a balanced equation, including state symbols, for the formation of ethanol by fermentation. [2]

       ii Fermentation only occurs in the presence of yeast. State <u>two</u> other essential conditions. [2]

       iii How would you know when fermentation of glucose is complete? [1]

    b Propan-2-ol is flammable and readily burns.

       Write a balanced equation for the complete combustion of propan-2-ol. [2]

Hint

Hint

continued

c Compound **D**, shown below, can be used as a solvent for plastics and fats and is also used in perfumery.

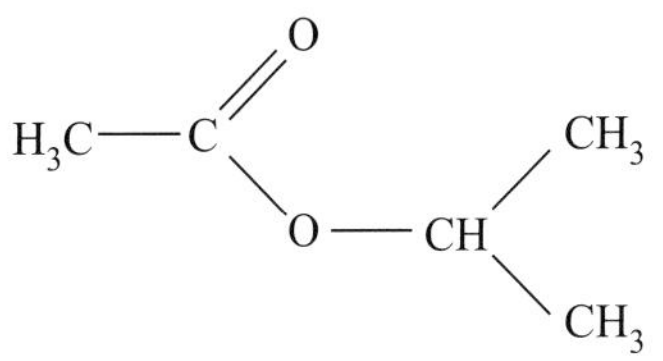

compound **D**

Compound **D** can be prepared from propan-2-ol and another organic compound. Identify this other compound. [1]

OCR Chemistry AS (2812) Jan 2007 [Total 8]

Answer

3 Compound **A** is an alcohol with the following percentage composition by mass: C, 60.0%; H, 13.3%; O, 26.7%.
  a i Calculate the empirical formula of compound **A**. [2]
  ii The relative molecular mass of compound **A** is 60. Show that this agrees with the molecular formula C_3H_8O. [1]
  b Compound **A** is one of two possible isomers that are both alcohols. Draw the structure of each isomer. [2]
  c Compound **A** was refluxed with an excess of an oxidising agent to form compound **B**, a ketone.
  i State a suitable oxidising agent for this reaction. [1]
  ii State the colour change you would see during this oxidation.
      From........................ to............................ [2]
  iii Explain what is meant by the term *reflux*. [1]
  iv Which isomer of compound **A** from part **b** was oxidised? Explain your answer. [1]
  v Write a balanced equation for the oxidation of compound **A** to form compound **B**.
      Use [O] to represent the oxidising agent. [2]

OCR Chemistry AS (2812) June 2003 [Total 12]

Answer

continued

4 A student was given the following instructions for the oxidation of an alcohol, C_3H_8O.
*To 20 cm³ of water in a flask, carefully add 6 cm³ of concentrated sulfuric acid, and set
up the apparatus as shown below.*

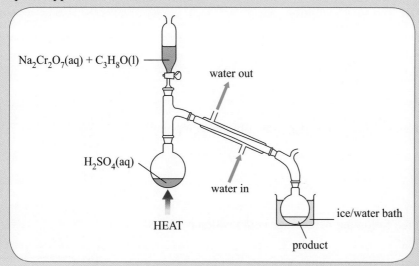

*Make up a solution containing 39.3 g of sodium dichromate(VI), $Na_2Cr_2O_7$, in 15 cm³
of water, add 18.0 g of the alcohol, C_3H_8O, and pour this mixture into the
dropping funnel.*
*Boil the acid in the flask. Add the mixture from the dropping funnel at such a rate that
the product is slowly collected.*
Re-distil the product and collect the fraction that boils between 48 °C and 50 °C.

a Identify the possible isomers of the alcohol C_3H_8O. [2]

b The balanced equation for the reaction is:

$$3C_3H_8O + Na_2Cr_2O_7 + 4H_2SO_4 \longrightarrow 3C_3H_6O + Na_2SO_4 + Cr_2(SO_4)_3 + 7H_2O$$

   i The mass of $Na_2Cr_2O_7$ used was 39.3 g. Calculate how many moles of
     $Na_2Cr_2O_7$ were used. (The molar mass of $Na_2Cr_2O_7$ is 262 g mol⁻¹.) [1]

   ii The amount of C_3H_8O used was 0.300 mol. Explain whether C_3H_8O or
     $Na_2Cr_2O_7$ was in excess. [1]

   iii State the colour change that the student would observe during the reaction. [2]

c The student obtained 5.22 g of the carbonyl compound, C_3H_6O.

   i Calculate how many moles of C_3H_6O were produced in the experiment. [2]

   ii The theoretical yield of C_3H_6O is 0.300 mol. Calculate the percentage yield of
     C_3H_6O obtained by the student. [1]

Hint

Halogenoalkanes

e-Learning

Objectives

This chapter deals with the properties and reactions of the simple halogenoalkanes. These have the general formula $C_nH_{2n+1}X$, where X is a halogen atom: one of F, Cl, Br or I. They are named by prefixing the name of the alkane with fluoro, chloro, bromo or iodo and a number to indicate the position of the halogen on the hydrocarbon chain. For example, $CH_3CH_2CHClCH_3$ is 2-chlorobutane.

SAQ

1 Name the following compounds.

Hint

   a $CH_3CH_2CH_2I$
   b $CH_3CHBrCH_3$
   c $CBrF_2CBrF_2$

Answer

The classification of halogenoalkanes

Halogenoalkanes are classified according to their structures (Figure 14.1) in a similar way to the classification of alcohols (see Chapter 13).

● In a *primary halogenoalkane* such as 1-chlorobutane, the halogen atom is covalently bonded to a carbon atom which, in turn, has a covalent bond to just *one* other carbon atom (or none in the case of a halogenomethane).

● In a *secondary halogenoalkane* such as 2-chlorobutane, the halogen atom is covalently bonded to a carbon atom which, in turn, has covalent bonds to *two* other carbon atoms.

● In a *tertiary halogenoalkane* such as 2-chloro-2-methylpropane, the halogen atom is covalently bonded to a carbon atom which, in turn, has covalent bonds to *three* other carbon atoms.

SAQ

2 a Which type of isomerism is shown by the compounds in Figure 14.1?

   b Draw the structural formula of one further isomer of C_4H_9Cl. Is this a primary, a secondary or a tertiary chloroalkane?

Answer

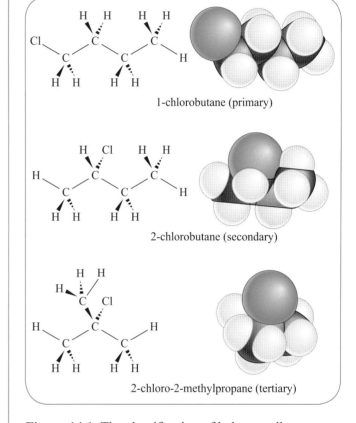

1-chlorobutane (primary)

2-chlorobutane (secondary)

2-chloro-2-methylpropane (tertiary)

Figure 14.1 The classification of halogenoalkanes as primary, secondary or tertiary.

Physical properties

Typically, halogenoalkanes and halogenoarenes are volatile liquids that do not mix with water.

SAQ

3 a Explain why 1-chloropropane, C_3H_7Cl, is a liquid at room temperature (boiling point = 46.7 °C) whereas butane, C_4H_{10}, is a gas (boiling point = 0 °C).

Hint

   b Why is it that halogen compounds such as 1-chloropropane do not mix with water?

Answer

Nucleophilic substitution

The predominant type of chemical reaction shown by halogenoalkanes involves **substitution** of the halogen by a variety of other groups. As the halogen atom is more electronegative than carbon, the carbon–halogen bond is polar:

In a substitution reaction, the halogen atom will leave as a halide ion. This means that the atom or group of atoms replacing the halogen atom must possess a lone-pair of electrons. This lone-pair is donated to the slightly positive, $\delta+$, carbon atom, and a new covalent bond is formed. A chemical that can donate a pair of electrons, with the subsequent formation of a covalent bond, is called a **nucleophile** (see Chapter 10).

The mechanism for the nucleophilic substitution of bromine in bromomethane by a hydroxide ion is:

Nucleophilic attack is followed by loss of the bromine atom as a bromide ion. A new covalent bond between the nucleophile and carbon is formed. Overall a substitution reaction has occurred.

Some nucleophiles possess an overall negative charge (i.e. are a negatively charged ion), but this is not necessary for nucleophilic behaviour. Nucleophiles which will substitute for the halogen atom in halogenoalkanes include the hydroxide ion, water and ammonia.

Hydrolysis

As the halogenoalkanes do not mix well with water, they are mixed with ethanol (which acts as a solvent for water and the halogenoalkane) before being treated with dilute aqueous sodium hydroxide. Warming the mixture causes a nucleophilic substitution to occur, producing an alcohol. The same **hydrolysis** reaction will occur more slowly without alkali, if the halogenoalkane is mixed with ethanol and water. The equation for the hydrolysis of bromoethane with alkali is:

$$CH_3CH_2Br + OH^- \longrightarrow CH_3CH_2OH + Br^-$$

The equation for the hydrolysis of bromoethane with water is:

$$CH_3CH_2Br + H_2O \longrightarrow CH_3CH_2OH + HBr$$

SAQ

4 Write a balanced equation for the alkaline hydrolysis of 2-bromo-2-methylpropane, using structural formulae for the organic compounds. Name the organic product.

> Hint

> Answer

The relative rates of hydrolysis of 1-halogenobutanes

Hydrolysis gets easier as you change the halogen from chlorine to bromine to iodine. At first sight this may seem strange, since the polarity of the carbon–halogen bond decreases from chlorine to iodine. You might expect that a less positively charged carbon atom would react less readily with the nucleophilic hydroxide ion.

However, examination of the carbon–halogen bond enthalpies (Table 14.1) shows that the strength of the bond decreases significantly from C–Cl to C–I.

Bond	Bond enthalpy /kJ mol^{-1}
C–F	467
C–Cl	340
C–Br	280
C–I	240

Table 14.1 Bond enthalpies of carbon–halogen bonds.

This suggests that the ease of breaking the carbon–halogen bond is more important than the size of the positive charge on the carbon atom. A nucleophile may be attracted strongly to the carbon atom but it will not displace the halogen unless the carbon–halogen bond breaks.

The carbon–fluorine bond does not undergo nucleophilic substitution because it is the strongest carbon–halogen bond. Despite the high polarity of the bond, no nucleophile will substitute for the fluorine. This accounts for the very high stability of the fluoroalkanes.

You can observe the relative rates of hydrolysis of halogenoalkanes by adding aqueous ethanolic silver nitrate to solutions of different halogenoalkanes in ethanol, and timing the first appearance of a silver halide precipitate (Figure 14.2). This will form as soon as sufficient halide ions have been formed by the hydrolysis of the halogenoalkane. For example:

$$Ag^+(aq) + Cl^-(aq) \longrightarrow AgCl(s)$$

- 1-chlorobutane slowly produces a faint white precipitate of silver chloride.
- 1-bromobutane produces a creamy white precipitate of silver bromide more rapidly.
- 1-iodobutane produces a yellow precipitate of silver iodide most rapidly.

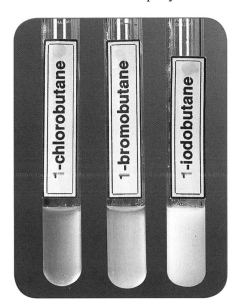

Figure 14.2 The hydrolysis of halogenoalkanes by aqueous ethanolic silver nitrate after 15 minutes. The silver nitrate produces an insoluble precipitate of a silver halide.

The uses of organic halogen compounds

As a functional group, the halogen atom provides chemists with useful routes to the synthesis of other compounds. This use of organic halogen compounds is more important than their usefulness as products. For example, the synthesis of a medicine such as ibuprofen requires alkyl groups to be joined to benzene. This is achieved by reactions between halogenoalkanes and benzene. Ibuprofen is an anti-inflammatory medicine that brings relief to many people suffering from rheumatoid arthritis (which causes painful inflammation of the joints).

Halogenoalkanes which do have direct applications include:
- the polymers poly(chloroethene), better known as PVC, and poly(tetrafluoroethene) (Figure 14.3)
- several CFCs, for example dichloro-difluoromethane or trichlorofluoromethane, have been used as refrigerants, aerosol propellants or blowing agents for producing foamed polymers (see the next section for the negative effects these have had on the environment and how chemists are developing alternative compounds)

Figure 14.3 Poly(tetrafluoroethene), PTFE, is used in the non-stick coating on saucepans and in waterproof clothing. *continued*

- CCl_2FCClF_2 as a dry cleaning solvent or degreasing agent for printed circuit boards
- firefighting compounds such as bromochlorodifluoromethane, known as BCF (Figure 14.4). BCF is used in some fire extinguishers. The high temperatures in fires break BCF down, producing free radicals such as $Br\bullet(g)$. These react rapidly with other free radicals produced during combustion, quenching the flames.

Figure 14.4 Bromochlorodifluoromethane, BCF, is very effective at extinguishing fires. However, it is not now in general use because the breakdown products are poisonous.

Chemists and the environment

Trouble in the ozone layer

Chlorofluorocarbons (CFCs) are regularly blamed for causing damage to our environment. They absorb much more infrared radiation per molecule than carbon dioxide. However, their contribution to the 'greenhouse effect' is very low due to their very low abundance in the atmosphere (carbon dioxide is the main cause of the 'greenhouse effect'). More importantly, CFCs are responsible for a thinning of the protective ozone layer (Figure 14.5) in the **stratosphere**. (High-level ozone absorbs significant quantities of harmful ultraviolet radiation which can cause skin cancer.)

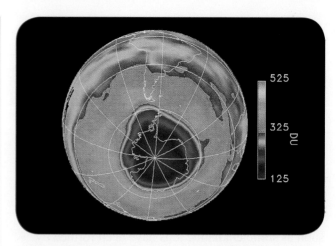

Figure 14.5 Representation of satellite measurements of the ozone 'hole' over Antarctica. Ozone concentration is measured in Dobson units (DU). The depletion of ozone reaches a maximum in October, the Antarctic spring, and is thought to be due mainly to the effects of chlorofluorocarbons (CFCs).

CFCs are still used in air conditioners and were formerly used as refrigerants and aerosol propellants. They were chosen for these purposes because they are gases that liquefy easily when compressed. They are also very unreactive, non-flammable and non-toxic.

The high stability of CFCs has been part of the cause of the problems in the ozone layer. This has enabled the concentration of CFCs to build up in the atmosphere. When they reach the stratosphere, CFCs absorb ultraviolet radiation, which causes *photodissociation* of carbon–chlorine bonds. For example:

$$CF_2Cl_2(g) \xrightarrow{\text{UV light}} \bullet CF_2Cl(g) + Cl\bullet(g)$$

Very reactive chlorine free radicals, $Cl\bullet(g)$, are formed.

These radicals catalyse the decomposition of ozone to oxygen. The overall reaction equation is:

$$2O_3(g) \longrightarrow 3O_2(g)$$

The developments of compounds such as CFCs and BCF (see box above) illustrate aspects of the work of chemists which benefit society and the environment. The development of BCF has helped to prevent unwanted fires which cause considerable economic and environmental damage.

Often, chemists respond to the needs of society by developing new, safer products. This happened in 1928 when Thomas Midgeley (an American engineer) was asked to find a safer alternative to the early refrigerants sulfur dioxide and ammonia, which are toxic. Leakage of ammonia or sulfur dioxide caused a number of deaths in the 1920s. As a result, some parts of the USA took the drastic measure of banning these early domestic refrigerators. Thomas Midgeley suggested the use of CF_2Cl_2 and demonstrated its lack of toxicity by inhaling the gas and blowing out a candle!

In recent years, we have learnt that the introduction of CFCs like CF_2Cl_2 was not without environmental consequences. These consequences have been identified by chemists and other scientists. An understanding of the processes involved has also helped in the search for safer replacements.

Nowadays, chemists are designing new 'ozone-friendly' chemicals to replace the destructive CFCs. The compound 1,1,1,2-tetrafluoroethane, CF_3CH_2F, is now being manufactured as an appropriate alternative. This and other HFCs (hydro-fluoro-carbons), plus hydrocarbons, are now used in aerosols. The presence of the hydrogen atoms in their compounds increases their reactivity relative to CFCs, so that they are broken down in the lower atmosphere much more rapidly, i.e. they are biodegradable alternatives, they don't reach the stratosphere. CFCs were also used as a 'blowing agent' in the manufacture of foam plastics. Now we use carbon dioxide as the blowing agent. The use of such alternatives will hopefully allow the ozone layer to recover, although the natural process may well be slow.

Summary

Glossary

- Halogenoalkanes have the general formula $C_nH_{2n+1}X$, where X is F, Cl, Br or I. They are named by prefixing the name of the alkene with fluoro, chloro, bromo or iodo and a number to indicate the position of the halogen on the hydrocarbon chain.

- Halogenoalkanes react with a wide range of nucleophiles. Nucleophiles possess a pair of electrons, which is donated to the positively charged carbon atom in a C–X bond. The halogen is substituted by the nucleophile, which forms a new covalent bond to the carbon atom attacked.

- Bromoethane is attacked by nucleophiles, e.g. when the nucleophile is water or aqueous alkali, ethanol will be the organic product formed.

- The reactivity of different halogenoalkanes depends on the relative strengths of the C–X bonds. The C–F bond is very unreactive due to its high bond energy. The rate of hydrolysis of the C–X bond increases from chlorine to iodine as the bond energy decreases.

- Poly(tetrafluoroethene) (a fluoroalkane) is an important polymer valued for its inertness, high melting point and smooth slippery nature. It is used for non-stick saucepans and waterproof clothing. PVC, poly(chloroethene), is a very widely used material.

- Chlorofluoroalkanes have been used extensively as they are inert, non-toxic, non-flammable compounds that have appropriate physical properties for use as propellants, refrigerants, blowing agents or cleaning solvents.

- Chemists play an important role, for example in the development of alternatives to CFCs to provide for the perceived needs of society and to minimise damage to the environment.

continued

- CFCs, which were used extensively in refrigerators and aerosol cans, are very unreactive. Their low reactivity means that they stay in the atmosphere for a long time. They are broken down by ultraviolet radiation to release chlorine free radicals, which have reduced the concentration of ozone in the stratosphere.

- CF_3CH_2F is being introduced as a replacement for various CFCs in refrigerants and aerosols.

- Bromochlorodifluoromethane (BCF), CF_2ClBr, has been used in some fire extinguishers. It is not now in general use because it produces poisonous breakdown products.

Questions

1 Halogenoalkanes are polar molecules and react with nucleophiles.
   a The displayed formula of chloromethane is shown below. Label the dipole on the C–Cl bond.

[1]

   b Chloromethane is hydrolysed by aqueous sodium hydroxide in a nucleophilic substitution reaction. An equation for this reaction is shown below.

$$CH_3Cl + OH^- \longrightarrow CH_3OH + Cl^-$$

      i What is meant by the term *nucleophile*? [1]
      ii Show, with the aid of curly arrows, the mechanism of this hydrolysis. [2]
   c i What would happen to the rate of hydrolysis if chloromethane were replaced
        by iodomethane? Explain your answer. [2]   Hint
      ii Suggest a reagent that could be used to compare the rate of the hydrolysis of
        both halogenoalkanes. State what you would see in each case. [5]
   d Compound **D** has the following composition by mass:
      C, 12.76%; H, 2.13%; Br, 85.11%.
      i Calculate the empirical formula of **D**. Show your working. [2]
      ii Compound **D** has a relative molecular mass of 187.8.
        What is the molecular formula of **D**? Show your working. [2]
      iii Identify two possible structural isomers of **D**. [2]
   e Complete hydrolysis of **D** forms ethane-1,2-diol, which is used as antifreeze in cars.
      i Draw the displayed formula of ethane-1,2-diol. [1]
      ii Which of the isomers in **d iii** is **D**? Explain your answer. [1]   Hint
      iii Write a balanced equation for the complete hydrolysis of **D**. [2]

OCR Chemistry AS (2812) Jan 2002   [Total 21]

Answer

continued

2 Halogenoalkanes undergo hydrolysis with hot aqueous NaOH to form alcohols.

  a **i** Write a balanced equation for the hydrolysis of 1-bromobutane, C_4H_9Br. [1]

     ii Describe, with the aid of curly arrows and relevant dipoles, the mechanism for the hydrolysis of 1-bromobutane with hydroxide ions, OH^-. [3]

  b The rates of hydrolysis of 1-chlorobutane, 1-bromobutane and 1-iodobutane were compared by an experiment using $AgNO_3$ solution. Aqueous $AgNO_3$ was added to each halogenoalkane. The rates of hydrolysis were measured by timing the appearance of the silver halide precipitate.

    i Place the three halogenoalkanes in the order of their rates of hydrolysis with the compound with the fastest rate first. [1]

    ii Explain the order in **i**. [1]

OCR Chemistry AS (2812) June 2003   [Total 6]

Answer

3 **a** Trifluorochloromethane, CF_3Cl, is an example of a chlorofluorocarbon, CFC, that was commonly used as a propellant in aerosols. Nowadays, CFCs have limited use because of the damage caused to the ozone layer.

    i Draw a diagram to show the shape of a molecule of CF_3Cl. [1]

    ii Predict an approximate value for the bond angles in a molecule of CF_3Cl. [1]

    iii Suggest a property that made CF_3Cl suitable as a propellant in an aerosol. [1]

  b **i** When CFCs are exposed to strong ultraviolet radiation in the upper atmosphere, homolytic fission takes place to produce free radicals. Explain what is meant by the term *homolytic fission*. [2]

    ii Suggest which bond is most likely to be broken when CF_3Cl is exposed to ultraviolet radiation. Explain your answer. [1]

Hint

    iii Identify the <u>two</u> free radicals most likely to be formed when CF_3Cl is exposed to ultraviolet radiation. [2]

OCR Chemistry AS (2812) Jan 2005   [Total 8]

Answer

Chapter 15

Modern analytical techniques

e-Learning

Objectives

Testing drivers

In Chapter 13 you saw how ethanol is the alcohol found in all alcoholic drinks and how some people abuse this substance. 'Drink driving' is a major factor in many fatal accidents. Innocent people can be killed because the driver of another car is 'over the limit'. The legal limit for driving in the UK is 80 mg of ethanol per 100 cm³ of blood.

If police suspect a driver has been drinking they conduct a road-side breathalyser test in which the driver blows into a small hand-held device (see Figure 15.1). This gives the police officer an approximate idea of the alcohol level of the suspect and indicates whether or not they might be over the limit.

A more accurate test is performed at a police station where an infrared spectrometer is used to test the driver's breath. If the breath sample test is positive, the driver can then opt for a further test on their blood to ensure there is no error.

Figure 15.1 This initial breathalyser test tells the police officer whether a more accurate test using infrared spectroscopy is needed at a police station.

continued

You can see the infrared spectrum of ethanol in Figure 15.2. Infrared radiation passes through a sample of ethanol in the machine, causing the bonds in the ethanol molecule to vibrate (e.g. bend or stretch more vigorously). The ethanol absorbs the infrared radiation when this happens. The troughs in the spectrum correspond to the absorption of different frequencies of infrared radiation by different bonds.

Water in the breath can interfere with the O–H bond absorbance (at 3340 cm⁻¹) but the sharp trough due to a C–H bond (at 2950 cm⁻¹) can be used by a computer to calculate the percentage of ethanol present.

Structural identification using infrared spectroscopy

In a modern infrared spectrometer, a beam of infrared radiation is passed through a sample of the chemical to be identified. Computer analysis enables the absorbance of radiation to be measured at different frequencies. Study of the resulting spectrum enables the presence (or absence) of particular functional groups to be established.

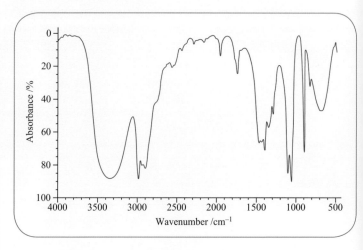

Figure 15.2 The infrared spectrum of ethanol. Notice that absorbance increases in a downward direction.

An unusual unit is used to measure frequency, the wavenumber or cm^{-1}. Table 15.1 shows the absorption frequencies which we shall use in this unit.

Functional group	Location	Wave-number / cm^{-1}	Absorb-ance
O–H	alcohols	3230–3550	strong, broad
O–H	carboxylic acids	2500–3500	medium, very broad
C=O	aldehydes, ketones, acids and esters	1680–1750	strong, sharp

Table 15.1 Infrared absorption frequencies of some functional groups.

Look again at the infrared spectrum of ethanol in Figure 15.2. Most of the absorptions are sharp and some overlap. The absorption of interest is the strong, broad absorption at about 3420 cm^{-1}, which shows the presence of the O–H group. (The O–H absorptions are usually broadened by the effect of hydrogen bonding between molecules.)

If ethanol is warmed with ethanoic acid in the presence of a few drops of concentrated sulfuric acid, ethyl ethanoate (see Chapter 13) is formed:

$$CH_3 - C \overset{\displaystyle O}{\underset{\displaystyle O - CH_2CH_3}{\big\langle}}$$

How do we know that the ester is present? The infrared spectrum of a pure sample of ethyl ethanoate is shown in Figure 15.3.

Note the absence of the strong, broad absorption from the O–H group in ethanol. Instead, there is a strong, sharp absorption at 1720 cm^{-1} which arises from the C=O group.

When ethanol is refluxed with an excess of potassium dichromate and dilute sulfuric acid,

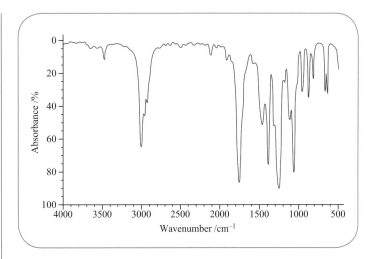

Figure 15.3 The infrared spectrum of ethyl ethanoate.

ethanoic acid is formed (see Chapter 13). Infrared spectroscopy again helps us to distinguish the product from ethanol. The infrared spectrum of a pure sample of ethanoic acid is shown in Figure 15.4.

The strong, very broad absorption between 2500 and 3500 cm^{-1} is partly due to the O–H group in the acid. (Although groups containing C–H bonds also absorb in this region, they are of little help in identification as such bonds are present in all organic compounds.) Compare the spectrum of ethanoic acid with that of ethanol. There are clear differences between the two spectra.

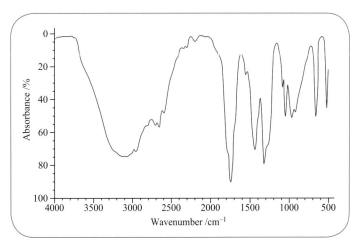

Figure 15.4 The infrared spectrum of ethanoic acid.

SAQ

1 a Draw the structural formula of ethanoic acid. _Hint_

b Apart from the O–H group, which other group can be identified in the spectrum for ethanoic acid? Use Figure 15.4 to record the absorbance and frequency of this bond. _Answer_

- Oxidation of a primary alcohol under milder conditions produces an aldehyde, for example ethanal, CH_3CHO, from ethanol, C_2H_5OH.
- Oxidation of secondary alcohols produces ketones, for example propanone, CH_3COCH_3, from propan-2-ol, $CH_3CHOHCH_3$ (see Chapter 13).

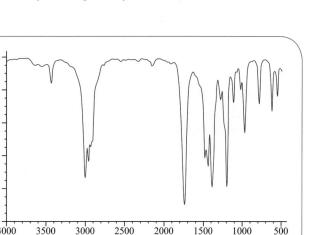

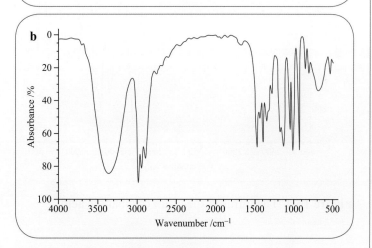

Figure 15.5 The infrared spectra of butanone and butan-2-ol, for use with SAQ 2.

SAQ

2 The infrared spectra of butan-2-ol and butanone are shown in Figure 15.5. _Hint_

Identify which of **a** and **b** is butanone. Explain your reasoning. _Answer_

Mass spectrometry

Determination of A_r from mass spectra

You have used relative atomic masses and relative isotopic masses in Chapter 2. You may have wondered how tables of relative atomic masses have been obtained. An instrument called a **mass spectrometer** is used for this purpose; such instruments are too expensive to be found in most schools or colleges. Academic or industrial chemical laboratories may have one or two, depending on their needs and resources. Mass spectrometers have even been sent into space (Figure 15.11).

When a mass spectrometer is used to find the relative atomic mass of an element it does two things. It measures the mass of each different isotope of the element and it measures the relative abundance of each isotope. A vaporised sample of the element is injected into the mass spectrometer. Then the atoms of the element are turned into ions with a positive charge. This is done by electron bombardment. Then the mass spectrometer separates the atoms with different masses, measures the masses of the atoms and counts them.

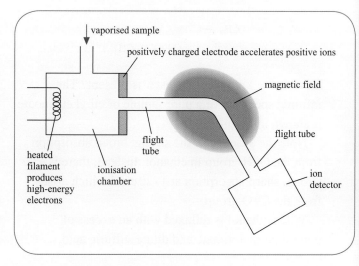

Figure 15.6 Simplified diagram of a mass spectrometer.

A simplified diagram of a mass spectrometer is shown in Figure 15.6. (NB You will not be tested on the details of how a mass spectrometer works.)

The results from a mass spectrometer are displayed on a computer monitor as a chart of abundance against mass (see Figure 15.7 for example, the mass spectrum of germanium). The abundance is on the vertical axis. To be strictly correct the horizontal axis displays the mass-to-charge ratio (m/e), but you will often see this axis simply labelled 'mass'.

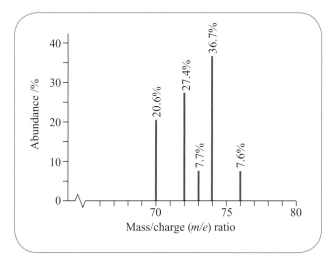

Figure 15.7 The mass spectrum of germanium, Ge.

We can use the data from the mass spectrum to calculate the relative atomic mass, A_r, of germanium. Each peak in the mass spectrum corresponds to a different isotope of germanium. So in the sample we have 20.6% ^{70}Ge, 27.4% ^{72}Ge, 7.7% ^{73}Ge, 36.7% ^{74}Ge and 7.6% ^{76}Ge. Remember that the relative atomic mass takes into account the different proportions of naturally occurring isotopes in an element (see Chapter 1). Therefore we have to calculate the weighted mean of the relative masses of the isotopes identified:

A_r of germanium =

$$\frac{(20.6 \times 70) + (27.4 \times 72) + (7.7 \times 73) + (36.7 \times 74) + (7.6 \times 76)}{100}$$

$= 72.7$

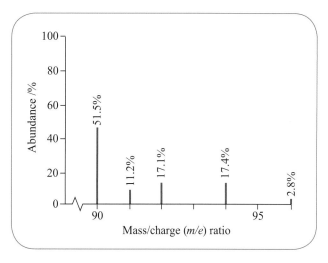

Figure 15.8 The mass spectrum of zirconium, Zr.

SAQ

3 a Look at the mass spectrum of the element zirconium in Figure 15.8.
List the isotopes present in zirconium.

b Use the percentage abundance of each isotope to calculate the relative atomic mass of zirconium.

Answer

Some modern mass spectrometers can be set up to determine isotopic masses to four or five decimal places. Figure 15.9 below shows a photograph of such a spectrometer.

Using mass spectra to identify unknown substances

The mass spectrometer can also be used to identify unknown organic substances, such as in testing athletes for prohibited drugs. A vaporised sample of the unknown substance is injected into the mass

Figure 15.9 A high-resolution mass spectrometer. It measures isotopic masses and their abundances very accurately.

The Viking space probe

Figure 15.11 The surface of Mars, seen from the *Viking 2* space probe.

When the two *Viking* space probes were launched by NASA in 1975, they carried mass spectrometers. The purpose of these spectrometers was to look for traces of organic compounds on the surface of Mars. Scientists had put forward the hypothesis that living organisms would have left behind traces of organic compounds. However, the soil sampled on Mars showed no trace of organic compounds.

spectrometer. The molecules of the substance are then changed into singly charged positive ions with the same relative molecular mass as the molecules of the unknown compound. These are called the 'molecular ions'. On the mass spectrum these will appear as the peak at the highest mass. So this peak gives us the relative molecular mass of the unknown compound.

This greatly reduces the number of possible compounds that the unknown could be. However, there will still be alternatives, e.g. if the relative molecular mass is 45, the molecular formula could be C_2H_7N or CH_3NO or any other molecule with this mass.

Organic molecules can also be broken up by the bombarding electrons in the mass spectrometer, to produce positively charged 'fragments' of the original molecule. Each fragment has a lower mass than the molecular ion. Look at the mass spectrum of ethanol in Figure 15.10.

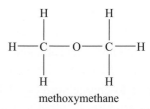
Every new compound discovered has its own characteristic mass spectrum which is stored in a computer database. The computer then matches the mass spectrum of the compound to be identified with the correct one from its database. This is called '*fingerprinting*'. Infrared spectra are also used for fingerprinting. The fragments formed in a mass spectrometer also can help chemists work out the structure of newly discovered compounds.

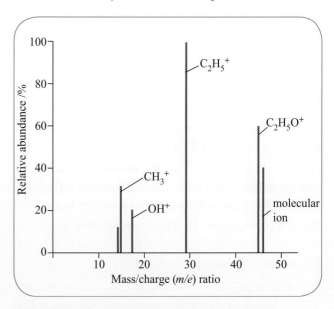

Figure 15.10 The mass spectrum of ethanol, C_2H_5OH. The main fragments have been labelled. Can you see how fragmentation of the ethanol molecule can give rise to each of these ions?

Summary

Glossary

- Infrared spectrometers analyse the radiation (energy) absorbed by different covalent bonds. This energy causes the bonds to vibrate in a number of different ways, e.g. bending or stretching. Different bonds absorb different parts of the infrared spectrum so they can be identified as being present in a sample.

- Infrared spectroscopy assists in the identification of alcohols, aldehydes and ketones, carboxylic acids and esters by the presence or absence of O–H and C=O absorption frequencies in their spectra.

- Infrared spectroscopy is used to measure the ethanol content of a driver's breath when they are suspected of drink driving.

- Mass spectra of elements enable isotopic abundances to be found and relative atomic masses to be calculated.

- We can also use mass spectroscopy to identify unknown organic compounds by 'fingerprinting' (matching the spectrum to other known spectra). The fragmentation peaks give us clues about the structure of the original molecule.

Questions

Use these characteristic infrared absorptions in organic molecules to help you answer questions **1** to **3**.

Bond	Location	Wavenumber /cm^{-1}
C–O	alcohols, esters	1000–1300
C=O	aldehydes, ketones, carboxylic acids, esters	1680–1750
O–H	hydrogen bonded in carboxylic acids	2500–3300 (broad)
N–H	primary amines	3100–3500
O–H	hydrogen bonded in alcohols, phenols	3230–3550
O–H	free	3580–3670

1 The infrared spectrum of compound **B** is shown below.
Identify:

- **a** the absorption responsible for peak **X** [1]
- **b** the absorption responsible for peak **Y** [1]
- **c** the functional group present in compound **B**. [1]

OCR Chemistry AS (2812) June 2003 [Total 3]

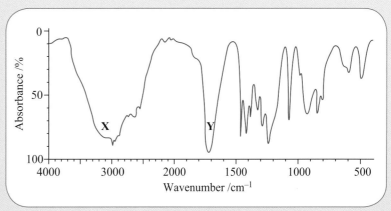

Answer

continued

2 A student refluxed a solution of chlorocyclohexane and NaOH.
The organic product was separated and analysed by infrared spectroscopy.
The infrared spectrum of the organic product is shown below.

Hint

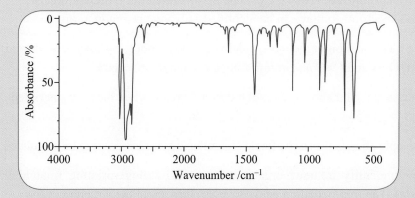

The student used the infrared spectrum to conclude that cyclohexene had been produced.
Use the table on page 177 to explain why the student was justified in ruling out
cyclohexanol as the organic product. [1]

OCR Chemistry AS (2812) June 2006 [Total 1]

Answer

3 Ethanal was analysed by infrared spectroscopy. Use the table on page 177 to justify
which of the three spectra shown below is most likely to be that of ethanal. Give
three reasons. [3]

Hint

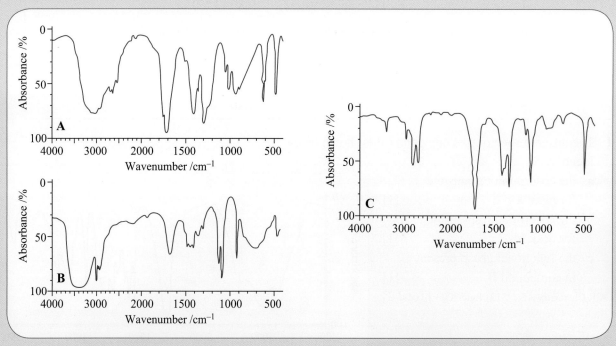

OCR Chemistry AS (2812) Jan 2005 [Total 3]

Answer

continued

4 A sample of gallium was analysed in a mass spectrometer to produce this mass spectrum. The relative atomic mass of gallium can be calculated from this mass spectrum.

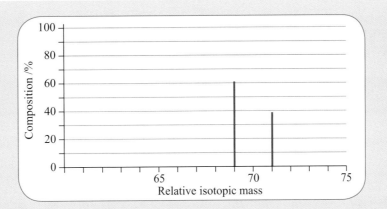

Hint

    a Estimate the percentage composition of each isotope present in the sample. [1]

    b Calculate the relative atomic mass of this sample of gallium. Your answer should be given to three significant figures. [2]

OCR Chemistry AS (2811) Jan 2003 [Total 3]

Answer

5 Rubidium, atomic number 37, was discovered in 1861 by Bunsen and Kirchoff. Rubidium is in Group 1 of the Periodic Table and the element has two natural isotopes, ^{85}Rb and ^{87}Rb.

    a Explain the term *isotopes*. [1]

    b A sample of rubidium was analysed in a mass spectrometer to produce the mass spectrum below.

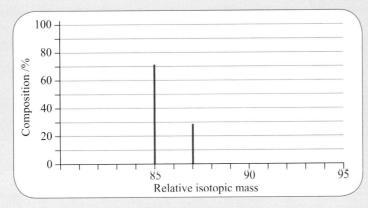

    i Use this mass spectrum to help you complete a copy of the table below.

Isotope	Percentage	Number of		
		protons	**neutrons**	**electrons**
^{85}Rb				
^{87}Rb				

[3]

    ii Calculate the relative atomic mass (A_r) of this rubidium sample. Give your answer to three significant figures. [2]

    c Which isotope is used as the standard against which the masses of the two rubidium isotopes are measured? [1]

OCR AS Chemistry 2811 Jan 2007 [Total 7]

Answer

Chapter 16

Enthalpy changes

Objectives

All chemical reactions involve change. In flames, for example, we can see the changes caused by very fast reactions between the chemicals in burning materials and oxygen from the atmosphere (Figure 16.1). There are new substances, new colours and changes of state, but the most obvious changes in these reactions are the transfers of energy as light and heat to the surroundings. All life on Earth depends on the transfer of energy in chemical reactions. Plants need the energy from the Sun for the production of carbohydrates by photosynthesis; animals gain energy from the oxidation of their food chemicals.

Figure 16.1 The chemical reactions in this fire are releasing large quantities of energy.

Energy transfer: exothermic and endothermic reactions

Most chemical reactions release energy to their surroundings. These reactions are described as **exothermic**. We recognise exothermic reactions most easily by detecting a rise in the temperature of the reaction mixture and the surroundings (the test-tube or beaker, the solvent, air, etc.). Examples of exothermic reactions include:

- important oxidation reactions such as the combustion of fuels
  e.g. $CH_4(g) + 2O_2(g) \longrightarrow CO_2(g) + 2H_2O(l)$ (+ energy)
  and respiration in plants and animals (involving oxidation of carbohydrates such as glucose)
  e.g. $C_6H_{12}O_6(aq) + 6O_2(g) \longrightarrow$
  $\qquad\qquad 6CO_2(g) + 6H_2O(l)$ (+ energy)

- acids with metals
  e.g. $Mg(s) + H_2SO_4(aq) \longrightarrow$
  $\qquad\qquad MgSO_4(aq) + H_2(g)$ (+ energy)

- water with 'quicklime' (calcium oxide) (see Chapter 8):
  $CaO(s) + H_2O(l) \longrightarrow Ca(OH)_2(aq)$ (+ energy)

Some chemical reactions occur only while energy is transferred to them *from* an external source. Reactions such as these, which require a heat input, are called **endothermic** reactions. The energy input may come from a flame, electricity, sunlight or the surroundings. Examples of endothermic reactions include:

- the decomposition of limestone by heating (Figure 16.2)
  $CaCO_3(s)$ (+ energy) $\longrightarrow CaO(s) + CO_2(g)$

- photosynthesis (Figure 16.3) – the energy is supplied to the reactions in the cells by sunlight
  $6CO_2(g) + 6H_2O(l)$ (+ energy) $\longrightarrow$
  $\qquad\qquad C_6H_{12}O_6(s) + 6O_2(g)$

- dissolving ammonium chloride in water, such as in a cold pack (Figure 16.4).
  $NH_4Cl(s) + water$ (+ energy) $\longrightarrow$
  $\qquad\qquad NH_4^+(aq) + Cl^-(aq)$

Figure 16.2 A modern lime kiln. Calcium carbonate, as limestone or chalk, has been converted to calcium oxide (quicklime) for centuries, by strong heating in lime-kilns.

Figure 16.3 Photosynthesis in green leaves is the most essential chemical reaction of all.

Figure 16.4 When the pack is kneaded, water and ammonium chloride crystals mix. As the crystals dissolve, energy is transferred from the surroundings, cooling the injury.

SAQ

1 Classify the following processes as exothermic or endothermic: evaporation; crystallisation; making magnesium oxide from magnesium and oxygen; making copper oxide from copper carbonate.

Answer

Energy is conserved

It is important to understand that energy is not being created by exothermic chemical reactions, and it is not destroyed in endothermic reactions. Energy is transferred from the reacting chemicals to the surroundings during an exothermic chemical reaction. Energy is transferred from the surroundings to the reacting chemicals during an endothermic chemical reaction. The total energy of the whole system of reacting chemicals and the surroundings remains *constant*. This applies to any energy transfer and is summarised in the *law of conservation of energy:* energy can neither be created nor destroyed.

You may also hear this universal law called the *first law of thermodynamics.* Thermodynamics is the science of transfer of energy.

Enthalpy and enthalpy changes

Measurements of the energy transferred during chemical reactions must be made under controlled conditions. A special name is given to the energy exchange with the surroundings when it takes place at constant pressure. This name is '**enthalpy** change'.

Enthalpy is the total energy content of the reacting materials. It is given the symbol H. Enthalpy cannot be measured as such, but it is possible to measure the enthalpy *change* when energy is transferred to or from a reaction system.

Enthalpy change is given the symbol ΔH. Δ is the upper case of the Greek letter δ, pronounced 'delta', and it is often used in mathematics as a symbol for change. The units are kilojoules per mole ($kJ\,mol^{-1}$).

We can illustrate enthalpy changes on enthalpy profile diagrams (Figure 16.5). An exothermic enthalpy change has a negative value, as the energy is lost from the reactants to the surroundings. It is shown in Figure 16.5a as:

$$\Delta H = -x\,kJ\,mol^{-1}$$

For example, when methane burns:
$$CH_4(g) + 2O_2(g) \longrightarrow CO_2(g) + 2H_2O(l);$$
$$\Delta H = -890.3\,kJ\,mol^{-1}$$

This means that when one mole of methane burns completely in oxygen, 890.3 kilojoules of energy are transferred to the surroundings (Figure 16.6a).

An endothermic enthalpy change has a positive value, as the energy is gained by the system from the surroundings. It is shown in Figure 16.5b as:

$$\Delta H = +y\,kJ\,mol^{-1}$$

For example, on heating calcium carbonate:
$$CaCO_3(s) \longrightarrow CaO(s) + CO_2(g); \Delta H = +572\,kJ\,mol^{-1}$$

This means that an input of 572 kilojoules of energy is needed to break down one mole of calcium carbonate to calcium oxide and carbon dioxide (Figure 16.6b).

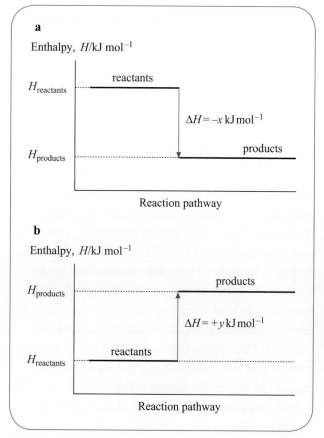

Figure 16.5 Enthalpy profile diagrams for **a** an exothermic reaction and **b** an endothermic reaction.

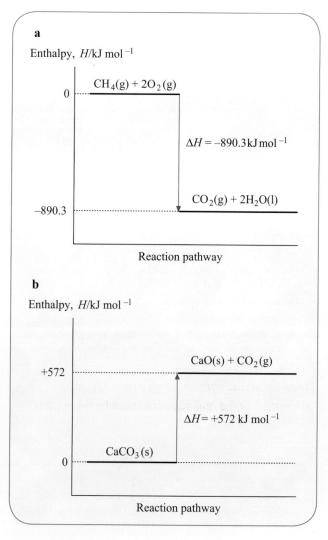

Figure 16.6 Enthalpy profile diagrams for **a** the combustion of methane and **b** the decomposition of calcium carbonate.

Standard enthalpy changes: standard conditions

When we compare the enthalpy changes of various reactions we must use standard conditions, such as known temperatures, pressures, amounts and concentrations of reactants or products. This allows us to compare the standard enthalpy changes for reactions.

A standard enthalpy change for a reaction takes place under these standard conditions:
- a pressure of 100 kilopascals (10^2 kPa)
- a temperature of 298 K (this is 25 °C).

The symbol for an enthalpy change measured under standard conditions is $\Delta H^{\ominus}$.

Standard enthalpy change of reaction $\Delta H_r^{\ominus}$

Definition: the *standard enthalpy change of reaction* is the enthalpy change when amounts of reactants, as shown in the reaction equation, react together under standard conditions to give products in their standard states.

It is necessary to make clear which reaction equation we are using when we quote a standard enthalpy change of reaction. For example, the equation for the reaction between hydrogen and oxygen can be written in two different ways and there are different values for $\Delta H_r^{\ominus}$ in each case:

equation **i**
$$2H_2(g) + O_2(g) \longrightarrow 2H_2O(l); \Delta H_r^{\ominus} = -572 \text{ kJ mol}^{-1}$$
equation **ii**
$$H_2(g) + \tfrac{1}{2}O_2(g) \longrightarrow H_2O(l); \Delta H_r^{\ominus} = -286 \text{ kJ mol}^{-1}$$
Note that the value of $\Delta H_r^{\ominus}$ in **ii** is half that of $\Delta H_r^{\ominus}$ in **i**.

Standard enthalpy change of formation $\Delta H_f^{\ominus}$

Definition: the *standard enthalpy change of formation* is the enthalpy change when one mole of a compound is formed from its elements under standard conditions; the compound and the elements must be in their standard states.

For example, water is formed in both equations **i** and **ii** above, but only in equation **ii** is one mole of water formed. Thus equation **ii** shows that the value of $\Delta H_f^{\ominus}(H_2O) = -286 \text{ kJ mol}^{-1}$ (Figure 16.7).

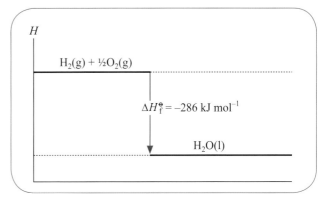

Figure 16.7 The standard enthalpy change of formation of water.

SAQ
2 a Write balanced equations for the formation of **i** ethane (C_2H_6) and **ii** aluminium oxide (Al_2O_3). Use a data book to add values for $\Delta H_f^{\ominus}$ in each case.
b Draw the enthalpy profile diagram for the enthalpy change of formation of ethane. Label your diagram fully.

Answer

Standard enthalpy change of combustion $\Delta H_c^{\ominus}$

Definition: the *standard enthalpy change of combustion* is the enthalpy change when one mole of an element or compound reacts completely with oxygen under standard conditions.

For example, the standard enthalpy change of combustion of hydrogen is given by equation **ii** above:
$$H_2(g) + \tfrac{1}{2}O_2(g) \longrightarrow H_2O(l); \Delta H_c^{\ominus} = -286 \text{ kJ mol}^{-1}$$
Another example is shown in Figure 16.8.

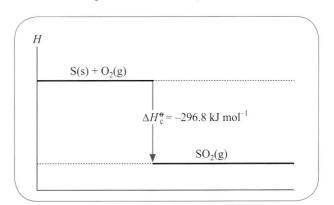

Figure 16.8 The standard enthalpy change of combustion of sulfur to form sulfur dioxide.

SAQ

3 a Which of the labels $\Delta H_r^\ominus$, $\Delta H_f^\ominus$, $\Delta H_c^\ominus$ could be used for the enthalpy changes shown in Figure 16.6?

   b What are the reaction equations for the combustion of **i** octane (C_8H_{18}) and **ii** ethanol (C_2H_5OH)? Include the values for $\Delta H_c^\ominus$ (use a data book).

   c Why is the $\Delta H_f^\ominus$ of water the same as the $\Delta H_c^\ominus$ of hydrogen?

   [Answer]

4 Define:

   [Hint]

   a standard enthalpy change of formation

   b standard enthalpy change of combustion.

   [Answer]

Bond making, bond breaking and enthalpy change

A typical combustion reaction, such as the burning of methane, is

$$CH_4(g) + 2O_2(g) \longrightarrow CO_2(g) + 2H_2O(l);$$
$$\Delta H_c^\ominus = -890.3\,\text{kJ}\,\text{mol}^{-1}$$

or, drawing the molecules to show the bonds:

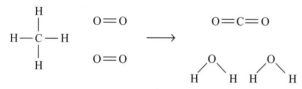

For this reaction to occur, some bonds must break and others form:

● bonds breaking $4 \times$ C–H and $2 \times$ O=O
● bonds forming $2 \times$ C=O and $4 \times$ H–O

To understand energy transfers during chemical reactions, remember the following rules.

When bonds break, energy is absorbed (endothermic process). When bonds form, energy is released (exothermic process).

If the energy released by the formation of bonds is greater than the energy absorbed by the breaking of bonds, energy will be transferred to the surroundings. The overall reaction will be exothermic.

If the energy released by bond formation is less than the energy absorbed by bond breaking then, overall, energy must be transferred from the surroundings. The reaction will be endothermic.

In the case of the combustion of methane, after all the bond breaking and bond formation, the energy transferred to the surroundings is 890.3 kJ for each mole of methane.

Bond enthalpy

It is useful to measure the amount of energy needed to break a covalent bond, as this indicates the strength of the bond. This amount of energy is called the **bond enthalpy**. The energy values are always quoted as bond enthalpy per mole (of bonds broken in the substance in the gaseous state).

Consider the example of oxygen gas, $O_2(g)$. The bond enthalpy of oxygen is the enthalpy change for the process:

$$O_2(g) \rightarrow 2O(g); \Delta H = +498\,\text{kJ}\,\text{mol}^{-1}$$

The symbol $E(X–Y)$ is often used for bond enthalpy per mole of X–Y bonds. Therefore $E(C–C)$ is the amount of energy needed to break one mole of carbon to carbon single bonds.

Typical values of bond enthalpies per mole are shown in Table 16.1.

Bond	$E(X–Y)$ /kJ mol^{-1}
H–H	+436
C–C	+347
C=C	+612
C–H	+410
O=O	+500
O–H	+465
C–O	+336
C=O	+805

Table 16.1 Some useful bond enthalpies.

● Bond enthalpies are all positive. The changes during breaking of bonds are endothermic (energy is absorbed). The same quantities of energy would be released in an exothermic change when the bonds form.
● Bond enthalpies are average values. The actual value of the bond enthalpy for a particular bond depends upon which molecule the bond is in. For example, the C–C bond has slightly different strengths in ethane C_2H_6 and in propane C_3H_8,

as it is affected by the other atoms and bonds in the molecules. The bond enthalpy quoted in data books for C–C is an average of the values from many different molecules.

● Bond enthalpies are very difficult to measure directly. They are usually calculated using data from measurements of enthalpy changes of combustion of several compounds.

SAQ

5 A book of data gives a value for the standard enthalpy change of combustion of hydrogen as $-285.8 \, kJ \, mol^{-1}$. A value for the enthalpy change of formation of water, calculated from bond energies, is $-283.1 \, kJ \, mol^{-1}$.
Suggest why these values are slightly different.

[Answer]

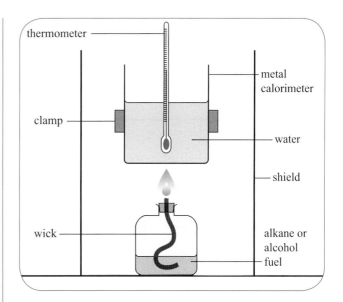

Figure 16.9 Apparatus used for approximate measurements of energy transferred by burning known masses of flammable liquids.

Measuring energy transfers and enthalpy changes

Simple laboratory experiments can give us estimates of the energy transferred during some reactions. Enthalpy changes may then be calculated.

Enthalpy changes of combustion

Measurements of $\Delta H_c^{\ominus}$ are important as they help to compare the energy available from the oxidation of different flammable liquids that may be used as fuels.

The type of apparatus used for a simple laboratory method is shown in Figure 16.9. A fuel, such as an alkane or an alcohol, burns at the wick. Measurements are made of:

● the mass of cold water in the metal calorimeter ($m \, g$)
● the temperature rise of the water ($\Delta T \, ^{\circ}C$)
● the loss in mass of the fuel ($y \, g$).

The energy required to raise $m \, g$ of water by $\Delta T \, ^{\circ}C$ is given by the general relationship:

energy transfer (as heating effect) $= mc\Delta T$ joules

where the symbol c is the specific heat capacity of a liquid, in this case water, and is the number of joules of energy required to raise the temperature of 1 g of the liquid by 1 °C. This value for water is $4.18 \, J \, g^{-1} \, ^{\circ}C^{-1}$.

Therefore, in the experiment, $m \times 4.18 \times \Delta T$ joules of energy are transferred during the burning of y grams of the fuel. Therefore, if one mole of the fuel has a

mass of M grams, $m \times 4.18 \times \Delta T \times M/y$ joules of energy are transferred when one mole of the fuel burns. The answer will give an approximate value of the enthalpy change of combustion of the fuel in joules per mole ($J \, mol^{-1}$). Dividing this answer by 1000 gives the value for $\Delta H_c^{\ominus}$ in kilojoules per mole ($kJ \, mol^{-1}$).

We shall now look at an example. In an experiment using the simple apparatus above to find the enthalpy change of combustion of propanol (C_3H_7OH), the following measurements were made:

mass of water in the calorimeter (m)   = 100 g
temperature rise of the water (ΔT)   = 21.5 °C
loss in mass of propanol fuel (y)   = 0.28 g

Data needed: A_r (H) = 1.0, A_r (C) = 12.0, A_r (O) = 16.0; and specific heat capacity of water c = $4.18 \, J \, g^{-1} \, ^{\circ}C^{-1}$. The energy transferred as heat from the burning propanol is:

$mc\Delta T = 100 \times 4.18 \times 21.5 \, J$
$= 8987 \, J$

This is the energy transferred (heat produced) by burning 0.28 g of propanol. The mass of one mole of propanol is 60 g. Therefore energy transferred by burning one mole of propanol is:

$\Delta H_c^{\ominus} = 8987 \times 60/0.28 \, J \, mol^{-1}$
$= 1\,926\,000 \, J \, mol^{-1}$
$= 1926 \, kJ \, mol^{-1}$

From this experiment, the value for $\Delta H_c^{\ominus}$ (C_3H_7OH) = $-1926 \, kJ \, mol^{-1}$.

SAQ

6 The value for $\Delta H_c^\ominus$ (C_3H_7OH) in a book of data is given as $-2010\,kJ\,mol^{-1}$. Suggest why the value calculated from the experimental results above is so much lower.

Answer

Measuring enthalpy changes of other reactions

The experiments outlined above involved burning fuels. Experiments can also be performed to find the enthalpy changes from reactions between chemicals in solutions. The enthalpy change of neutralisation in the reaction between an acid and an alkali is an example:

e.g. hydrochloric acid plus sodium hydroxide solution:

$$HCl(aq) + NaOH(aq) \longrightarrow Na^+(aq) + Cl^-(aq) + H_2O(l)$$

The reaction that produces the enthalpy change here is shown more simply as:

$$H^+(aq) + OH^-(aq) \longrightarrow H_2O(l)$$

$Na^+(aq)$ and $Cl^-(aq)$ are *spectator ions* and take no part in the reaction producing the enthalpy change.

To perform such an experiment you would:

- use a heat-insulated vessel, such as a vacuum flask or a thick polystyrene cup (Figure 16.10), and stir the reactants
- use known amounts of all reactants and known volumes of liquids – if one reactant is a solid, make sure you have an excess of solvent or other liquid reactant, so that all the solid dissolves or reacts
- measure the temperature change by a thermometer reading to at least 0.2°C accuracy

Figure 16.10 A simple apparatus used in school laboratories to measure enthalpy changes for reactions in aqueous solutions.

- calculate the energy transfers using the relationship
  energy transferred (joules) = $mc\Delta T$
  = mass of liquid (g) × sp. heat cap. of aq. soln.
  ($J\,g^{-1}\,°C^{-1}$) × temp. rise (°C).

We shall now work through some typical results from an experiment. When $50\,cm^3$ of HCl(aq) are added to $50\,cm^3$ of NaOH(aq), both of concentration $1\,mol\,dm^{-3}$, in an insulated beaker, the temperature rises by 6.2°C. The acid and alkali are completely neutralised.

We can calculate the molar enthalpy change of neutralisation (for the reaction between hydrochloric acid and sodium hydroxide) as follows:

$$50\,cm^3\ HCl(aq) + 50\,cm^3\ NaOH(aq) = 100\,cm^3$$
solution

mass of this solution (m)   = 100 g
change in temperature (ΔT)   = 6.2°C

We assume that the specific heat capacity (c) of the solution is the same as that for water ($4.18\,J\,g^{-1}\,°C^{-1}$). Therefore the energy transferred (heat produced) by the reaction is:

$$mc\Delta T = 100 \times 4.18 \times 6.2$$
$$= 2592\,J$$

$50\,cm^3$ of HCl(aq) and $50\,cm^3$ of NaOH(aq), both of concentration $1\,mol\,dm^{-3}$ contain 50/1000 moles = 5×10^{-2} moles of HCl and 5×10^{-2} moles of NaOH. So, the molar enthalpy change of neutralisation, for the reaction between 1 mole of HCl and 1 mole of NaOH to give 1 mole of NaCl, is given by:

$$\Delta H_r^\ominus = \frac{-2592}{5 \times 10^{-2}}\,J\,mol^{-1}$$
$$= -51\,840\,J\,mol^{-1}$$
$$= -51.84\,kJ\,mol^{-1}$$

The data book value for the molar enthalpy change of neutralisation is $\Delta H_r^\ominus = -57.1\,kJ\,mol^{-1}$. In the above experiment, some heat is lost to the surroundings. So the result obtained is less exothermic than the data book value.

SAQ

7 Suggest why the molar enthalpy changes of neutralisation for hydrochloric acid, sulfuric acid or nitric acid, reacting with alkalis such as aqueous sodium hydroxide or potassium hydroxide are all very similar in value, at about $-57.2\,kJ\,mol^{-1}$.

Hint

Answer

The enthalpy change of solution of sodium hydroxide

In this experiment, the temperature is measured over time and a graph is plotted. Extrapolation of the curve obtained as the mixture cools allows a correction to be made to the estimated temperature rise. The corrected figure makes allowance for cooling losses to the surroundings.

Method

1 Weigh a polystyrene cup of the kind shown in Figure 16.10.
2 Weigh 100 g of distilled water into the polystyrene cup. (How could you work out roughly how much to add in advance?)
3 Measure the temperature of the water in the cup. Keep a check on it until the temperature is steady. Record this temperature.
4 Add a few pellets of solid sodium hydroxide straight from a previously sealed container. (Solid sodium hydroxide absorbs water from the air, so it gets heavier if you leave it standing. *Take care* – it is also very *corrosive*. Wash it off immediately with water if you get it on your skin, and report to your teacher.)
5 Stir the mixture immediately, and start a stopwatch. Keep stirring with the thermometer, and record the temperature every 30 seconds.
6 The temperature will reach a maximum, and then it will start to fall. When it has fallen for five minutes, you can stop taking readings.
7 Weigh the cup + solution to calculate the mass of sodium hydroxide you dissolved.
8 Plot a graph of temperature against time, and work out the maximum temperature the mixture might have reached (see graph in Figure 16.11).

9 Calculate the amount of heat input to the solution, and calculate the amount of heat given out by the sodium hydroxide and water.
10 Scale the result to tell you how much heat energy would have been released on dissolving one mole (40.0 g) of sodium hydroxide to make the same strength of solution.

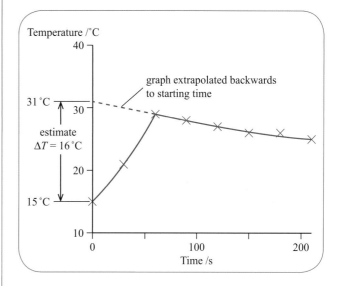

Figure 16.11 The results of the example experiment.

A typical set of results

The example below will help you to understand the procedure and calculations:

mass of polystyrene cup	= 8.00 g
mass of polystyrene cup + distilled water	= 108.15 g
mass of distilled water used	= 100.15 g
mass of cup + water + sodium hydroxide	= 114.35 g
mass of sodium hydroxide that dissolved	= 6.20 g
initial temperature of water in the cup	= 15.0 °C

Table showing temperature at fixed times after mixing:

Time /s	0	30	60	90	120	150	180	210
Temperature /°C	15.0	21.0	29.0	28.0	27.0	26.0	26.0	25.0

Calculation

Estimated temperature rise caused by 6.20 g NaOH is $\Delta T = 16\,°C$.

Energy transferred $= m \times c \times \Delta T$

where m = mass of water, c = specific heat capacity of water ($4.18\,J\,g^{-1}\,°C^{-1}$), and ΔT = maximum temperature rise.

So: energy transferred $= 100.15\,g \times 4.18\,J\,g^{-1}\,°C^{-1} \times 16\,°C$
$= 6.70\,kJ$

6.20 g NaOH releases 6.70 kJ energy on dissolving in water.

Therefore 1 mole of NaOH (40 g) releases

$\dfrac{40}{6.20} \times 6.70\,kJ\,mol^{-1}$ on dissolving in water

$= 43.2\,kJ\,mol^{-1}$.

Enthalpy changes by different routes: Hess's law

When we write a chemical equation, we usually show only the beginning and the end, that is the reactants and products. But there may be many different ways that the reaction actually occurs in between. The reactants may be able to change into the products by more than one route.

For example, consider a reaction system, with initial reactants A + B and final products C + D, in which two different routes (1 and 2) between A + B and C + D are possible (Figure 16.12). What can be said about the enthalpy changes for the two different routes? Are they different too?

The answer to this question was first summarised in 1840 by Germain Hess and is now called **Hess's law**.

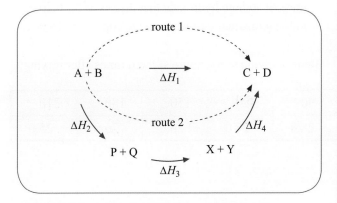

Figure 16.12 Two different routes between reactants and products. Hess's law tells us that $\Delta H_1 = \Delta H_2 + \Delta H_3 + \Delta H_4$.

Hess's law: The total enthalpy change for a chemical reaction is *independent* of the route by which the reaction takes place, provided the initial and final conditions are the same.

In the case of our example above, Hess's law tells us that the enthalpy change for route 1 would equal the total of the enthalpy changes for route 2; that is

$\Delta H_1 = \Delta H_2 + \Delta H_3 + \Delta H_4$

The overall enthalpy change is affected only by the initial reactants and the final products, not by what happens in between.

Hess's law seems fairly obvious in the light of the more universal first law of thermodynamics (law of conservation of energy – see page 181). If different routes between the same reactants and products were able to transfer different amounts of energy, energy would be being created or destroyed and we could make 'perpetual motion' machines (Figure 16.13) and gain free energy forever! Unfortunately, as in most aspects of life, you cannot get something for nothing.

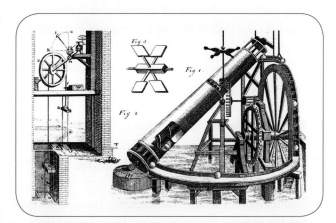

Figure 16.13 An attempt to design a mechanical perpetual motion machine. The heavy spheres cause the wheel to rotate. This operates the 'screw', which lifts the spheres back to the top of the wheel. Why does it not work?

Using Hess's law: enthalpy cycles

Chemists often use an **enthalpy cycle** to calculate the enthalpy change for a reaction which cannot easily be measured directly. We shall look at three examples. The first example makes use of bond enthalpies; the second, enthalpy changes of formation; and the third, enthalpy changes of combustion.

Calculating $\Delta H_r^\ominus$ from bond enthalpy data

An important reaction which is covered in more detail in Chapter 17 is the reaction for the Haber process for the synthesis of ammonia:

$$N_2(g) + 3H_2(g) \longrightarrow 2NH_3(g)$$

Bond enthalpy data is shown in Table 16.2.

Bond	Bond enthalpy /kJ mol^{-1}
N≡N	945
H–H	436
N–H	391

Table 16.2 Bond enthalpies important for the Haber process.

When calculating $\Delta H_r^\ominus$ from bond enthalpy data, you will find it helpful to draw the enthalpy cycle showing the bonds present in the reactants and products. Remember that bond enthalpy is the enthalpy change for breaking one mole of the bonds.

The enthalpy cycle for the Haber process is shown in Figure 16.14. Look at the equation for route 1. The following bonds are broken:

- one N≡N triple bond
- three H–H single bonds.

And the following bonds are formed:

- three N–H single bonds per molecule of ammonia – a total of six because two ammonia molecules are produced from one nitrogen and three hydrogen molecules.

Route 2 shows the bond breaking and bond forming as two separate steps. By Hess's law:

enthalpy change for route 1 =
                    total enthalpy change for route 2

$\Delta H_r^\ominus$ = enthalpy change for bonds broken +
                    enthalpy change for bonds formed

$= E(N≡N) + 3 \times E(H–H) + 6 \times -E(N–H)$

$= 945 + (3 \times 436) + (6 \times -391)$

$= 2253 - 2346$

$= -93 \, \text{kJ mol}^{-1}$

Remember that bond breaking is endothermic, bond formation is exothermic (hence the negative sign before E(N–H) and before 391).

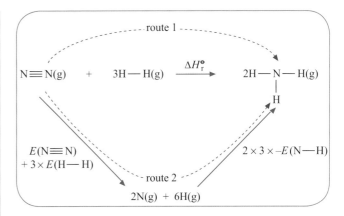

Figure 16.14 The enthalpy cycle for the production of ammonia by the Haber process.

Figure 16.15 will help you to visualise which bonds are being broken and which formed with the help of molecular models.

Bond enthalpy calculations lend themselves to use of a spreadsheet on a computer. Table 16.3 shows such a spreadsheet for the Haber process.

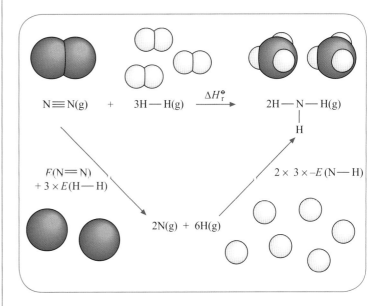

Figure 16.15 Bond breaking and bond formation in the enthalpy cycle for the Haber process.

	A	B	C	D	E	F
1	**Enthalpy of formation of ammonia from bond enthalpies**					
2						
3						
4	Bond	Bond enthalpy /kJ mol^{-1}	Number of bonds broken	Enthalpy input /kJ mol^{-1}	Number of bonds formed	Enthalpy output /kJ mol^{-1}
5						
6	N≡N	945	1	945		0
7	H–H	436	3	1308		0
8	N–H	391		0	6	−2346
9						
10		Total enthalpy input:		2253		
11		Total enthalpy output:				−2346
12						
13				Enthalpy of reaction /kJ mol^{-1}:		−93

Table 16.3 Bond enthalpy calculation for the Haber process.

SAQ

8 A balanced chemical equation for the complete combustion of methane is:

$$CH_4(g) + 2O_2(g) \longrightarrow CO_2(g) + 2H_2O(g)$$

a Re-write this equation to show the bonds present in each molecule.

b Using your equation, draw an enthalpy cycle showing bonds broken in the reactants, forming gaseous atoms, and bonds formed in the products from these gaseous atoms.

c Use the bond enthalpies given on page 184 to calculate the enthalpy change of combustion for methane (use a spreadsheet if possible).

d Compare the value for $\Delta H_c^{\ominus}$ with the experimentally determined value of −890 kJ mol^{-1}. Which value is more accurate?

Answer

Enthalpy change of reaction from enthalpy changes of formation

Enthalpy changes of formation of many compounds have been determined experimentally under carefully controlled conditions. Often these have been found indirectly from other experimental enthalpy changes such as enthalpy changes of combustion (see page 183). Many data books provide detailed tables of enthalpy changes of formation. Such data can be used, with Hess's law, to calculate the enthalpy change of a reaction.

These enthalpy figures help chemists and chemical engineers to design a new chemical production plant. In such a plant, an exothermic process may release sufficient energy to cause a fire or explosion. Careful design of the plant allows the release of energy to be controlled and even re-used for heating or electricity generation.

Imagine that you are building a plant to make slaked lime from quicklime (see Chapter 8). The equation for this exothermic reaction is:

$$CaO(s) + H_2O(l) \longrightarrow Ca(OH)_2(s); \Delta H_r^{\ominus} = ?$$

The enthalpy changes of formation for each of the reactants and the product are:

$$Ca(s) + \tfrac{1}{2}O_2(g) \longrightarrow CaO(s);$$
$$\Delta H_f^{\ominus}[CaO(s)] = -635.1\,kJ\,mol^{-1}$$

$$H_2(g) + \tfrac{1}{2}O_2(g) \longrightarrow H_2O(l);$$
$$\Delta H_f^{\ominus}[H_2O(l)] = -285.8\,kJ\,mol^{-1}$$

$$Ca(s) + H_2(g) + O_2(g) \longrightarrow Ca(OH)_2(s);$$
$$\Delta H_f^{\ominus}[Ca(OH)_2(s)] = -986.1\,kJ\,mol^{-1}$$

We can now draw the enthalpy cycle (Figure 16.16).
- As both $CaO(s)$ and $H_2O(l)$ appear in the top left-hand corner of the cycle, we must add the enthalpy changes of formation of each compound.
- Route 1 contains the $\Delta H_r^{\ominus}$ which we wish to determine.

The total enthalpy change for route 1 is:

$$\Delta H_f^{\ominus}[CaO(s)] + \Delta H_f^{\ominus}[H_2O(l)] + \Delta H_r^{\ominus}\,kJ\,mol^{-1}$$
$$= (-635.1) + (-285.8) + \Delta H_r^{\ominus}\,kJ\,mol^{-1}$$

and for route 2 is:

$$\Delta H_f^{\ominus}[Ca(OH)_2(s)] = -986.1\,kJ\,mol^{-1}$$

By Hess's law, the enthalpy change for route 1 = enthalpy change for route 2.

So:
$$(-635.1) + (-285.8) + \Delta H_r^{\ominus} = -986.1\,kJ\,mol^{-1}$$
$$\text{or } \Delta H_r^{\ominus} = (-986.1) - [(-635.1) + (-285.8)]\,kJ\,mol^{-1}$$
$$= -65.2\,kJ\,mol^{-1}$$

Notice the brackets inserted round each enthalpy change figure. These help ensure that you do not make a mistake over the signs!

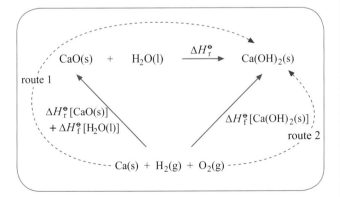

Figure 16.16 The enthalpy change for the reaction of quicklime with water.

SAQ

9 The balanced equation for the decomposition of magnesium carbonate is:
$$MgCO_3(s) \longrightarrow MgO(s) + CO_2(g)$$
   a Draw an enthalpy cycle showing the formation of each of the reactants and products from their elements in their standard states.
   b Use the following enthalpy changes of formation to calculate the enthalpy change for the decomposition of magnesium carbonate.

	$MgCO_3(s)$	$MgO(s)$	$CO_2(g)$
$\Delta H_f^{\ominus}\,(kJ\,mol^{-1})$	−1096	−602	−394

Answer

Enthalpy change of formation from enthalpy changes of combustion

Consider the formation of methane from carbon and hydrogen:
$$C(s) + 2H_2(g) \longrightarrow CH_4(g);\ \Delta H_f^{\ominus} = ?$$
This could be a very useful reaction for making methane gas starting with a plentiful supply of carbon such as coal or wood charcoal. Scientists are trying to find ways of making it occur directly and need to know the value of the enthalpy change of formation of methane. The best way is to use enthalpy of combustion data, and Hess's law.

Carbon, hydrogen, and methane all burn in oxygen and the enthalpy changes of combustion of each can be measured.

$$C(s) + O_2(g) \longrightarrow CO_2(g);\ \Delta H_c^{\ominus} = -393.5\,kJ\,mol^{-1}$$

$$H_2(g) + \tfrac{1}{2}O_2(g) \longrightarrow H_2O(l);$$
$$\Delta H_c^{\ominus} = -285.8\,kJ\,mol^{-1}$$

$$CH_4(g) + 2O_2(g) \longrightarrow CO_2(g) + 2H_2O(l);$$
$$\Delta H_c^{\ominus} = -890.3\,kJ\,mol^{-1}$$

Next we must draw an enthalpy cycle (Figure 16.17):
- As in previous enthalpy cycles, the balanced equation for the enthalpy change we wish to calculate is written at the top of the cycle.

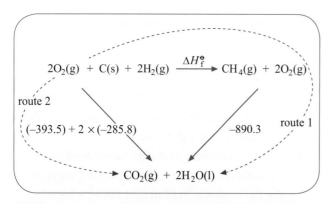

Figure 16.17 An enthalpy cycle used to calculate the enthalpy change of formation of methane.

- In this example the common products of combustion are written at the bottom of the cycle. We add oxygen to both sides of the top equation to balance the equations for the downward pointing arrows. This does not alter the reaction in the top equation.
- Note that, in route 2, one mole of carbon is oxidised by combustion together with two moles of hydrogen.
  The enthalpy change for route 1:
  $$= \Delta H_f^\ominus + (-890.3)\,\text{kJ mol}^{-1}$$
  The enthalpy change for route 2:
  $$= (-393.5) + 2 \times (-285.8)\,\text{kJ mol}^{-1}$$
  Applying Hess's law:
  $$\Delta H_f^\ominus + (-890.3) = (-393.5) + 2 \times (-285.8)$$
  Hence:
  $$\Delta H_f^\ominus = (-393.5) + 2 \times (-285.8) - (-890.3)$$
  $$= -74.8\,\text{kJ mol}^{-1}$$

SAQ

10 a Write the balanced chemical equation for the enthalpy change of formation of ethane, $C_2H_6(g)$. [Hint]

b Using your equation, draw an enthalpy cycle. The enthalpy change of combustion of ethane is $-1560\,\text{kJ mol}^{-1}$.

c Using the enthalpy changes of combustion of carbon and hydrogen in the above example, calculate the enthalpy change of formation of ethane. [Answer]

Summary

- Chemical reactions are often accompanied by transfers of energy to or from the surroundings, mainly as heat. In exothermic reactions, energy is transferred away from the reacting chemicals; in endothermic reactions, energy is gained by the reacting chemicals.

- Changes in energy of reacting chemicals at constant pressure are known as enthalpy changes (ΔH). Exothermic enthalpy changes are shown as negative values (−) and endothermic enthalpy changes are shown as positive values (+).

- Standard enthalpy changes are compared under standard conditions of pressure and temperature.

- Standard enthalpy changes of formation, $\Delta H_f^\ominus$, and of combustion, $\Delta H_c^\ominus$, are defined in terms of one mole of compound formed or of one mole of element or compound reacting completely with oxygen; their units are $kJ\,mol^{-1}$.

- Bond breaking is an endothermic process; bond making is an exothermic process.

- Bond enthalpy is a measure of the energy required to break a bond. The values quoted are usually average bond enthalpies, as the strength of a bond between two particular atoms is different in different molecules.

- Enthalpy changes may be calculated from measurements involving temperature change in a liquid, using the relationship:
    enthalpy change = mass of liquid × specific heat capacity of liquid × temperature change.
    $\Delta H \qquad = mc\Delta T$

- Hess's law states that 'the total enthalpy change for a chemical reaction is independent of the route by which the reaction takes place, provided the inital and final conditions are the same'.

- The principle of Hess's law may be used to calculate enthalpy changes for reactions that do not occur directly or cannot be found by experiment.

- Hess's law can be used to calculate the approximate enthalpy change of a reaction using the average bond cnthalpies of the reactants and of the products.

- Hess's law can be used to calculate the enthalpy change of a reaction using the enthalpy changes of formation of the reactants and products.

- Hess's law can be used to calculate the enthalpy change of formation of an organic compound using the enthalpy changes of combustion of carbon, hydrogen and the organic compound.

Questions

1 a Energy changes during reactions can be considered using several different enthalpy changes. These include average bond enthalpies and enthalpy changes of combustion.

   The table below shows the values of some average bond enthalpies.

Bond	Average bond enthalpy /kJ mol^{-1}
C–H	+410
O–H	+465
O=O	+500
C=O	+805
C–O	+336

   i   Why do bond enthalpies have positive values? [1]

   ii Define the term *bond enthalpy*. [2]

b i   The equation below shows the combustion of methanol, CH_3OH, in the gaseous state.

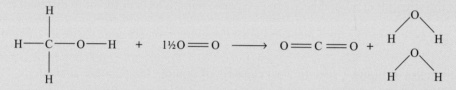

   Use the average bond enthalpies from the table above to calculate the enthalpy change of combustion of gaseous methanol, ΔH_c in kJ mol^{-1}. [3]

   ii Suggest <u>two</u> reasons why the *standard* enthalpy change of combustion of methanol will be different to that calculated in part **b i**. [2]

OCR Chemistry AS (2813) June 2006 [Total 8]

Answer

2 a The changes in energy during reactions are often considered using enthalpy changes of reaction. One such enthalpy change is the standard enthalpy change of formation.

   i   Define the term *standard enthalpy change of formation*. [3]

   ii Write the equation for the reaction corresponding to the standard enthalpy change of formation of magnesium nitrate, $Mg(NO_3)_2$. Include state symbols. [2]

b When magnesium nitrate is heated, it decomposes to give magnesium oxide, nitrogen dioxide and oxygen.

   Hint

   Use the standard enthalpy changes of formation in the table below to find the enthalpy change of reaction for this decomposition.

Substance	Standard enthalpy change of formation /kJ mol^{-1}
$Mg(NO_3)_2$	−791
MgO	−602
NO_2	−33

continued

The equation for this reaction is shown below.

$$Mg(NO_3)_2(s) \longrightarrow MgO(s) + 2NO_2(g) + \tfrac{1}{2}O_2(g)$$

Give the enthalpy change in $kJ\,mol^{-1}$. [3]

OCR Chemistry AS (2813) June 2006 [Total 8]

Answer

3 There are several oxides of lead. This question is about the enthalpy changes that occur during the reactions of some of these oxides.

  a What are the standard conditions of temperature and pressure used in enthalpy calculations? [1]

  b Write an equation, including state symbols, representing the standard enthalpy change of formation of PbO. [2]

  c Metal priming paints often contain 'red lead', Pb_3O_4. Red lead can be made by heating PbO in the presence of air.

$$3PbO(s) + \tfrac{1}{2}O_2(g) \longrightarrow Pb_3O_4(s)$$

    i Use the $\Delta H_f^{\ominus}$ values in the table below to calculate the standard enthalpy change for the above reaction, in $kJ\,mol^{-1}$.

Hint

Compound	$\Delta H_f^{\ominus}$ /kJ mol^{-1}
PbO(s)	−217
Pb_3O_4(s)	−718

[3]

    ii Red lead can also be obtained by reacting PbO_2 with PbO.

$$PbO_2(s) + 2PbO(s) \longrightarrow Pb_3O_4(s); \Delta H^{\ominus} = -10\,kJ\,mol^{-1}$$

    Use the value of $\Delta H^{\ominus}$ for this reaction, together with the values of $\Delta H_f^{\ominus}$ in the table, to calculate a value for the enthalpy change of formation of PbO_2(s) in $kJ\,mol^{-1}$. [3]

OCR Chemistry AS (2813) Jan 2004 [Total 9]

Answer

4 Butane, C_4H_{10}, is a gas at room temperature. It is used as a fuel for portable gas cookers.

  a The combustion of butane is shown in the equation below.

$$C_4H_{10}(g) + 6\tfrac{1}{2}O_2(g) \longrightarrow 4CO_2(g) + 5H_2O(l)$$

    i The standard enthalpy change of combustion of butane is $-2877\,kJ\,mol^{-1}$. What does *standard* mean in this context? [1]

    ii Define the term *enthalpy change of combustion*. [2]

    iii Copy and complete the enthalpy profile diagram for the combustion of butane. Label the activation energy, E_a, and the enthalpy change, ΔH. [3]

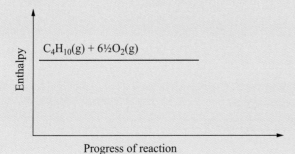

continued

b   Enthalpy changes of combustion can be used to determine enthalpy changes of formation.

    i   Write the equation for the standard enthalpy change of formation of butane, C_4H_{10}. Include state symbols in your answer. [2]

    ii Use the data in the table below to calculate the standard enthalpy change of formation of butane. Give your answer in $kJ\,mol^{-1}$.

Hint

	Standard enthalpy change of combustion /$kJ\,mol^{-1}$
carbon	−394
hydrogen	−286
butane	−2877

[3]

OCR Chemistry AS (2813) Jan 2005

[Total 11]

Answer

5 This question is about the relationship between bond enthalpies and enthalpy changes of combustion.

  a Methane burns in air according to the following equation.

$$CH_4(g) + 2O_2(g) \longrightarrow CO_2(g) + 2H_2O(g)$$

    i   Use the average bond enthalpies given in the table below to calculate a value for the enthalpy change of combustion of methane, ΔH_c. Give your answer in $kJ\,mol^{-1}$.

Bond	Average bond enthalpy / $kJ\,mol^{-1}$
C–H	+410
O–H	+465
O=O	+500
C=O	+805

[3]

    ii The standard enthalpy change of combustion of methane is $-890\,kJ\,mol^{-1}$.

       Suggest why your calculated value differs from this value. [1]

  b The table below gives values for the standard enthalpy changes of combustion for the first three alkanes.

Alkane	Formula	$\Delta H_c^{\ominus}$/$kJ\,mol^{-1}$
methane	CH_4	−890
ethane	C_2H_6	−1560
propane	C_3H_8	−2220

[3]

    i   Write a balanced equation, including state symbols, illustrating the standard enthalpy change of combustion of propane gas. [2]

    ii As the chain length in the alkanes increases, the value of $\Delta H_c^{\ominus}$ becomes more negative. Use your understanding of bond breaking and bond making to explain this trend. [2]

    iii Propane can be cracked to form ethene and methane.

$$C_3H_8(g) \longrightarrow C_2H_4(g) + CH_4(g)$$

continued

The standard enthalpy change of combustion of ethene, $\Delta H_c^{\ominus} = -1410 \text{ kJ mol}^{-1}$.

Use this value together with relevant values from the table above to calculate the enthalpy change $\Delta H^{\ominus}$ of this reaction. Give your answer in kJ mol^{-1}. [3]

OCR Chemistry AS (2813) June 2004 [Total 11]

Answer

6 Propane, C_3H_8, is a gas at room temperature and pressure. It is used in blow-torches to melt the bitumen needed to apply felt to flat roofs.

a Write the equation for the complete combustion of propane. [2]

b Define the term *standard enthalpy change of combustion*. [3]

c A blow-torch was used to determine the enthalpy change of combustion of propane. The apparatus is shown below.

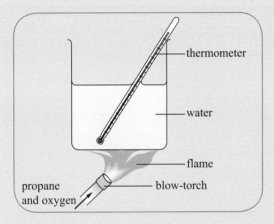

In the experiment, 200 g of water were used. The temperature of the water changed from 18.0 °C to 68.3 °C when 1.00 g of propane was burnt.

i Calculate the energy produced in kJ. The specific heat capacity of water is 4.18 J g^{-1} °C^{-1}. [2]

ii Calculate the number of moles of C_3H_8 burnt during the experiment. [1]

iii Deduce the enthalpy change of combustion, in kJ mol^{-1}, of C_3H_8. [2]

d Values of enthalpy changes of combustion can be used to calculate enthalpy changes of formation. The enthalpy change for the reaction in the equation below is the enthalpy change of formation of propane.

$$3C(s) + 4H_2(g) \longrightarrow C_3H_8(g)$$

The table below shows the enthalpy changes of combustion of carbon, hydrogen and propane.

	Enthalpy change of combustion /kJ mol^{-1}
carbon	−394
hydrogen	−286
propane	−2219

i Use these data to calculate the enthalpy change of formation of propane in kJ mol^{-1}. [3]

ii Suggest why the enthalpy change of formation of propane cannot be measured directly. [1]

OCR AS Chemistry 2813 Jan 2007 [Total 14]

Answer

Chapter 17

Rates and equilibrium

Background

e-Learning

Objectives

Speed, rates and reactions

The speed of the winner of a Formula 1 Grand Prix is much higher than that of the Olympic 100 m. Different chemical reactions also proceed at very different speeds. Some reactions are very fast whilst others may be very slow (Figure 17.1).

The speed of a chemical reaction may be affected by changing the conditions. For example, glucose will burn rapidly in air, but when used as an energy source in our bodies, it is oxidised much more slowly. In both cases the products are the same (carbon dioxide and water).

Figure 17.1 a A variety of rapid combustion reactions take place following the ignition of fireworks.
b Fortunately, rusting is a very slow reaction.

For chemical reactions we use the term *rate* instead of speed to describe how fast a reaction proceeds. The rate of a reaction is found by measuring the amount in moles of a reactant used up or product formed in a given time. The study of rates of reactions is referred to as *chemical kinetics*.

Rates of reaction – why bother?

There are many reasons why chemists study reaction rates, for example:

- to improve the rate of production of a chemical
- to help understand the processes going on in our bodies or in the environment
- to gain an insight into the mechanism of a reaction.

During the manufacture of a chemical such as a fertiliser or a medicine, the reaction rate is one of the factors which determine the overall rate of production. An understanding of how fast a reaction proceeds helps chemists and chemical engineers to choose the conditions used in the manufacture of a particular chemical.

Understanding reaction rates can also be important for chemists seeking to manage environmental issues. For example, the rate of formation of ozone in the stratosphere is dependent on the intensity of ultraviolet (UV) radiation reaching the Earth from the Sun. In Chapter 11, Chapter 12 and Chapter 14, you learnt how ozone depletion has been caused by chlorine free radicals. The ozone layer normally helps to filter out UV radiation from sunlight, but its destruction leads to high levels of UV radiation reaching the Earth's surface where it causes problems such as skin cancer. The chlorine free radicals are formed by the action of UV radiation on chlorofluorocarbons (see Chapter 14).

continued

A knowledge of the rates of these various reactions has enabled chemists to contribute much to an understanding of this environmental problem, highlighting the urgent need to control the use of chlorofluorocarbons.

Also in Chapter 11, Chapter 12 and Chapter 14, you met several organic reaction mechanisms. Many of these mechanisms have been discovered by a study of reaction rates. It is the slowest step in a mechanism which determines the overall **rate of reaction**. The slowest step in a reaction mechanism is called the *rate-determining step*. In the formation of ozone in the stratosphere, the slowest step involves the photodissociation (breakdown by light) of oxygen molecules by high-energy ultraviolet radiation into oxygen atoms. Environmental chemists have studied the rates of many of the reactions which take place in the atmosphere. Such research has contributed much to our understanding of these reactions and of the effects of pollutant gases. This work has both demonstrated the need for the control of artificial pollutants and led to the development of more environmentally friendly products (Figure 17.2).

Figure 17.2 Chemists are developing new environmentally friendly products.

Factors that affect the rate of a reaction

Several factors may affect the rate of a chemical reaction.

1 *Concentration of reactants*: Increasing the concentration of hydrochloric acid in the reaction of magnesium with the acid (see Chapter 8), will cause the reaction rate to increase. This will be seen in the more vigorous evolution of hydrogen gas. For reactions involving gases, an increase in pressure will increase the reaction rate, as pressure is proportional to concentration. The Haber process is operated under high pressure in order to increase the rate of reaction (see page 213).

2 *Temperature*: Nearly all reactions show an increase in rate as the temperature is increased. In general, an increase of 10 °C causes the rate of many reactions to approximately double.

3 *Catalysts*: A catalyst speeds up a chemical reaction while remaining chemically unchanged itself at the end of the reaction. For example, a nickel catalyst is used to speed up the hydrogenation of vegetable oils in the manufacture of margarine (see Chapter 12). Enzymes are biological catalysts.

4 *Surface area*: If a solid reactant is changed from lumps into a powder its surface area is increased and the reaction rate will be increased. For example, powdered magnesium produces hydrogen more rapidly than magnesium ribbon when added to hydrochloric acid. Catalysts also work more effectively if they have a large surface area.

SAQ

1 Hydrogen peroxide solution decomposes very slowly to oxygen and water. How can the rate of decomposition of hydrogen peroxide solution be speeded up? Try to include a reference to all of points 1 to 4 above.

Answer

The collision theory of reactivity

Collisions occur between billiard balls in a game of snooker. There are a few stories from the second half of the nineteenth century of explosions occurring when two billiard balls collided with exceptional force. One story describes how such an explosion set off a gunfight in a Colorado saloon.

Why should billiard balls explode? At the time billiard balls were made from celluloid (a mixture of nitrocellulose and camphor). Sometimes the balls were varnished with a nitrocellulose paint. If two such billiard balls collided with sufficient energy they might conceivably explode. As modern billiard balls are no longer made of celluloid, this is not something we are likely to experience, however many hours we spend watching snooker on television!

The 'exploding billiard balls' story enables us to visualise the collision theory of reactivity. Collision theory helps to provide explanations for the following experimental observations, made by measuring rates of reaction. The measurements show that the rate of reaction can be increased by:

- increasing the concentration of a reactant
- increasing the pressure of a gaseous reactant
- increasing the temperature
- using a catalyst.

An example of a reaction, well known to welders, is the combustion of ethyne in oxygen when using an oxyacetylene torch. (Acetylene is a more traditional name for ethyne. This gas is an example of an alkyne. Alkynes contain a C≡C triple bond.) This gaseous reaction involving two reactants is shown in Figure 17.3 and Figure 17.4. The reactant molecules are moving around and, occasionally, random collisions will occur.

SAQ

2   Write the balanced equation for the reaction in Figure 17.3.

Answer

It is not hard to imagine that, like the exploding billiard balls, collision of an ethyne and an oxygen molecule can result in a reaction. However, again like our nineteenth century billiard balls, only a few of these collisions result in a reaction. Not all the collisions are effective; a collision is not necessarily followed by a reaction. Effective collisions occur when the kinetic energy of the colliding molecules provides sufficient energy for reaction. We shall explore this aspect later (see page 203). However, a reaction certainly *cannot* occur if the molecules don't collide.

This simple notion is the basis of the *collision theory of reactivity*. When there are more balls on a billiard table, more collisions are likely to occur. If we increase the pressure of a gas, the molecules are closer together and more collisions will occur. Increasing the number of collisions in a given time will increase the number of effective collisions and so the reaction proceeds at a faster rate (Figure 17.5).

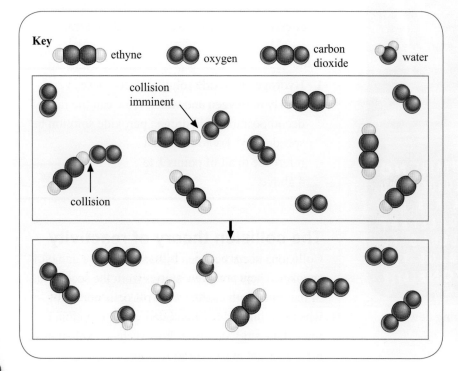

Key
ethyne   oxygen   carbon dioxide   water

collision imminent

collision

Figure 17.3 Molecules of ethyne, C_2H_2(g), and oxygen, O_2(g), can collide. If the collision is big enough (sufficiently energetic), chemical bonds are broken. They are re-formed when the fragments combine to make new molecules: carbon dioxide, CO_2(g), and steam, H_2O(g).

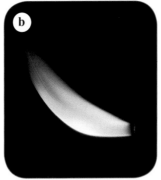

Figure 17.4

a Ethyne, mixed with oxygen, is used in an oxyacetylene torch. Here the gas is not ignited, and you can see it bubbling through water.

b The ethyne is now ignited, but is not completely combusting because the yellow flame indicates the presence of carbon: the temperature of the flame is relatively low.

c The ethyne is now being completely converted into carbon dioxide and water: the temperature of the flame is much higher.

The theory also generally applies to reactions in solutions. The reactants in solution behave rather like those in a gas – in each case the reactant molecules are spread apart. An increase in pressure forces gas molecules closer together. This increases the number of gas molecules in a given volume, which means the concentration is increased. When we increase the concentration of reactants in solution the reactant molecules become closer together. The rate of reaction also increases.

In studying the influence of concentration on rate, we have to be careful to keep temperature constant, because a change in temperature will alter the reaction rate. An increase in temperature increases reaction rate, and using the simple collision theory model it is not hard to see why. If the temperature of a substance is increased the average kinetic energy of its molecules increases. This means the average speed of the molecules increases also. The increased speeds of molecules will lead to more molecules gaining sufficient energy to react on collision.

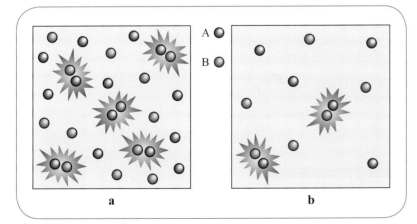

Figure 17.5 The molecules in **a** are closer together than the molecules in **b**. This leads to more frequent collisions between molecules and a faster reaction rate.

We can summarise all this as follows.

- Molecules will react only if they collide with each other.
- Reactions will occur only if there is enough energy in the collision.
- Increased concentration of molecules increases the frequency of collision, which increases reaction rate.
- Increased temperature increases the proportion of molecules with sufficient energy to react which increases reaction rate.

SAQ

3 Explain why a 2 cm strip of magnesium ribbon reacts more quickly with $50 \, cm^3$ of $1.0 \, mol \, dm^{-3}$ hydrochloric acid at $50\,°C$ than it does with $50 \, cm^3$ of $0.2 \, mol \, dm^{-3}$ hydrochloric acid at $20\,°C$.

Hint

Answer

The Boltzmann distribution

In any mixture of moving molecules, the energy of each molecule varies enormously. Like bumper cars at a fairground, some are belting along at high speeds while others are virtually at a standstill. The situation changes moment by moment: a particle travelling at a fairly gentle pace can get a shunt from behind and speed off with much greater energy than before; the fast particle that caused the collision will slow down during the collision.

The Boltzmann distribution represents the numbers of particles with particular energies in a sample of gas, where there are billions and billions of molecules in constant random motion. A few molecules are almost motionless. The majority have speeds around an average value. A minority have momentary speeds far in excess of the average. This is illustrated by the graph shown in Figure 17.6.

This average molecular energy will increase if the temperature of the entire collection of molecules

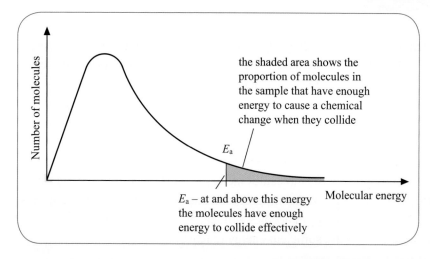

the shaded area shows the proportion of molecules in the sample that have enough energy to cause a chemical change when they collide

E_a

E_a – at and above this energy the molecules have enough energy to collide effectively

Molecular energy

Figure 17.6 The Boltzmann distribution for molecular energies in a sample of gas. Since the mass of each molecule is the same, the difference in energies is due to a difference in speed. Note that the curve is *not* symmetrical. The activation energy of the reaction is E_a.

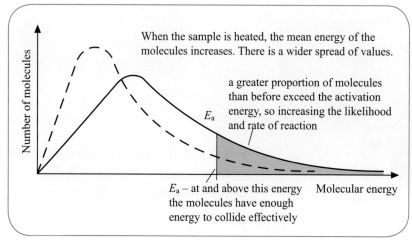

When the sample is heated, the mean energy of the molecules increases. There is a wider spread of values.

a greater proportion of molecules than before exceed the activation energy, so increasing the likelihood and rate of reaction

E_a

E_a – at and above this energy the molecules have enough energy to collide effectively

Molecular energy

Figure 17.7 Note how the Boltzmann distribution flattens and shifts to the right at the higher temperature. The areas under both curves are the same – they represent the total number of molecules in the sample, and this should not change before a reaction occurs.

is increased. Some molecules will still be almost immobile, but at any one time there is a greater number at a higher speed than before. The new distribution is shown in Figure 17.7.

The effect of this shift in the distribution is to increase the proportion of molecules with sufficient energy to react. This energy value is called the **activation energy**.

Activation energy

Just as two cars with effective bumpers may collide at low speed with no real damage being done (apart from frayed tempers), so low-energy collisions will not result in reaction. The molecules will bounce apart unchanged (Figure 17.8).

Figure 17.8 These collisions, frequent as they are, are not effective. They do not, we hope, result in permanent damage.

On the other hand, a high speed collision between one car and another will result in permanent damage, and the configuration of each vehicle will be drastically altered (and the same may go for the drivers). In the same sort of way, molecules have to collide with a certain minimum energy, called E_a, for there to be a chance of reaction. E_a is referred to as the activation energy for the reaction. Like other energy changes, activation energy has units of $kJ\,mol^{-1}$.

But why should we have to surmount this energy barrier E_a to bring about reaction? After all, as we saw in Chapter 16, if a reaction is exothermic the sum of the bond energies in the product molecules is less than the sum of the bond energies in the reactant molecules. Why doesn't a reaction, such

as the combustion of methane in oxygen, flow spontaneously downhill to give carbon dioxide and water (a less energetic, more stable, state) as illustrated in Figure 17.9?

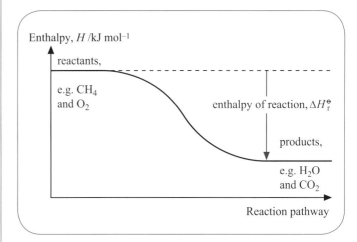

Figure 17.9 A 'downhill-all-the-way' reaction. Fortunately, it does not happen for methane and oxygen at normal temperatures and pressures. For an explanation of enthalpy, see Chapter 16.

Before we consider the answer to this question, it must be pointed out that such a situation would be inconvenient, if not catastrophic. Methane (or other hydrocarbons) would ignite spontaneously on contact with air! The equation for the complete combustion of methane is:

$$CH_4(g) + 2O_2(g) \longrightarrow CO_2(g) + 2H_2O(l)$$

We have to ignite the methane; that is, we must give it sufficient energy for the reaction to get started. There is no reaction between the two gases (methane and oxygen) before ignition, and without this boost they sit together quite contentedly for an indefinite length of time. This is because, as the methane and oxygen molecules approach one another, the outer electrons of one molecule repel the outer electrons of the other. It's only if this repulsion can be overcome by a substantial input of energy that bonds can be broken and the attractive forces (between the electrons of one molecule and the positive nuclear charge of the other) can take over. The redistribution of electrons that occurs results in the bond-breaking and bond-making processes – it sets off a molecular reaction. Once the reaction has started, enough heat energy is produced to keep the reaction going (it is

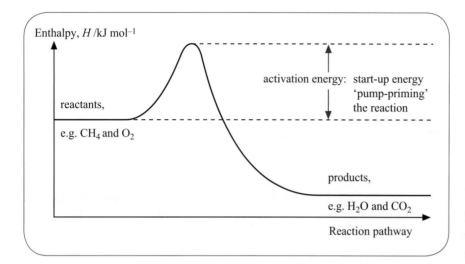

Figure 17.10 A reaction pathway diagram, showing the activation energy. This is an exothermic reaction.

self-sustaining). Figure 17.10 shows the situation diagrammatically. Overall the reaction pathway (or coordinate) lies downhill, but initially the path lies uphill.

SAQ

4 In the case of the reaction between methane and oxygen, where could the activation energy come from?

[Answer]

The reaction pathway for an endothermic reaction is shown in Figure 17.11.

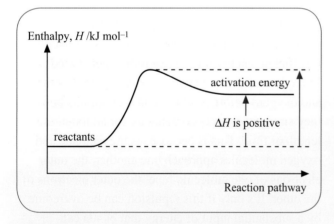

Figure 17.11 A reaction pathway diagram, showing the activation energy for an endothermic reaction.

Catalysis

A catalyst is something added to a reaction that increases its rate, but does not itself change in concentration: the same amount remains after the reaction as before.

Catalysts work by providing a different reaction pathway (route or mechanism) for the reaction. A reactant (in some case more than one reactant) will combine weakly with the catalyst to form an activated complex. This activated complex will undergo further reaction to form the products, releasing the catalyst for re-use. The catalyst takes part in the reaction but is restored at the end of the reaction.

The reaction rate increases because the catalysed reaction pathway has a lower activation energy than that of the uncatalysed reaction. This is shown in Figure 17.12.

The Boltzmann distribution in Figure 17.13 shows how the lower activation energy for the catalysed reaction increases the number of molecules that will react on collision.

Many economically important industrial processes involve the use of catalysts. Here are examples.

1 The production of poly(ethene) and other addition polymers using organometallic catalysts (these compounds contain metals and organic parts in their molecules, e.g. triethylaluminium) and other metal compounds (such as titanium(IV) chloride). These are called Ziegler–Natta catalysts (see Chapter 12). They enable chemists to control the shape of the polymer chains produced. An example is the production of high density poly(ethene) or HDPE. Without a Ziegler–Natta catalyst the chains produced are branched and cannot pack closely together. This is called low density poly(ethene). However, with the Ziegler–Natta catalyst the chains formed are straight and can pack tightly together, producing a stronger plastic, high density poly(ethene).

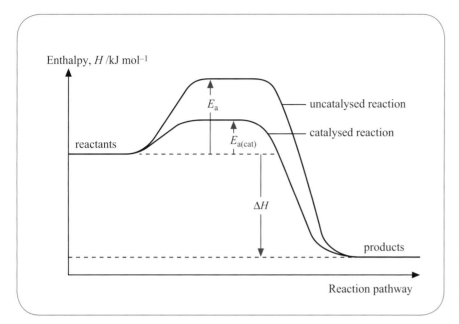

Figure 17.12 The catalysed reaction follows a different route (pathway) with a lower activation energy, $E_{a(cat)}$.

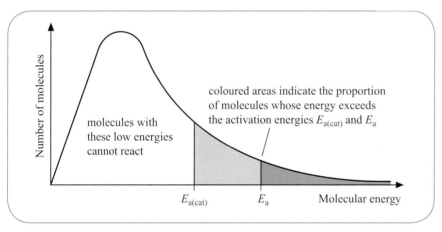

Figure 17.13 The route with the lower activation energy does not alter the Boltzmann distribution; however, it does increase the number of molecules with energies above the activation energy.

2 Hydrogenation of vegetable oils to produce margarine using a nickel catalyst (see Chapter 12): An unsaturated fat is converted to a saturated fat by the addition of a hydrogen molecule to each of the carbon–carbon double bonds in the fat:

$$-CH=CH- + H_2 \xrightarrow{\text{nickel}} -CH_2-CH_2-$$

The nickel catalyst must be finely divided to provide a large surface area for reaction.

3 The production of ammonia by the Haber process: Before the German chemist Fritz Haber developed this process, much nitrogen was converted to ammonia using an expensive electrical discharge process. The Haber process uses a gas-phase reaction between nitrogen and hydrogen:

$$N_2(g) + 3H_2(g) \rightleftharpoons 2NH_3(g)$$

Most of the ammonia produced is used to manufacture fertilisers for increasing the yield of food crops. It is often said that, without such fertilisers, a far greater proportion of the world's human population would have suffered starvation during the twentieth century. Fritz Haber's discovery gained him the 1918 Nobel Prize for Chemistry.

The Haber process reaction is catalysed by contact with a finely divided iron catalyst. Without a catalyst, the activation energy needed to break the very strong N≡N triple bond is extremely high. In the presence of the iron catalyst, molecules of nitrogen are weakly adsorbed on to iron atoms at the surface. This process weakens the nitrogen triple bond sufficiently for reaction to take place. We will look further at the Haber process later in this chapter. Notice that the reactions in points **1–3** here all have an atom economy of 100% (see Chapter 10). Points 1 and 2 are addition reactions and in the Haber process there is only one product formed in the reaction (unreacted nitrogen and hydrogen are continuously recycled so there is no waste at the synthesis stage). So we can thank the catalysts discovered for helping to reduce waste products that alternative reactions might well produce.

4 The use of enzymes in biotechnology: Enzymes are the biological catalysts found in all living things. They usually catalyse specific reactions and work best close to room temperature and pressure, saving on energy costs in the biotechnology industry. Also the pollution that is produced when generating high temperatures and pressures in traditional chemical processes is reduced.

We have already seen how enzymes in yeast are used to make ethanol by the fermentation of glucose (see Chapter 13). This is essential in the brewing, baking and dairy industries, as well as the manufacture of ethanol as a biofuel. But enzymes also have many other uses. It was in the 1950s that enzyme technology really started but progress has accelerated over the last 20 years. There are great benefits in using enzymes as catalysts. They are over 10 000 times more efficient than other catalysts used in industry. One enzyme molecule can catalyse 10 million reactions in a single second! An example from the food industry is the conversion of glucose into fructose, which is a sweeter sugar. Manufacturers of reduced-calorie meals can use less fructose to get the same taste as more glucose and help slimmers to lose weight.

We have looked at the advantages of using enzymes, but scientists have also had to overcome some problems. For example, if you want to catalyse one particular reaction you need a *pure enzyme* (not the mixture of enzymes found in cells). Their separation is difficult, which makes pure enzymes expensive. It is also desirable to use expensive enzymes over and over again. However, it is hard to remove enzymes from liquid products. So biotechnologists developed a technique in which the enzymes are stuck to plastic beads. They can be trapped in tiny pores inside inert structures. We say that they are *immobilised*. They do not get washed out with the product, allowing more economical continuous processes.

5 The use of platinum / palladium / rhodium catalysts in car exhausts: These expensive metals are helping to reduce pollution from cars. They are recycled when they are changed on a car. You can read more about these in Chapter 18.

SAQ

5 Why is the iron catalyst used in the Haber process finely divided?

Hint

Answer

Chemists are engaged in studying the surfaces of catalysts to find out just how they work, with the aim of developing new catalysts or improving existing ones. Improving the rates of large-scale chemical processes leads to savings in energy and other costs such as that of the chemical plant. In recent years progress has been more rapid due to new techniques such as scanning probe microscopy (SPM). This technique enables the positions of gaseous molecules or atoms to be seen on a metal surface. SPM provides powerful evidence to support reaction pathways such as adsorption of reactants, breaking of covalent bonds in reactant molecules and the presence of atoms on catalyst surfaces (Figure 17.14).

Chemists are also involved in the safe disposal of traditional metal catalysts such as vanadium, chromium, lead and nickel. These are all toxic. If they are simply buried in landfill sites the metals can dissolve as ions in water and soak away into waterways, harming wildlife and contaminating our water supplies. Treating the waste with lime helps to reduce this.

Research has found that heating metal waste with clay binds it into the clay. The product can then be used to make bricks, concrete or ceramics. The properties of these new materials are not adversely affected compared with traditional products and there is very little leaching of the metals.

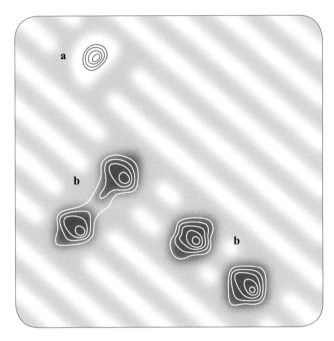

Figure 17.14 A scanning probe microscope (SPM) picture of oxygen on a copper surface. The diagonal rows coloured yellow are copper atoms. **a** is an oxygen molecule, O_2, adsorbed on the surface. **b** are four O^- ions. The distance between these is about 0.80 nm, which is large enough to show they are not bonded together.

Equilibrium

Reversible reactions

You will be familiar with a number of reversible physical processes. For example, if you decrease the temperature of water below $0\,^\circ C$, ice forms. Allow the ice to warm to room temperature and it soon melts. The process can be represented as follows:

$$H_2O(s) \rightleftharpoons H_2O(l)$$

The $\rightleftharpoons$ sign in this equation is used to indicate that the process is reversible.

Another reversible physical process is the dissolving of carbon dioxide in water. You will have met aqueous carbon dioxide in the form of fizzy drinks such as cola. Carbon dioxide is dissolved in the drink under pressure. When the drink is poured, the carbon dioxide escapes as bubbles of gas, producing a pleasant sensation when the cola is consumed. An equation for this reversible change is:

$$CO_2(aq) \rightleftharpoons CO_2(g)$$

The solubility of carbon dioxide in water is enhanced by chemical reaction with water, producing hydrogen ions, $H^+(aq)$, and hydrogencarbonate ions, $HCO_3^-(aq)$:

$$H_2O(l) + CO_2(aq) \rightleftharpoons H^+(aq) + HCO_3^-(aq)$$

This reaction is also easily reversed. Boiling the water will decompose the hydrogencarbonate ions and drive off carbon dioxide.

Many other chemical reactions are reversible. An environmentally important reversible reaction is the formation of ozone, $O_3(g)$, from oxygen. Ultraviolet light is needed to form ozone; chlorine atoms (from CFCs) have the overall effect of reversing the reaction, causing damage to the ozone layer:

$$3O_2(g) \overset{\text{UV light}}{\underset{\text{CFCs}}{\rightleftharpoons}} 2O_3(g)$$

In the rest of this chapter we shall explore the nature of reversible reactions in more detail.

Equilibrium – a state of balanced change

The notion of a system being in equilibrium is a familiar one. You can stir salt into water until no more will dissolve. At this point the solution is described as a saturated solution and is in equilibrium with the undissolved solid. Although the concentration of the saturated solution stays the same, the ions in the solid and the solution are in a constant state of exchange. We describe this as a **dynamic equilibrium**. Ions in the crystal lattice of the undissolved solid continue to go into solution. However, they are immediately replaced elsewhere in the lattice by the same numbers and kinds of ion from the solution. The dynamic nature of the equilibrium is only observable at the ionic or molecular level. The situation is one of continued but balanced change (Figure 17.15a).

A similar situation exists when you close the tap on a cylinder of butane gas, in a camping gas stove for example. Evaporation and condensation go on until the liquid and gas phases are in equilibrium with one another. Again the equilibrium is dynamic. At equilibrium some of the molecules of liquid butane are evaporating, but only at the same rate as molecules of gaseous butane are condensing (Figure 17.15b).

In general, in a dynamic equilibrium, the rate of reaction in the *forwards* direction equals the rate of reaction in the *reverse* direction.

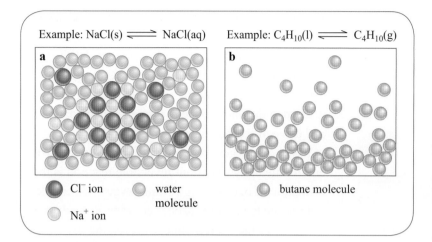

Example: $NaCl(s) \rightleftharpoons NaCl(aq)$
Example: $C_4H_{10}(l) \rightleftharpoons C_4H_{10}(g)$

a

b

Cl^- ion   water molecule

Na^+ ion

butane molecule

Figure 17.15 Two physical equilibria. In both situations there is a constant interchange of particles, which maintains a steady balance. In **a** ions leave a crystal structure, while others join it. In **b** molecules escape from the crush in a liquid to relative isolation in a gas, while others leave the gas to join the liquid.

Equilibrium and chemical change

If you heat calcium carbonate it decomposes, forming calcium oxide and carbon dioxide:

$$CaCO_3(s) \longrightarrow CaO(s) + CO_2(g)$$

On the other hand, if you leave calcium oxide in an atmosphere of carbon dioxide, the reverse reaction occurs, and calcium carbonate forms:

$$CaO(s) + CO_2(g) \longrightarrow CaCO_3(s)$$

If these substances are put in a sealed container at a high temperature (say $500\,^{\circ}C$) and left to get on with it, an equilibrium is set up. Both of the above reactions occur until a balance is reached. At this point the rate of formation of calcium carbonate equals its rate of decomposition.

All chemical reactions can reach equilibrium, a situation where the reactants are in equilibrium with the products. Again, these are dynamic equilibria: reagents are constantly being converted to products, and vice versa. At equilibrium the rate of the forward process is the same as that of the backward one. The idea that *all* chemical reactions can reach equilibrium seems to conflict with experience, e.g. the burning of magnesium in air. In many cases the degree of conversion of reactants to products is so large that, at the conclusion of the reaction, no reactants can be detected by normal analytical means. At other times two reagents, e.g. the nitrogen and oxygen in the air, do not seem to react at all. Such reactions are often considered to be irreversible one-way reactions under those conditions.

Suppose, for example, we mix an equal number of molecules of hydrogen and bromine. We then provide some energy to start the reaction. The orange colour of the bromine disappears, and we are left with hydrogen bromide. The reverse reaction is so minimal that the reaction appears to go to completion as indicated by the equation:

$$H_2(g) + Br_2(g) \longrightarrow 2HBr(g)$$

If we do the same with hydrogen gas and iodine vapour, however, we find that the violet colour of the iodine persists. There is an equilibrium set up between the three components in which all three are present in significant amounts, as shown in the equation below and in Figure 17.16.

$$H_2(g) + I_2(g) \rightleftharpoons 2HI(g)$$

The equation tells us that when a molecule of hydrogen reacts with a molecule of iodine, two molecules of hydrogen iodide are formed. It also enables us to examine the reaction in reverse. If two molecules of hydrogen iodide dissociate (i.e. split apart), then a molecule each of hydrogen and iodine are formed.

When you cook on a camping gas stove, butane gas is released and burned. The liquid butane will evaporate to maintain the gas supply. However, equilibrium will not be restored in the cylinder unless the gas is turned off at the tap. We must have a **closed system** to achieve a dynamic equilibrium. A closed system is a system in which the substances cannot leave and cannot enter.

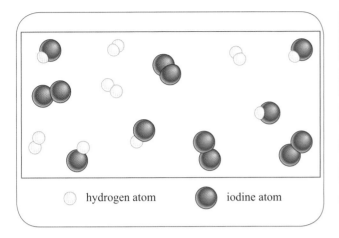

Figure 17.16 A snapshot of the dynamic equilibrium between hydrogen gas, iodine gas and hydrogen iodide gas.

SAQ

6   A beaker contains saturated aqueous sodium chloride in contact with undissolved solid sodium chloride. Is this a closed system? Explain your answer.

[Hint]

[Answer]

When we view a system at equilibrium, we are not aware that constant change is taking place. From our viewpoint the system looks static because all the dynamic change is occurring at the molecular or ionic level. The properties that we can see or measure remain constant. We call these macroscopic properties. In a camping gas cylinder with the tap closed, the volume of liquid and the pressure do not change once dynamic equilibrium is achieved. The concentration of salt in a saturated solution is also constant at equilibrium. Another feature of a dynamic equilibrium is this *constancy of macroscopic properties.*

In summary the characteristic features of an equilibrium are:
● it is dynamic at the molecular or ionic level
● both forward and reverse processes occur at equal rates
● a closed system is required
● macroscopic properties remain constant.

Changing conditions: Le Chatelier's principle

Now that we have established the characteristic features of an equilibrium, we can ask the question "What happens to the equilibrium if we change the conditions in some way?" We could, for example, alter the temperature or change the concentration of a reactant.

Suppose we add more water to the equilibrium mixture of solid sodium chloride and saturated aqueous sodium chloride we saw in Figure 17.15. The mixture will no longer be in equilibrium. However, more of the solid will dissolve and, providing there is sufficient solid, the solution will again become saturated. The system readjusts to restore equilibrium.

Raising the temperature of our sealed container of calcium carbonate in equilibrium with calcium oxide and carbon dioxide provides the energy that allows more calcium carbonate to decompose. Again, the system adjusts to restore equilibrium. Initially, more calcium oxide and carbon dioxide is formed. Equilibrium is restored when the rate of formation of the calcium oxide and carbon dioxide is the same as the rate of the reverse reaction to form calcium carbonate.

Observations of this type led the French chemist Henri Louis Le Chatelier in 1884 to put forward an important principle. The essence of **Le Chatelier's principle** is that:

> when any of the conditions affecting the position of a dynamic equilibrium are changed, then the position of that equilibrium will shift to minimise that change.

'The position of equilibrium' refers to the relative amounts of reactants and products present. We can predict the effect of changing temperature, pressure or concentration on the equilibrium position using Le Chatelier's principle.

The effect of temperature on the position of equilibrium

We know that calcium carbonate decomposes to calcium oxide and carbon dioxide at a high temperature. At room temperature, no change is seen. The white cliffs of Dover are still calcium carbonate, as they were when first seen by Julius Caesar!

When calcium carbonate is heated in a closed system (see Chapter 8), an equilibrium mixture containing both reactant and products results. The reaction is endothermic and, on raising the temperature, the equilibrium shifts towards the formation of calcium oxide and carbon dioxide. In the closed system, the higher the temperature, the greater is the proportion of products at equilibrium.

The dissociation of hydrogen iodide is an example of a homogeneous endothermic reaction:

$$2HI(g) \rightleftharpoons H_2(g) + I_2(g)$$

The effect of different temperatures on the equilibrium concentration of hydrogen can be seen in Table 17.1 and Figure 17.17. As the temperature rises, the equilibrium concentration of hydrogen rises. The position of equilibrium in this gas-phase reaction shifts towards the formation of hydrogen and iodine at higher temperature.

When we raise the temperature of an endothermic reaction, there is an increase in the enthalpy in the system. According to Le Chatelier's principle, the equilibrium position should shift towards the products in order to compensate for the additional enthalpy input.

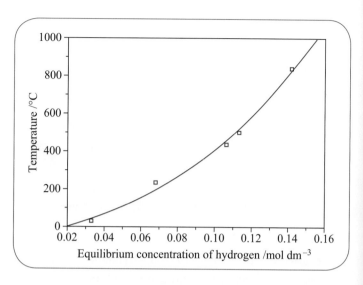

Figure 17.17 In an endothermic reaction such as the dissociation of hydrogen iodide, as the temperature is increased the equilibrium concentration of the products increases. The graph shows the increase in concentration of hydrogen with increasing temperature.

Temperature /°C	Equilibrium concentration of hydrogen iodide /mol dm^{-3}	Equilibrium concentration of hydrogen (or iodine) /mol dm^{-3}
25	0.934	0.033
230	0.864	0.068
430	0.786	0.107
490	0.773	0.114
830	0.714	0.143

Table 17.1 The dissociation of hydrogen iodide, HI(g), at various temperatures.

Example	Endothermic reaction $2HI(g) \rightleftharpoons H_2(g) + I_2(g)$	Exothermic reaction $2SO_2(g) + O_2(g) \rightleftharpoons 2SO_3(g)$
temperature increase	equilibrium position shifts towards products: more hydrogen and iodine form	equilibrium position shifts towards reactants: more sulfur dioxide and oxygen form
temperature decrease	equilibrium position shifts towards reactant: more hydrogen iodide forms	equilibrium position shifts towards product: more sulfur trioxide forms

Table 17.2 The effect of temperature change on the equilibrium positions of reactions involving gases.

Suppose we consider an increase in temperature for an exothermic reaction. The reverse reaction will be endothermic, so Le Chatelier's principle tells us that the equilibrium will shift towards the reactants to compensate for the extra enthalpy input.

Table 17.2 summarises the effects of temperature changes on the equilibrium position for exothermic and endothermic reactions.

The effect of changes in concentration on the equilibrium position

We will consider the formation of an ester, such as ethyl ethanoate. When ethanol is warmed with ethanoic acid in the presence of a few drops of concentrated sulfuric acid, ethyl ethanoate is formed (see Chapter 13)

The sulfuric acid catalyses this reaction. An equilibrium is soon established in the reaction mixture, with significant concentrations of both products and reactants present. Suppose we increase the concentration of ethanol in the mixture. The position of equilibrium is disturbed. Applying Le Chatelier's principle, more ethyl ethanoate and water will form and the concentration of ethanol and ethanoic acid will fall until a new position of equilibrium is established. The position of equilibrium moves towards the products.

SAQ

7 Consider the equilibrium involved in the formation of ethyl ethanoate.
   a Use Le Chatelier's principle to predict how the position of equilibrium would change on adding more water to the mixture.
   b Ethyl ethanoate is a useful solvent. It is used, for example, in nail varnish remover. Suggest how a chemical company might optimise the conversion of ethanol and ethanoic acid to ethyl ethanoate.

Answer

The effect of pressure on equilibria

Pressure has virtually no effect on the chemistry of solids and liquids. As shown in Figure 17.18, pressure does not affect the concentration of solids and liquids – the molecules concerned are already in contact and it is difficult to push them closer together.

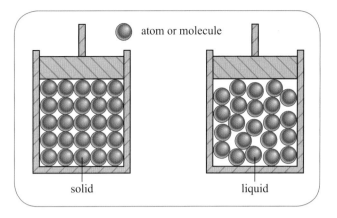

Figure 17.18 Pressure has little, if any, effect on the concentrations of solids and liquids.

Pressure does have significant effects on the chemistry of reacting gases. As Figure 17.19 shows, the concentrations of gases increase with an increase in pressure, and decrease with a decrease in pressure. Since chemical equilibria are influenced by concentration changes, pressure changes also have an effect on equilibria where one or more of the reagents is a gas.

Again we can apply Le Chatelier's principle to predict how pressure change will affect an equilibrium. Imagine a reaction in the gaseous phase where two molecules, A and B, combine to form a single molecule, C:

$$A(g) + B(g) \rightleftharpoons C(g)$$

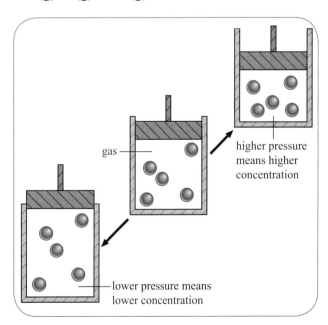

Figure 17.19 Pressure has a considerable effect on the concentrations of gases.

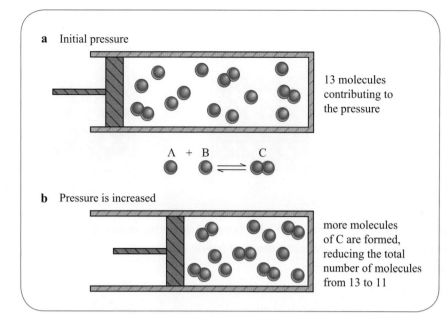

a Initial pressure

13 molecules contributing to the pressure

A + B ⇌ C

b Pressure is increased

more molecules of C are formed, reducing the total number of molecules from 13 to 11

Figure 17.20 An increase in pressure in this case causes the equilibrium to shift to the right, to produce more molecules of C than before, but fewer molecules in the reaction vessel overall.

The situation is illustrated in Figure 17.20.

It helps to remember that the pressure of a gas depends on the number of molecules in a given volume of the gas. The greater the number of molecules, the greater the number of collisions per second, and hence the greater the pressure of the gas. In the reaction above, when the pressure is increased *the equilibrium shifts to minimise this increase*, that is to reduce the pressure overall. Therefore, there must be fewer molecules present than before. This can only happen if A and B molecules react to make more molecules of C.

This reasoning is summarised in Table 17.3, which summarises the effects of increasing and decreasing the pressure on reactions in which either there are fewer molecules on the right of the equilibrium, or there are more molecules on the right of the equilibrium.

SAQ

8 Predict the effect of increasing the pressure on the equilibrium position of the following reactions.

  a $2HI(g) \rightleftharpoons H_2(g) + I_2(g)$

  b $N_2(g) + 3H_2(g) \rightleftharpoons 2NH_3(g)$

Hint

Answer

The effect of catalysts on equilibria

As you know, a catalyst reduces the activation energy of a reaction and so speeds it up. A catalyst affects the rate of reaction, but does not feature in the overall equation for the reaction. More catalyst could mean a faster reaction, one in which an equilibrium was established more quickly, *but does not change the equilibrium concentration of reactants or products*.

Example	Fewer molecules on right $2SO_2(g) + O_2(g) \rightleftharpoons 2SO_3(g)$	More molecules on right $N_2O_4(g) \rightleftharpoons 2NO_2(g)$
pressure increase	equilibrium position shifts towards products: more SO_3 forms	equilibrium position shifts towards reactants: more N_2O_4 forms
pressure decrease	equilibrium position shifts towards reactants: more SO_2 and O_2 form	equilibrium position shifts towards products: more NO_2 forms

Table 17.3 The effect of pressure change on the equilibrium position of reactions involving gases.

An equilibrium of importance: the Haber process

We will now gather together the ideas covered in this chapter using a reaction that is important from the theoretical, the practical and the industrial points of view. This is the Haber process for the 'fixation' of atmospheric nitrogen. We need large amounts of nitrogen compounds, particularly for fertilisers. Air is 80% nitrogen, so atmospheric nitrogen is the most plentiful and readily available source. At the same time, it cannot be used directly in the gaseous form; it needs to be 'fixed' into a chemically combined form to make a useful compound. One possible conversion might be to ammonia. Compared with nitrogen, ammonia is much more reactive. It is readily soluble in water, it is readily convertible to ammonium salts, and it can be converted to nitric acid by oxidation.

The equation for the reaction to form ammonia is:
$$N_2(g) + 3H_2(g) \rightleftharpoons 2NH_3(g); \Delta H = -93 \, kJ \, mol^{-1}$$
ΔH refers to the enthalpy change of reaction (see Chapter 16). The unreactive nature of nitrogen is, of course, a problem. Although the reaction is exothermic, the triple bond within nitrogen molecules lends them great strength, so the reaction has a high activation energy. This makes the reaction extremely slow. How can an industrial plant be set up to give a good yield of ammonia at an acceptable rate? This problem was solved in the early 1900s by the German chemist Fritz Haber, and the process that he developed is still in use today.

The obvious way to speed things up would be to increase the temperature. However, by using Le Chatelier's principle, since the reaction is exothermic, increasing the temperature will drive the equilibrium to the left (Table 17.2). This effect is quite dramatic (Table 17.4). 91.7% of the nitrogen and hydrogen is converted to ammonia at 100 °C and 25 atm. At 700 °C and 25 atm this percentage drops to 0.9%.

Pressure is another variable. We can reason that an increase of pressure will drive the equilibrium to the right. The proportions by volume of the gases are as follows:
$$N_2(g) + 3H_2(g) \rightleftharpoons 2NH_3(g); \Delta H = -93 \, kJ \, mol^{-1}$$

| 1 volume | 3 volumes | 2 volumes |

According to Le Chatelier's principle, an increase in pressure should drive the equilibrium to the right, since this will result in a decrease in volume. This is found in practice, and Table 17.4 gives the relevant figures.

Imagine you were asked to design a chemical production plant for the manufacture of ammonia. From our discussion we might consider using the highest possible pressure with a suitably low temperature, for example about 500 atm and 20 °C. However, such a choice would create difficulties. The high activation enthalpy means that the rate of reaction is effectively zero at 20 °C. Very high pressures increase the rate of reaction but dramatically increase the cost of the plant. The cost of labour for running the plant also increases as the type of pumps required for maintaining high pressure require more maintenance.

The problem of the very low rate of reaction can be partially overcome by the choice of a suitable catalyst such as porous iron (Figure 17.21). Small amounts of the oxides of potassium, magnesium, aluminium and silicon improve the efficiency of the catalyst. The catalyst enables the reaction to proceed by a different route with a lower activation energy (see page 203). However, it does not affect the percentage of ammonia in an equilibrium mixture.

Temperature /°C	Percentage of ammonia at equilibrium			
	25 atm	50 atm	100 atm	200 atm
100	91.7	94.5	96.7	98.4
300	27.4	39.6	53.1	66.7
500	2.9	5.6	10.5	18.3
700	0.9	1.2	3.4	8.7

Table 17.4 Percentage of ammonia in the equilibrium mixture at various temperatures and pressures.

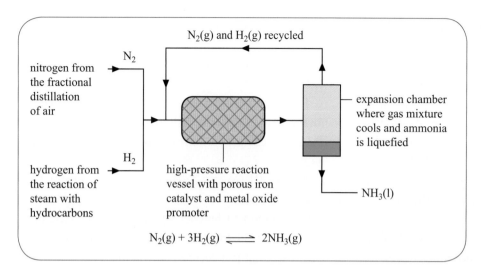

Figure 17.21 The production of ammonia by the Haber process.

The rate can also be increased by raising the temperature and accepting a reduced equilibrium percentage of ammonia in the mixture. A temperature is chosen to give a reasonable rate, and a reasonable yield. It is described as a *compromise* temperature.

Modern plants operate at high pressures, but these are not as high as the pressures used by older plants. The money saved by operating the plant at lower pressure compensates for the reduction in the equilibrium percentage of ammonia. Such compromises reduce the overall production costs sufficiently to justify replacing an old plant.

There are two more ways in which the efficiency of an ammonia plant may be improved.

- Ammonia is removed as it is formed so that the reaction mixture is not allowed to reach equilibrium. This means that the reaction rate stays reasonably high. If the reaction is allowed to approach equilibrium, the reaction rate decreases as concentrations of the reactants decrease.
- The plant operates continuously and, after passing through the reaction vessel, the reaction mixture is passed into an expansion chamber where rapid expansion cools the mixture and allows ammonia to liquefy (Figure 17.21). The liquefied ammonia is run off to pressurised storage vessels and the unreacted nitrogen and hydrogen is recirculated over the catalyst in the reaction vessel.

Modern Haber process plants (Figure 17.22):
- are highly efficient in conversion of nitrogen and hydrogen to ammonia

- have a low energy consumption (around 35 MJ per kg of nitrogen converted to ammonia)
- are smaller, so they are less expensive to build
- have less environmental impact
- may be sited where they are needed, reducing transport costs.

The conditions used in modern plants are:
- a pressure between 25 and 150 atm
- a temperature between 400 and 500 °C
- a finely divided or porous iron catalyst with metal oxide promoters.

Figure 17.22 This modern ammonia plant uses the chemical reaction devised by Fritz Haber in the early 1900s.

SAQ

9 A modern Haber process plant may operate at a temperature of 450 °C, a pressure of 100 atm, and using an iron catalyst.

   a Why don't they use a temperature above 450 °C?

   b Why don't they use a temperature below 450 °C?

   c Why don't they use a pressure above 100 atm?

   d Why don't they use a pressure below 100 atm?

Answer

Fertilisers

80% of ammonia production goes into making fertilisers such as ammonium sulfate. The increasing human population requires ever more intensive agricultural production techniques. In order to grow crops repeatedly in the same soil, artificial fertilisers are added to maintain fertility.

However, the solubility of these ammonia-based fertilisers has caused pollution of waterways by a process called eutrophication (see Figure 17.23). The fertilisers can reach streams and rivers through surface run-off from fields or by soaking through the ground.

In the water they encourage aquatic plants and algae to grow rapidly. Plants and algae not only produce oxygen during photosynthesis, they use it (as do aquatic animals) during respiration.

continued

Rapid growth of plants at the surface blocks light to organisms below. So plants growing deeper in the water stop photosynthesising and the oxygen concentration in the water starts to fall. Eventually it falls low enough that these plants start to die.

Bacteria feed on the decaying plants and increase rapidly in number. They use oxygen from the water for respiration, and the oxygen concentration falls to very low levels. At these levels, few other organisms can survive, so most of the plants and animals die, providing more food for the bacteria. Eutrophication can lead to the loss of most of the life in a river.

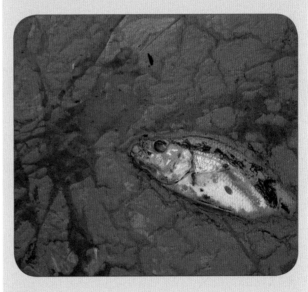

Figure 17.23 A river affected by eutrophication.

Summary

- The rate of a chemical reaction is measured by the amount (in moles) of a reactant used up in a given time. Chemical kinetics is the study of rates of chemical reactions.

- Chemists study rates of reaction to:
  - improve the rate of production of a chemical
  - help understand the processes going on in our bodies or in the environment
  - gain an insight into the mechanism of a reaction.

- The factors that affect the rate of a chemical reaction are:
  - concentration (or pressure of gases)
  - temperature
  - surface area
  - catalysts.

- At higher concentration (or pressure), more frequent collisions occur between reactant molecules. This increases reaction rate.

- At higher temperature, molecules have more kinetic energy, so a higher percentage of effective collisions occurs between reactant molecules.

- The activation energy of a reaction is the minimum energy required for reaction to occur. Enthalpy profile diagrams show how the activation energy provides a barrier to reaction.

- The Boltzmann distribution represents the numbers of molecules in a sample with particular energies. The change in the Boltzmann distribution as temperature is increased shows how more molecules have a kinetic energy higher than the activation energy. This leads to an increase in reaction rate.

- A catalyst increases the rate of a reaction by providing an alternative reaction pathway with a lower activation energy. More molecules have sufficient energy to react, so the rate of reaction is increased.

- A reversible reaction is a reaction that may proceed in either direction (forward or reverse), depending on the applied conditions.

- Dynamic equilibrium occurs when the rate for the forward reaction is equal to the rate of the reverse reaction. Products are formed and reactants are re-formed at the same rate. The equilibrium is dynamic because it is maintained by continual changes occurring between molecules.

- A closed system is needed for equilibrium to be established in a chemical reaction. Equilibria are characterised by the constancy of macroscopic properties such as concentration.

- Le Chatelier's principle states that the equilibrium will shift so as to minimise the effect of a change in concentration, pressure or temperature.

- A catalyst may accelerate the rate at which the reaction achieves equilibrium.

- The conditions used in the Haber process for the production of ammonia are a compromise between the position of the equilibrium and the rate of reaction whilst controlling the expense and minimising the energy needed to produce those conditions. The key requirements are to maximise ammonia production and minimise costs.

Questions

1 A group of students investigated the effect of concentration on the rate of a reaction. They used the reaction between magnesium carbonate and dilute hydrochloric acid and measured the rate at which the gas was collected.

   a An incomplete equation for this reaction is given below. Copy and complete the equation by balancing it and inserting state symbols.

      $MgCO_3 + HCl \longrightarrow MgCl_2 + CO_2 + H_2O$ [2]

   b The students added dilute hydrochloric acid to some magnesium carbonate. The students collected the gas and measured the volume, at regular intervals, until after the reaction was complete. They then plotted a graph of their results.

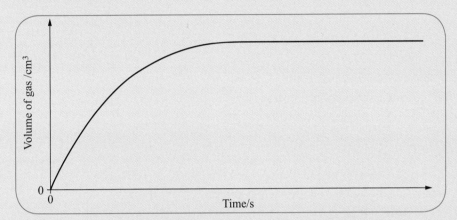

      Use collision theory to explain the changes in the rate of the reaction as it proceeds.   [3]

   c The students repeated the experiment using a weak acid instead of hydrochloric acid. Assume the concentration of both acids and all other conditions are the same.   Hint

     i Copy the axes in **b**, and sketch the graph the students obtained from this experiment.   [2]

     ii State and explain what effect changing the acid has on the rate of reaction.   [2]

OCR Chemistry AS (2813) Jan 2007   [Total 9]

Answer

2 **a** The diagram below shows the energy distribution of reactant molecules at a particular temperature. E_a represents the activation energy of the reaction.

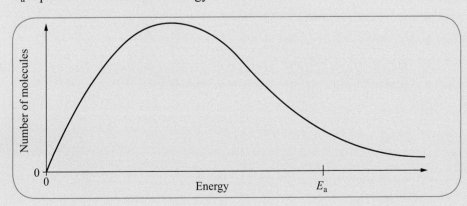

continued

> **i** Sketch the diagram and draw a second curve to represent the energy distribution of the same number of molecules at a higher temperature. [2]
>
> **ii** Use your completed diagram to explain how an increase in temperature can cause an increase in the rate of reaction. [2]
>
> **b** The rate of reaction between hydrogen and oxygen depends on the pressure as well as the temperature.
> $$2H_2(g)+O_2(g) \longrightarrow 2H_2O(g); \Delta H=-286\,kJ\,mol^{-1}$$
>
> **i** Describe and explain the effect of increasing the pressure on the rate of this reaction. [2]
>
> **ii** A sudden rapid increase in the rate of a reaction causes an explosion to occur.
> Suggest why highly exothermic reactions such as this one are more likely to explode than other reactions. [2]

OCR Chemistry AS (2813) June 2004 [Total 8]

Answer

3 a State Le Chatelier's principle. [2]

b State <u>two</u> characteristics of a dynamic equilibrium. [2]

c The following equation represents an equilibrium reaction.
$$Cr_2O_7^{2-}(aq)+H_2O(l) \rightleftharpoons 2CrO_4^{2-}(aq)+2H^+(aq)$$
orange   yellow

Use Le Chatelier's principle to describe and explain the colour change (if any) that might take place when dilute HCl(aq) is added to a solution containing K_2CrO_4(aq). [2]

Hint

d The following equation represents another equilibrium reaction.
$$2NO_2(g) \rightleftharpoons N_2O_4(g); \Delta H^\ominus=-58\,kJ\,mol^{-1}$$
brown   colourless

Use Le Chatelier's principle to describe and explain the colour change (if any) that might take place when

i a mixture of NO_2(g) and N_2O_4(g) is compressed at constant temperature, [2]

ii a mixture of NO_2(g) and N_2O_4(g) is heated at constant pressure. [2]

OCR Chemistry AS (2813) Jan 2004 [Total 10]

Answer

4 Esters are used as flavourings. They are made by a reversible reaction between a carboxylic acid and an alcohol.
carboxylic acid+alcohol $\rightleftharpoons$ ester+water

a The production of esters is catalysed by the presence of acids.

i What is meant by a catalyst? [1]

ii Using the fact that acids are needed to catalyse this reaction, deduce the formula of the ion that acts as the catalyst. [1]

iii Catalysts do <u>not</u> affect the position of an equilibrium. Explain why not. [1]

b i Draw and label a graph of the Boltzmann distribution to show the energies of molecules in a gas at a fixed temperature. [2]

ii Use the graph to explain the effect of a catalyst on the rate of a reaction. [2]

OCR Chemistry AS (2813) Jan 2007 [Total 7]

Answer

continued

5   A chemical **C** is made by reacting chemical **A** with chemical **B** in a reversible reaction. **A**, **B** and **C** are all gases under the reaction conditions.

Research chemists wanted to know the optimum conditions to use in the manufacture of **C**. They carried out a series of reactions under different conditions of temperature and pressure. The percentage conversion of **A** at equilibrium is shown in the table below.

pressure /MPa	temperature /°C	% A converted
10	350	8
	450	12
	550	16
20	350	11
	450	21
	550	29
40	350	18
	450	
	550	49

a   Suggest the percentage of **A** that is converted at 450 °C and 40 MPa.   [1]

b   i   Use the data in the table to state the effect of increasing pressure on the percentage of **A** converted.   [1]

ii   What can be deduced, from this change, about the total number of moles of reactants **A** and **B** compared with the number of moles of product **C** in the equation for the reaction? Explain how you reached your conclusion.   [2]

〔Hint〕

c   Use the data in the table to deduce whether the reaction between **A** and **B** is exothermic or endothermic. Explain how you reached your conclusion.   [2]

d   It was found necessary to use a catalyst in the production of **C**.

i   What is meant by a *catalyst*?   [2]

ii   Suggest and explain <u>two</u> reasons why catalysts are used in industrial processes.   [2]

e   Conditions were used that should have given a conversion of **A** of 39%. In the manufacture, using these conditions, it was found that only 20% conversion was achieved.

Suggest why the conversion was much less than the theory suggested.   [1]

f   Give <u>two</u> economically important processes that use catalysts.   [2]

OCR AS Chemistry 2813 Jan 2006   [Total 13]

〔Answer〕

Chapter 18

Chemistry of the air and 'Green Chemistry'

e-Learning

Objectives

The Earth's atmosphere is essential for life (see Figure 18.1). Oxygen is required for respiration by animals and plants, carbon dioxide is needed for photosynthesis, nitrogen is used for making proteins, and ozone protects us from the Sun's harmful rays.

Figure 18.1 View of the Earth from space.

The atmosphere extends roughly 2000 km above the Earth's surface and becomes less dense the higher you go.

The dry atmosphere at sea level contains 78.09% nitrogen, 20.94% oxygen, 0.93% argon, 0.03% carbon dioxide and traces of other gases. It is these traces of other gases, and increasing levels of carbon dioxide, that are causing environmental concerns.

Atmospheric pollution

Emissions into the atmosphere are in the form of either gases or particulate material (Figure 18.2). The term particulate material refers to dusts and liquid droplets.

Material emitted into the atmosphere is diluted with air and transported both vertically and horizontally. It also undergoes chemical and physical changes. Pollutants are either primary, emitted directly into the atmosphere, or secondary, formed in chemical reactions in the atmosphere.

Figure 18.2 Both human and natural activities cause pollutants to be emitted into the atmosphere.

Emissions can cause problems in various ways. Some emissions accumulate in the atmosphere, for example carbon dioxide. Others, that are unreactive in the lower atmosphere (called the **troposphere**), travel into the upper atmosphere (the **stratosphere**). Once there they participate in chemical reactions (for example CFCs, Chapter 14).

There are many different air pollutants, but the main ones are carbon monoxide, carbon dioxide, sulfur dioxide, oxides of nitrogen and low-level ozone. All of these have damaging effects on human and animal health, vegetation and building materials.

The greenhouse effect

Carbon dioxide in the atmosphere

The concentration of carbon dioxide in the atmosphere depends on the various processes involved in the carbon cycle (Figure 18.3). Carbon dioxide is removed from the atmosphere in photosynthesis:

$$CO_2(g) + H_2O(l) \longrightarrow [CH_2O](aq) + O_2(g)$$

([CH_2O] is an empirical formula for a carbohydrate.)

Photosynthesis is carried out by green plants and is an energy-storing process.

Carbon dioxide is returned to the atmosphere by respiration in plants and animals:

$$[CH_2O](aq) + O_2(g) \longrightarrow CO_2(g) + H_2O(l)$$

Respiration is an energy-liberating process.

Some carbon dioxide is also produced in the atmosphere by the oxidation of hydrocarbons, mainly methane. This methane is given off by microorganisms in marshes and by animals, such as cattle, due to the microorganisms in their intestines.

Atmospheric carbon dioxide is also in dynamic equilibrium with carbon dioxide dissolved in surface water. Dynamic equilibrium means that the concentrations of gaseous and aqueous (dissolved) carbon dioxide remain constant even though there is constant movement of individual molecules between the atmosphere and surface water:

$$CO_2(g) \rightleftharpoons CO_2(aq)$$

In general, the concentrations of carbon are in balance in the different sections of the cycle, but variations in the carbon dioxide content of the air do occur. For example, there is more carbon dioxide at night and in the winter. Carbon dioxide concentration is increased by the combustion of fossil fuels and the removal of large areas of tropical rainforest, where photosynthesis is particularly rapid. The effect of human intervention on the carbon cycle is discussed further in the next section.

SAQ

1   Explain why carbon dioxide levels in the air increase in winter, compared to summer, and increase at night, compared to daytime.

| Answer |

2   Predict how the destruction of large areas of tropical rainforest will affect the atmosphere.

| Hint |

| Answer |

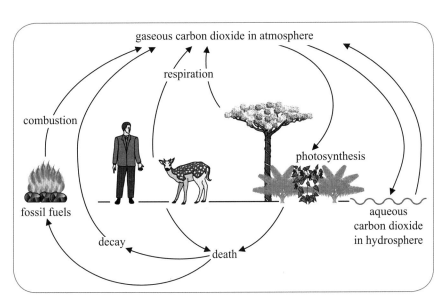

Figure 18.3 The carbon cycle.

You've never had it so hot in 1,000 years

by Anthony Browne
Environmental Correspondent

THIS YEAR is set to be the hottest of the millennium. With only six weeks Office predicts December will be 'warmer than normal'.

The prediction is a dramatic reinforcement of evidence that the whole Research unit at the University of East Anglia, a leading member of the intergovernmental Panel on Climate Change said: "There's a good chance particularly shocking since this year has not had a notable heatwave. Not one month in 1999 has broken the record, but every month apart and we had a sort of Indian summer'.

Met Office figures show that, up to 11 November, tempetatures have been 1.25 degrees centigrade

Figure 18.4 Part of an article from the front page of *The Observer* newspaper, 14 November 1999. UK temperature records were broken again in 2005.

The greenhouse effect and climate change

Look at the headline in Figure 18.4. There is obvious concern about the rate at which the Earth's surface is heating up, and about the extreme weather events. For example, hurricanes and cyclones have trebled in number over the last 30 years.

Climate depends on the global heat balance. A temperature change of 2–3 °C would have a pronounced effect on global and local climate. Of the energy which enters the Earth's atmosphere, 47% reaches the Earth's surface. Incoming energy from the Sun is in the ultraviolet, visible and infrared regions of the electromagnetic spectrum. Because the surface of the Earth is at a much lower temperature than the Sun, the radiation re-emitted from the Earth is of lower energy and is in the infrared region. Some of this infrared radiation is absorbed by water vapour, carbon dioxide and methane in the air.

These molecules absorb the infrared radiation because the O–H bonds in water, the C=O bonds in carbon dioxide and the C–H bonds in methane vibrate at frequencies in the infrared range. Therefore when exposed to infrared radiation the bonds stretch, twist or bend more vigorously, absorbing energy. We saw how this is used in infrared spectroscopy to identify compounds. Scientists use this technique to monitor levels of pollutants in the atmosphere, sometimes using infrared spectrometers stationed on satellites.

Having absorbed energy from the infrared radiation, the molecules will then re-emit the energy. The average temperature of the Earth's surface is maintained at 14 °C by the portion of this re-emitted energy which is returned to Earth. If it was not for

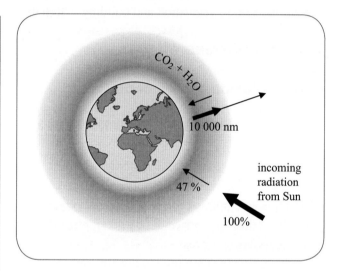

Figure 18.5 The 'greenhouse effect'. Some of the infrared radiation emitted by the Earth at a wavelength of 10 000 nm is absorbed by carbon dioxide, water vapour and other 'greenhouse gases', such as methane, in the atmosphere. Some of this radiation is re-emitted back to the Earth, keeping the surface of the Earth relatively warm.

this re-emitted infrared radiation from water and carbon dioxide, the temperature of the Earth would be –20 °C to –40 °C at its surface.

This effect, which keeps the Earth's surface relatively warm, is called the **'greenhouse effect'**. It resembles the way in which a greenhouse retains the Sun's warmth by internal reflection (see Figure 18.5). The carbon dioxide, water vapour and methane in the atmosphere play the same role as the glass in the greenhouse.

The greenhouse effect creates a steady-state system in which the rate at which the Earth is absorbing energy is equal to the rate at which energy is being radiated back into space. It is delicately balanced.

This delicate balance is being disturbed by emissions to the atmosphere. There is increasing concern that the rising levels of carbon dioxide in the atmosphere, together with other greenhouse gases discussed later, will lead to **global warming** with potentially disastrous climatic effects.

Levels of carbon dioxide in the atmosphere have shown a steady increase since around 1870. There are seasonal variations in carbon dioxide concentration: in the northern hemisphere it peaks in April and is at its lowest in September/October. This seasonal variation is in the most part caused by photosynthesis in the mid-latitude forests. Destruction of the rainforests could have a serious effect in raising the levels of carbon dioxide in the atmosphere.

As part of the International Geophysical Year in 1957, atmospheric monitoring stations were established at the South Pole and at Mauna Loa in Hawaii. Figures for carbon dioxide concentration from Mauna Loa are given in Figure 18.6. The concentration of CO_2 is measured here in parts per million by volume, ppmv. This measure is frequently used where concentrations are very small. Here it is the number of particles of carbon dioxide per million molecules of air; 350 ppmv corresponds to a concentration of 0.035%.

How do we know how our atmosphere has changed?

Scientists in Greenland have spent seven years drilling a hollow pipe down into the ice. On reaching the rock beneath the ice sheet, they had drilled out a cylinder of ice over 3 kilometres long!

When you shine light on the ice core sample, you see bands of light and dark. The dark bands are from winter snowfall, the light from summer. So it's a bit like looking a tree rings. Each band can provide us with evidence about the conditions on Earth at the time the snow fell.

There are tiny bubbles of air trapped in the ice core. Scientists can analyse the air to find how the composition of the gases in the atmosphere has changed over time. They can find the amounts of greenhouse gases, such as carbon dioxide and methane, in different bands. Initial analysis has shown that our current levels of carbon dioxide are higher than at any time in the last 440 000 years.

Figure 18.7 shows how the level of carbon dioxide and the Earth's temperature are related to each other.

continued

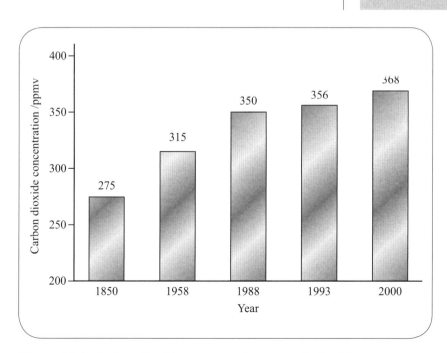

Figure 18.6 Carbon dioxide concentrations recorded at Mauna Loa, Hawaii, with earlier data from 1850 for comparison (see text for explanation).

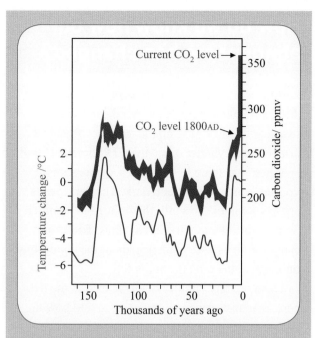

Figure 18.7 Changes in temperature (red) and the concentration of CO_2 in the atmosphere (blue) in the last 150 000 years.

Evidence like this is used to argue against those who think that our current warm phase is not the result of increased levels of greenhouse gases but just natural variations that have happened throughout the history of the Earth.

Although a link can be established, it does not show conclusively which came first. Does the carbon dioxide level rise because the temperature rises (possibly caused by some other effect)? Or does the increased carbon dioxide cause the temperature to rise?

By studying these results, the scientists hope that they will be able to predict the effect of human influences on climate in the future. They use complex computer models which attempt to simulate the many varying factors that affect climate throughout the world. These models predict how varying degrees of global warming will affect different parts of our planet (see page 225).

We do know that in the last 800 000 years there have been twelve ice ages and that the periods of warmer conditions between them are relatively short. Latest predictions say that the next ice age will be due in about 15 000 years.

How can we possibly know what the carbon dioxide concentration was in 1850, just over 100 years before the station at Mauna Loa was set up? This 1850 figure has been obtained very ingeniously from measurements on bubbles of ancient air trapped in the ice sheets of Antarctica and Greenland.

It is interesting to note the average annual rise in CO_2 levels in recent decades. In the 1960s the average annual rise in carbon dioxide concentration was 0.7 ppmv, in the 1970s it was 1.3 ppmv, in the 1980s it was 1.6 ppmv and in the 1990s it was 1.4 ppmv. It has been suggested that the lower figure for the 1990s is due to the economic and industrial problems in the former Soviet Union during this decade. As mentioned earlier, there is a seasonal variation, which is about 7 ppmv at Mauna Loa.

Most of the carbon dioxide put into the atmosphere by human activity comes from the burning of fossil fuels. However, natural sources have led to substantial rises in carbon dioxide levels in the past, long before the appearance of artificial sources.

Greenhouses gases

As mentioned earlier, some molecules will absorb infrared energy because of the way their bonds vibrate. Molecules containing three or more atoms absorb strongly in the infrared region because of asymmetric vibrations that affect the dipole moment of the molecule. There are several pollutant gases in the atmosphere that will absorb infrared radiation in this way. Gases which allow most of the Sun's ultraviolet and visible radiation to enter but prevent some of the Earth's infrared radiation leaving are called greenhouse gases. Apart from carbon dioxide, CO_2, water, H_2O, and methane, CH_4, other greenhouse gases include dinitrogen oxide, N_2O, CFCs and tropospheric (low-level) ozone, O_3.

The greenhouse effect of a given gas is dependent both on its concentration and its ability to absorb infrared radiation. The extent to which an atmospheric gas absorbs infrared radiation relative to the same amount of carbon dioxide is called its 'greenhouse factor'. Carbon dioxide is given a value of 1.

The contribution of various gases to the greenhouse effect is shown in Table 18.1. The more modes of bond vibration that are possible, the higher the absorption of infrared radiation.

Gas	Greenhouse factor	Concentration in troposphere /ppmv	Overall contribution /%
CO_2	1	358	60
CH_4	30	1.7	15–20
N_2O	150	0.3	} 20–25
O_3	2000	0.1 (varies)	
CFCs	10 000– 25 000	0.004	

Table 18.1 Greenhouse factors of various gases, relative to carbon dioxide, and their contribution to the 'enhanced' greenhouse effect – water vapour is not included.

From Table 18.1 you can see that some CFCs are 25 000 times more efficient at absorbing infrared radiation than carbon dioxide is. Fortunately, their concentration in the lower atmosphere is very low and is unlikely to increase. The production of CFCs in the developed world has now fallen to very low levels as a result of the Montreal Protocol and subsequent international agreements (page 230).

Global temperature change

It has been predicted that the build-up of greenhouse gases in the atmosphere will lead to an increase in the Earth's surface temperature. This is called global warming. The Earth's average temperature in 2000 was 1 °C above its 1850 value (1850 is chosen as the baseline as this precedes the polluting effects of the Industrial Revolution). The rise from 1970 to 2000 has been 0.4 °C. Comparing these values gives a feel for the accelerating rate of global warming. The average rise from 1850 to 1970 works out as 0.005 °C per year but for 1970 to 2000 the average rise is 0.013 °C per year, over twice the 1850–1970 value.

Making predictions is very difficult as it involves so many variable factors. Information is fed into very powerful computers which use mathematical models to predict climate changes.

Computer predictions suggest the climate may be heating up far faster than originally predicted. So fast, in fact, that natural systems would be unable to adapt and world food production would be affected.

Century of natural disasters predicted

New study shows global warming is overwhelming nature's defences

Paul Brown
Environment Correspondent

The world's climate is heating up far faster than predicted, with the average temperatures in Europe soaring in the next century, according to new super-computer predictions.

The increase will be faster than at any time in the world's history. It will be far too fast for natural systems to adapt and will threaten world food production.

ging their feet over the last decade, the cuts proposed in carbon dioxide emissions are just not enough. It is going to hurt most the poorest people in the world, the sub-sistance farmers who will no longer be able to feed themselves."

The computer modelling leader, Peter Cox, who had spent more than five years on the die-back of forest and release of carbon dioxide from soil, said the previous internationally accepted cal-

Figure 18.8 Part of an article in *The Guardian* newspaper, 4 November 1999.

Higher temperatures would cause die-back in tropical rainforests and would increase microbial activity in soils. Most of the Amazon rainforest is predicted to disappear by this model due to die-back. Carbon in the wood would be released as carbon dioxide, putting millions of extra tonnes into the atmosphere as well as fewer trees to absorb carbon dioxide during photosynthesis. At the same time, the increased warmth of the soil would cause microbes to release more carbon dioxide.

In 1989, predictions of the rise in the Earth's average temperature due to build-up of greenhouse gases were between 1.5 and 2.5 °C by 2100, but later predictions suggest that the rise may be as much as 8 °C above the 1850 level by 2100 for land masses, including Europe. Extreme weather events are already occurring more regularly (see Figure 18.8).

The cuts in carbon dioxide emissions proposed by the Climate Change Convention are not enough to prevent the changes predicted.

Internationally accepted calculations had put the level of carbon dioxide in 2100 at twice its current value if no attempt was made to reduce emissions. The new research, which takes into account die-back of rainforests and increased soil microbe activity, now puts the 2100 value at three times the current value if there is no reduction in carbon dioxide emissions. It had been thought that increased levels of carbon dioxide in the atmosphere would increase forest growth because of increased photosynthesis, thus removing some carbon dioxide from the atmosphere. However, the new research suggests that increased temperatures will increase respiration in the forests and lead to the die-back of healthy growth.

Worldwide, it is thought that the 2010 values for greenhouse gas emissions will be 8% above the 1990 values. This increase will be the result of population growth and increased economic activity.

If we accept these predictions, global warming needs to be tackled urgently. If we don't there will be profound adverse effects on agriculture, sea levels, ecosystems, water resources, human health and weather.

SAQ

3 The United Nations Environment Project, UNEP, states in its July 1999 report that if no new policies are adopted to reduce carbon dioxide emissions, the amount of fuel burned will rise from 7 billion tonnes of carbon per year in 1990 to 20 billion tonnes of carbon per year in 2100. What extra mass of carbon dioxide will this release to the atmosphere every year? What assumption do you make in your calculation?

Hint

Answer

4 Explain why the percentage increase in the concentration of carbon dioxide in the lower atmosphere (troposphere), as measured at the Mauna Loa laboratory, is actually less than the percentage increase in the concentration of carbon dioxide released to the atmosphere from the burning of fossil fuels.

Answer

Helping to solve the problem

Some ways to decrease the amount of carbon dioxide being released into the atmosphere are as follows.

- Use sources of energy that do not involve the combustion of fossil fuels, e.g. renewable sources of energy such as solar, wind, wave, tidal and geothermal energy. More nuclear power stations would help reduce carbon dioxide levels but would also bring other environmental concerns.
- Conserve energy so that less fossil fuel is burned.
- Capture and store carbon dioxide (CCS – **carbon capture** and storage) released from power stations or large-scale industrial processes. Research is being carried out to find out if carbon dioxide can be prevented from escaping into the atmosphere from major polluters. This would involve piping the gas deep underground and storing it within rock formations. Oil and gas fields reaching the end of their productive lives are one promising option already being explored by a Norwegian oil company. Pumping down CO_2 makes it easier to get diminishing supplies of oil and gas to the surface. The carbon dioxide can react with metal oxides in the underground rock formations, forming solid carbonates. It can also be dissolved in the salty water trapped underground in the rock formations near oil and gas deposits (see Figure 18.9). The UK has many oil and gas wells that will be suitable for CCS as North Sea supplies run out.

The idea of injecting liquid carbon dioxide into the deep oceans is also being investigated. It can take centuries for the deep waters to circulate to the surface. However, scientists are concerned that this method will eventually acidify the oceans, damaging the ecosystem.

Any CCS scheme will inevitably cost the power industry money, using extra energy to retrieve, liquefy and pump the CO_2 to its storage reservoir. This cost will ultimately be passed on to consumers, e.g. in increased electricity prices. However, some estimates put this at only 1p or 2p per unit of electricity used.

SAQ

5 Why will carbon capture and storage increase the cost of electricity generated by burning fossil fuels?

Answer

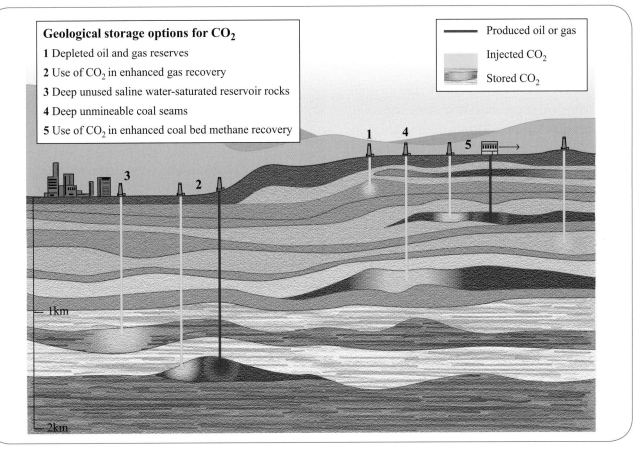

Figure 18.9 Carbon capture and storage is one way to prevent CO_2 escaping into the atmosphere.

There have also been international treaties on climate change where countries meet to set targets for the reduction of their carbon dioxide emissions. As of December 2006, 169 countries had signed the Kyoto Protocol. However some of the largest producers of carbon dioxide, such as the USA and Australia, have refused to sign, and other countries such as China and India, which are seen as still developing their industrial bases, do not have to reduce their emissions. Chemists will have a significant role to play in monitoring such agreements, using techniques such as infrared spectroscopy (see Chapter 15).

Chemical balance in the stratosphere

Ozone

Ozone, O_3, is an important gas in the upper part of the atmosphere called the stratosphere. Ozone is a form of oxygen with three oxygen atoms. It is produced in the stratosphere by the following **photochemical reactions**:

$$O_2(g) \xrightarrow{\text{UV}} O(g) + O(g)$$
$$O(g) + O_2(g) \longrightarrow O_3(g)$$

The ozone produced absorbs ultraviolet (UV) radiation to re-form diatomic oxygen. The two reactions are:

$$O_3(g) \xrightarrow{\text{UV}} O_2(g) + O(g)$$

followed by:

$$O_3(g) + O(g) \longrightarrow 2O_2(g)$$

Combining the two reactions above gives the equation for the overall destruction of ozone as:

$$2O_3(g) \longrightarrow 3O_2(g)$$

You can see from these reactions that ozone is being made and destroyed all the time. Over millions of years a steady state has been reached in which the rate of formation of ozone is equal to the rate of its destruction.

Until very recently the thickness of the ozone layer in the stratosphere remained relatively constant. The situation has altered over the last 50 years as a result of chemicals being released into the atmosphere.

It is the absorption of ultraviolet radiation in the stratosphere which prevents most of the radiation of wavelengths less than 340 nm from reaching the Earth. In this way, plants and animals are protected from this damaging radiation.

There are other reactions which remove ozone. The most important of these are free radical reactions with oxides of nitrogen, which occur at a height of roughly 25 km:

$$NO\bullet(g) + O_3(g) \longrightarrow NO_2\bullet(g) + O_2(g)$$
$$NO_2\bullet(g) + O(g) \longrightarrow NO\bullet(g) + O_2(g)$$
$$NO_2\bullet(g) \xrightarrow{UV} NO\bullet(g) + O(g)$$

The dot ($\bullet$) after a formula refers to an unpaired electron. A species such as NO$\bullet$ that has an unpaired electron is called a **radical** or a free radical. Note that the oxygen atom is recycled and nitrogen monoxide, NO$\bullet$, is regenerated. This means that one molecule of a nitrogen oxide can remove many ozone molecules.

Oxides of nitrogen enter the atmosphere naturally, e.g. from thunderstorms, and they can also be formed during the combustion of fossil fuels, e.g. from aircraft travelling at high altitudes.

SAQ

6 State the reactions which **a** produce ozone and **b** remove ozone by reaction with nitrogen oxides in the upper atmosphere (stratosphere).

$\boxed{\text{Answer}}$

Damage to the ozone layer in the stratosphere

The inter-related reactions for the production and removal of ozone in the stratosphere, described above, ensured that the amount of ozone in the stratosphere was sufficient to protect the Earth from excessive ultraviolet radiation.

From the 1930s until the late 1980s, chemicals called CFCs (chlorofluorocarbons) were used extensively in aerosol cans, refrigerators, air-conditioning systems and the production of plastics (see Chapter 14). They

seemed ideal for such purposes because of their lack of reactivity, low flammability and low toxicity. However, since the 1980s, there has been much concern about the effect of CFCs on the ozone in the stratosphere and this has led to a severe reduction in their use, except for approved specialised needs.

CFCs are unaffected by ultraviolet radiation in the lower atmosphere, but they are susceptible to attack by ultraviolet radiation in the stratosphere, releasing chlorine free radicals which react with ozone.

The problem now is the quantity of CFCs reaching the stratosphere and the length of time they remain there due to their unreactivity. This leads to more ozone being destroyed than is created, as can be seen in the following reactions for two typical CFCs, CCl_2F_2 and CCl_3F. (All species are in the gaseous state.)

- Initiation of ozone-destroying species:
$$CCl_2F_2 \xrightarrow{UV} \bullet CClF_2 + Cl\bullet$$
$$CCl_3F \longrightarrow \bullet CCl_2F + Cl\bullet$$

- It is the C–Cl bond rather than the C–F bond which breaks because the ultraviolet radiation from the Sun reaching the upper atmosphere has enough energy to break the weaker C–Cl bond but not the stronger C–F bond (see box on page 229).

- Then:
$$Cl\bullet + O_3 \longrightarrow ClO\bullet + O_2$$

- The chlorine free radical, Cl$\bullet$, is regenerated when ClO$\bullet$ reacts with an oxygen atom:
$$ClO\bullet + O \longrightarrow Cl\bullet + O_2$$

- This chlorine free radical can then react with more ozone. The reaction rate is fast and a few chlorine free radicals rapidly destroy many ozone molecules. It has been estimated that, during its lifetime, one chlorine free radical could destroy up to 100 000 ozone molecules.

The effects of ozone depletion

To predict the precise consequences of ozone depletion requires three-dimensional atmospheric modelling on an enormous scale. Very powerful computers using programs of immense complexity are used to simulate possible atmospheric changes as a result of pollution. Monitoring the problem is made more difficult by natural changes in the ozone layer. The thickness of the layer varies with time and in space. Holes in the ozone layer come and go and levels are lower in winter and at night, due to lack of sunshine.

- The British Antarctic Survey reported a 'hole' in the ozone layer over the Antarctic in 1985.
- In the early 1990s levels of ozone depletion in the Arctic were found to be 50 times greater than scientists expected on the basis of atmospheric modelling predictions.
- In the mid-latitudes of the Northern Hemisphere, the average ozone depletion has been 7% per decade since 1979.

The hole in the Antarctic ozone layer is caused by unusual conditions. Most ozone is created at the tropics and transported to the poles. Climatic conditions in the Antarctic winter effectively cut off a cone of air over the South Pole from the surrounding atmosphere.

The Antarctic hole was mapped using the Total Ozone Mapping Spectrometer (TOMS). TOMS was carried on the Nimbus-7 and Meteor-3 satellites between November 1978 and December 1994; during this time the ozone hole was discovered. The satellite 'Earth Probe' was launched on 2 July 1996 and now provides data on a near real-time basis (see Figure 18.10).

Thinning of the ozone layer results in more UV radiation of wavelengths below 320 nm reaching the Earth's surface. UV radiation in the wavelength range 290–320 nm is known as UV-B. In living tissue this is absorbed by chemicals in cells, such as nucleic acids, and may damage them, leading to increased levels of skin cancer in humans. Fair-skinned people are much more likely to develop skin cancers since they do not have the pigments

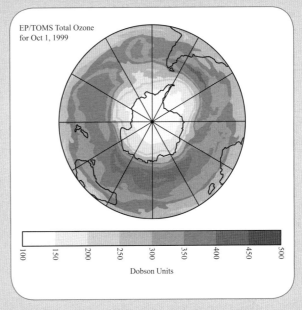

EP/TOMS Total Ozone for Oct 1, 1999

100 150 200 250 300 350 400 450 500
Dobson Units

Figure 18.10 Satellite map showing a severe depletion or 'hole' in the ozone layer over the Antarctic on October 1, 1999. The Antarctic ozone hole reaches a maximum in October each year, the Antarctic spring. Concern about the ozone layer in the Antarctic was growing in the early 1980s, with results being published in May 1985. (Ozone concentration is measured in Dobson units.)

that are present in dark skin that help to screen out the UV rays. Some predictions suggest that a 1% decrease in stratospheric ozone causes a 2% increase in UV-B and a 2–5% increase in levels of skin cancer. Another effect of UV-B is that it appears to prevent normal immune responses in the skin and other parts of the body. Large ozone depletions, and hence increased UV radiation, will also affect crop yields in plants due to cell damage.

Larvae of fish, shrimp and crab, zooplankton (tiny aquatic animals) and phytoplankton (microscopic aquatic plants) are particularly affected by UV radiation. Indeed, a significant increase in levels of UV radiation reaching the oceans could cause some microscopic life-forms to become extinct. Plankton are very important as they are the beginning of the food chain for

continued

animals living in the sea. Phytoplankton take in carbon dioxide and give out oxygen into the water and into the atmosphere. Phytoplankton need only water, dissolved carbon dioxide, salts and sunlight to make all their vital substances. Zooplankton feed on these microscopic plants and are in turn eaten by fish. The fish are then eaten by other animals and humans. If increased UV levels were damaging to plankton the effects would be passed up the food chain and would be profound. Destruction of phytoplankton would also lead to more carbon dioxide in the atmosphere, and so would contribute to an increased greenhouse effect and global warming (see page 225).

Fighting ozone depletion

There are now controls over the manufacture and use of CFCs in the Montreal Protocol, agreed in 1987 and tightened further in 1990 and in 1997. Over 60 nations agreed to phase out CFCs by 2000.

Other halogen-containing carbon compounds that cause damage to the stratospheric ozone layer are tetrachloromethane, CCl_4, and bromomethane, CH_3Br. Large-scale production of these chemicals has now ceased.

Typically a CFC molecule remains unreacted in the atmosphere for 50 to 80 years, so atmospheric modellers predict that it will take up to 100 years for existing CFCs to disperse. Meanwhile a search for substitutes is being made. The most likely candidates are the hydrofluorocarbons, HFCs, and hydrochlorofluorocarbons, HCFCs. These molecules contain C–H bonds, which are broken down in the lower atmosphere. This initiates the breakdown of the entire molecule, with the result that the chlorine in HCFCs is unable to reach the stratosphere. Unfortunately both these types of compound are potent 'greenhouse gases' and may contribute to global warming (see page 225). The use of HFCs and HCFCs is therefore only a stopgap substitute for CFCs. It is planned to phase out the use of these substitutes by 2015 in the EU and by 2030 in the rest of the world.

SAQ

7 Give the property of CFCs which led to their original widespread use and now leads to their long-term presence in the stratosphere.

Hint

Answer

8 What are the types of chemical now being used as replacements for CFCs? Explain, chemically, why they do not have such a detrimental effect upon the ozone layer in the upper atmosphere (stratosphere). Discuss the disadvantages of their use.

Hint

Answer

Controlling air pollution

Nitrogen oxides

The nitrogen oxides are serious pollutants. For example, together with unburnt hydrocarbons, they are responsible for the **photochemical smog** that occurs in cities such as Los Angeles (see Figure 18.11). Nitrogen oxides also contribute to the formation of acid rain.

Nitrogen can have a variety of oxidation numbers in its oxides.

NO_3^-, HNO_3	+5
$2NO_2 \rightleftharpoons N_2O_4$	+4
NO_2^-, HNO_2	+3
NO	+2
N_2O	+1

Figure 18.11 Photochemical smog in Los Angeles.

Nitrogen monoxide, NO, and nitrogen dioxide, NO_2, are the main pollutant oxides. They are often referred to as NO_x.

Nitrogen itself, N_2, has an oxidation number of 0. It is stable and has low reactivity. However, at high temperatures, such as in an internal combustion engine (see page 236), it will combine with oxygen to form nitrogen monoxide:

$$N_2(g) + O_2(g) \longrightarrow 2NO(g)$$

Nitrogen monoxide is a primary pollutant in the lower atmosphere.

Nitrogen dioxide is formed by reaction of nitrogen monoxide with oxygen. It is therefore a secondary pollutant:

$$2NO(g) + O_2(g) \longrightarrow 2NO_2(g)$$

Nitrogen monoxide, NO, plays a part in the removal of ozone in the stratosphere (see page 228). The nitrogen monoxide can be formed by the reaction of oxygen atoms and dinitrogen oxide and so is a secondary pollutant in the stratosphere.

It has been estimated that annual emissions of nitrogen oxides are 53 million tonnes from artificial sources and 1092 million tonnes from natural sources. You can see that larger amounts of nitrogen oxides come from natural processes (for example biological activity in soil, volcanoes and lightning) than come from human activities. The problem with artificial emissions is that they are concentrated in urban areas and can reach high concentrations. The biggest source is the combustion of oil and petrol, followed by the combustion of coal. Power stations and transport are by far the biggest contributors to artificial nitrogen oxide pollution. Urban levels follow seasonal variations: the concentration of nitrogen oxides is significantly higher in winter.

Ozone in the lower atmosphere

We saw earlier (page 229) that ozone in the stratosphere is essential in screening the Earth from dangerous ultraviolet radiation. In contrast, ozone present in the lower atmosphere (low-level ozone) is a dangerous pollutant.

Low-level ozone arises from photochemical reactions of primary pollutants originating from motor vehicle emissions and incomplete combustion of fossil fuels. Depending on concentration it can have a serious effect on human health, on vegetation, on synthetic polymers, and it plays a part in the complex series of reactions involved in forming photochemical smog.

The fluctuation of ozone levels in urban areas varies over a 24-hour period. Figure 18.12 shows that the nitrogen monoxide produced from vehicle exhausts in the early-morning rush hour leads to a rise in the level of ozone via the formation of nitrogen dioxide.

Seasonal variations in tropospheric ozone in the northern hemisphere are shown in Figure 18.13. High levels in June, July and August relate to longer hours of daylight and the greater intensity of sunshine.

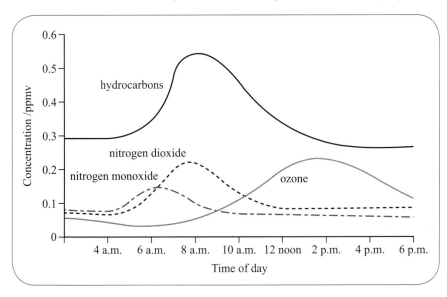

Figure 18.12 The variation in concentration of atmospheric pollutants during daylight hours in Los Angeles.

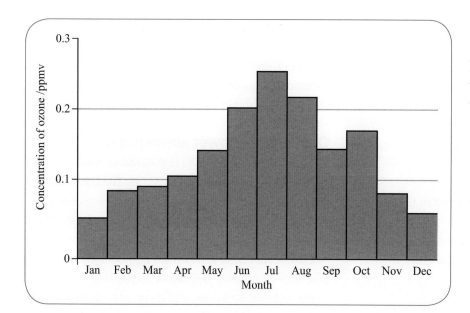

Figure 18.13 The seasonal variation in low-level ozone in the northern hemisphere.

Low-level ozone affects our health. It causes shortness of breath and irritation of eyes, nose and throat at low concentrations. It also has other detrimental effects. Ozone will add across C=C double bonds such as those found in synthetic materials, e.g. plastics, paints and dyes. Rubber is a natural polymer containing C=C double bonds; reaction with ozone causes rubber to crack, and so can damages car tyres.

The level at which low-level ozone is likely to cause damage to health has been defined in the US as exposure to a concentration of 120 ppbv (0.120 ppmv) for 1 hour. Levels in excess of this figure have been recorded in Britain during the summer months in urban, suburban and rural sites.

SAQ

9 Ozone is a called a secondary pollutant. Explain what this means.

$\boxed{\text{Answer}}$

Pollution effects of the internal combustion engine

Cars have revolutionised our lives. Unfortunately there is a price to pay in terms of atmospheric pollution.

Pollutants produced by an internal combustion engine are particulates, carbon monoxide, oxides of nitrogen, oxides of sulfur and unburnt hydrocarbons (Table 18.2). Most of these pollutants are emitted from the exhaust. However, a significant amount of hydrocarbon pollution comes from the crankcase, carburettor (where fitted) and fuel tank, not from the exhaust. (In some parts of the world lead is still emitted from older cars that need to use leaded petrol, but in the UK leaded petrol has been unavailable since 1 January 2000 and has been substituted by lead replacement petrol, LRP.)

In a car's engine a mixture of fuel and air burns in the combustion chamber. Then the hot gases produced by the burning fuel are expelled, eventually leaving the car as the exhaust. The air-to-fuel ratio plays an important part in determining the relative levels of emissions (Figure 18.14). The most common ratios used are between 12:1 and 15:1 for petrol engines.

Fuel	Carbon monoxide	Hydrocarbons	Nitrogen oxides	Sulfur dioxide	Black smoke
petrol	236	25	29	0.9	0.6
diesel	10	17	59	3.8	18.0

Table 18.2 Emission factors for motor vehicles, measured in grams of pollutant produced per kilogram of fuel.

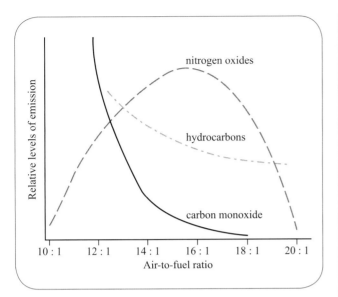

Figure 18.14 Exhaust emissions at various air-to-fuel ratios.

The stoichiometric ratio is 15:1. (Stoichiometric means the same ratio as in the chemical equation.) At this ratio it can be seen from Figure 18.14 that emissions of nitrogen oxides are relatively high, and emissions of carbon monoxide and hydrocarbons are relatively low.

To reduce nitrogen oxide emissions and to keep carbon monoxide and hydrocarbon emissions low, a high air-to-fuel ratio needs to be used. When there is more air than the stoichiometric ratio it is referred to as a 'lean' mixture. The trouble is that a lean mixture will lead to misfiring.

A richer mix (an air-to-fuel ratio lower than 15:1) reduces nitrogen oxide emissions but increases carbon monoxide and hydrocarbon emissions.

Modern engine technology has produced 'lean-burn' engines which use an air-to-fuel ratio of 18:1. These engines have specially designed combustion chambers and electronically controlled fuel injection to overcome the problem of misfiring.

A summary of pollutants from motor vehicles and their effects is given in Table 18.3.

Catalytic converters – a benefit of modern chemistry

By the application of careful research, chemists are enabling us to enjoy all the benefits of the car without the associated air pollution. Huge improvements in the reduction of harmful exhaust emissions are taking place in the EU, as is shown in Figure 18.15.

Name of gas	Formula	Origin	Effect
carbon monoxide	CO	incomplete combustion of hydrocarbons in the fuel	poisonous gas that combines with haemoglobin in the blood, and prevents oxygen from being carried instead
nitrogen dioxide	NO_2	N_2 and O_2 react at high temperature in the engine to form NO, then NO is oxidised in the atmosphere to form NO_2 (diesel engines produce lower levels of nitrogen oxides as they operate at lower temperatures than petrol engines)	nitrogen dioxide is involved in the formation of acid rain, photochemical smog and low-level ozone
hydrocarbons	C_xH_y	hydrocarbons in petrol fail to combust	toxic to humans, possibly carcinogenic (causing cancer), cause photochemical smog

Table 18.3 Pollutants in vehicle exhaust fumes.

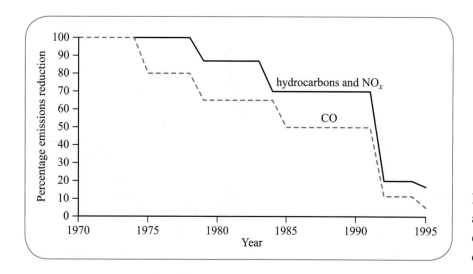

Figure 18.15 Reductions in allowed levels of pollutants in exhaust emissions; EU, 1970 onwards.

	Date effective	Petrol			Diesel			
		CO	HC	NO$_x$	CO	HC + NO$_x$	NO$_x$	PM
Stage 1	1992	2.72	0.97		2.72	0.97		0.14
Stage 2	1996	2.2	0.5		1.0	0.9		0.10
Stage 3	2000	2.3	0.2	0.15	0.64	0.56	0.50	0.05
Stage 4	2005	1.0	0.1	0.08	0.5	0.30	0.25	0.025

Table 18.4 Development of EU emission legislation. (The figures are in grams of pollutant per kilometre during a 20-minute test driving cycle. PM is particulate matter.)

Table 18.4 shows the progression of EU legislation on emissions from 1992 to 2005. The figures are in grams of pollutant per kilometre travelled and refer to a 20-minute test cycle on a rolling road (dynamometer). During this test cycle, the vehicle is driven according to a standard procedure which represents a typical European driving pattern. Each figure is a maximum amount – any car emitting more than this is not allowed on the road. In the UK we would say it had 'failed its MOT'.

SAQ

10 Calculate the reduction in the legal maximum number of grams of pollutant emitted per kilometre for carbon monoxide that have taken place between 1992 and 2005. Consider both petrol and diesel. Use data from Table 18.4.

Answer

Exhaust emissions can be controlled by the use of **catalytic converter**s fitted to exhaust systems (Figure 18.16). Hot exhaust gases are passed over a mixture of platinum, rhodium and palladium on a supporting honeycomb structure of cordierite (2MgO.2Al$_2$O$_3$.5SiO$_2$), which gives a very large surface area where the exhaust gases can come into contact with the catalyst.

Catalyst systems

There are two forms of catalyst system – oxidation catalysts and three-way catalysts.

- *Oxidation catalysts* can be used in conjunction with 'lean-burn' engines such as diesel to control carbon monoxide and hydrocarbon emissions. The exhaust gases of lean-burn engines are rich in oxygen. This enables unburnt hydrocarbons and carbon

Figure 18.16 A typical catalytic converter, cut away to show the open honeycomb structure of the ceramic support coated with platinum, rhodium and palladium.

monoxide to be rapidly oxidised on the surface of the catalyst to give carbon dioxide and water at lower temperatures than normal (200–250 °C):

$$2CO(g) + O_2(g) \longrightarrow 2CO_2(g)$$
$$C_8H_{18} + 12\tfrac{1}{2}O_2(g) \longrightarrow 8CO_2(g) + 9H_2O(g)$$

(Octane, C_8H_{18}, in the isomeric form 2,2,4-trimethylpentane, is a major constituent of petrol.)

Nitrogen oxides are not removed by this type of catalyst, but as you can see from Figure 18.15, lean-burn engines have a low nitrogen oxide content in their exhausts. Tighter legislation has now made it necessary to control nitrogen oxide emissions as well.

- *Three-way catalysts* work with conventional engines, controlling carbon monoxide, hydrocarbon and nitrogen oxide emissions. The nitrogen oxides are reduced to nitrogen:
  $$2NO(g) + 2CO(g) \longrightarrow N_2(g) + 2CO_2(g)$$
  Palladium and platinum catalyse the oxidation of carbon monoxide and unburnt hydrocarbons. Rhodium catalyses the reduction of nitrogen monoxide to nitrogen.

The presence of rhodium in the catalyst enables it to start working at temperatures as low as 150 °C. Most engines attain an exhaust gas temperature of 200 °C or more within 30 seconds from a cold start, so rhodium is an important factor in reducing emissions.

In a catalytic converter the metal catalysts catalyse the formation of nitrogen from nitrogen oxide, and carbon dioxide from carbon monoxide:

- the pollutant reactant molecules are *adsorbed* onto the metal atoms at the surface of the catalyst
- covalent bonds in the reactants (NO and CO) are weakened and broken
- new bonds form to give the product molecules (N_2 and CO_2) which are *desorbed* from the catalyst.

The catalyst has to form bonds with the molecules that are strong enough to weaken the bonds in the reactants but are weak enough to break and allow the products to leave.

Figure 18.17 shows the actual and the projected future reduction in carbon monoxide emissions in the UK resulting from the use of catalytic converters. You will notice that the projected reduction is roughly 6 000 000 tonnes per year by 2025.

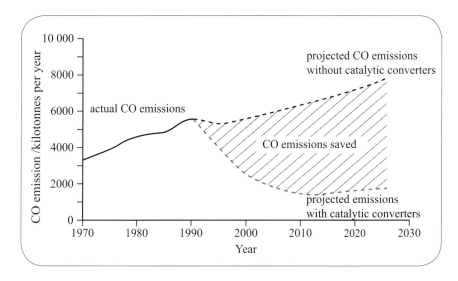

Figure 18.17 The savings, both actual and projected, in carbon monoxide emissions due to the use of catalytic converters; UK, 1970 onwards.

235

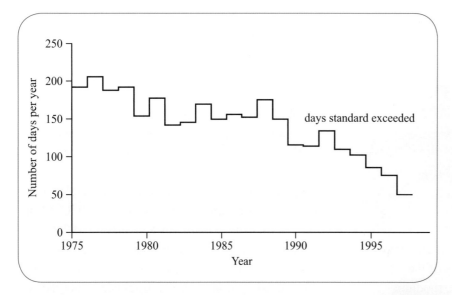

Figure 18.18 The number of days per year when photochemical smog, as measured by ozone levels, is a problem in Los Angeles has greatly reduced due to the use of catalytic converters.

Figure 18.18 shows the enormous effect that the introduction of catalytic converters in the late 1970s has had on urban ozone levels in and around Los Angeles.

Green Chemistry

The people who work in the chemical industry are becoming increasingly aware of the need to conserve the Earth's resources and to stop the damage we are causing to our environment. In the past, profits might have been the only consideration when making new materials but now industrialists have realised their responsibilities to the planet and its inhabitants. Hence the birth of 'Green Chemistry'. Green Chemistry enables us to maintain and improve living standards through sustainable development that will safeguard the Earth for future generations.

Five important principles of a greener chemical industry are as follows.

1 The design of processes to maximise the amount of raw material that is converted into product. Synthetic methods should be chosen to maximise the percentage incorporation of all materials used in the process into the final product. Catalytic reagents that are regenerated help to improve the **atom economy** of a process (see Chapter 10 and Chapter 17).

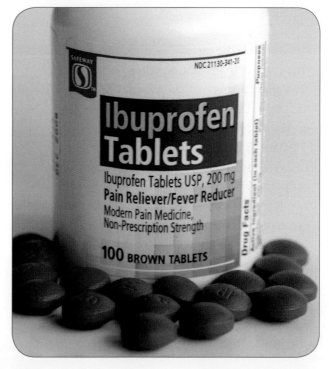

Figure 18.19 The production of ibuprofen has been made more efficient since its introduction in the 1980s.

Example: Ibuprofen is an 'over-the-counter' drug which came into use in the 1980s (see Figure 18.19). At that time the synthetic route used to produce it consisted of six steps. Its overall atom economy was just 40.1%. However, in the 1990s the Hoechst Celanese Corporation developed a new three-stage process. This had the effect of improving the atom economy to 77.4%, making more efficient use of the raw materials and creating less waste. This type of innovation relies on the creativity of research chemists to design new methods and the technologists who can put these into practice on an industrial scale.

2   The use of raw materials or feedstock that are renewable rather than finite wherever possible. However, technical and economic factors may make this difficult.

Example: The use of **biofuels**, such as biodiesel and ethanol, instead of fossil fuels (see Chapter 11). However, sometimes these initiatives have unforeseen consequences, as happened in 2007 with a shortage of corn for the food industry in the USA. Farmers who had always grown corn (called maize in the USA) were being encouraged to produce crops for biofuels instead (see Figure 18.20). This drove up the price of corn-based foodstuffs, and made it more difficult for poorer nations to import corn.

3   The use of safe, environmentally friendly solvents or no solvents at all where possible. The use of auxiliary substances (e.g. solvents, separation agents, etc.) should be made unnecessary wherever possible. If there is no way to carry out a process without the use of auxiliary substances, those chosen should be harmless. Wherever practicable, synthetic methods should be designed to use and generate substances that are non-toxic to humans and the environment.

Figure 18.20 This oilseed rape is being grown for the production of biodiesel. Biodiesel is potentially a 'carbon-neutral' fuel as the carbon dioxide given off when the biodiesel burns returns what was removed from the air during photosynthesis (see Chapter 11).

The substances, and the form of the substances, used in a chemical process should be selected to minimise the potential risk of chemical accidents, including emissions, explosions and fires.

Example: The use of *new foam* fire-fighting compounds and *aerosol sprays* that do not rely on halogenated hydrocarbons that contribute to ozone depletion in the atmosphere (see Chapter 14).

4   The design of energy-efficient processes; the energy requirements of chemical processes should be minimised to reduce their impact on the environment (and should also reduce costs). If possible, synthetic methods should be conducted at ambient temperature and pressure. Wherever possible the energy that is required should come from a renewable source, such as solar or wind generators, rather than a finite source such as a fossil fuel.

Example: The use of *enzyme-catalysed* reactions in the biotechnology industry which can operate at much lower temperatures and pressures than traditional industrial processes (see Chapter 17).

5 The consideration of waste reduction in the production process and at the end of a product's life cycle, aiming not to create waste in the first place! It is more efficient to prevent waste than to treat it or clean it up. Chemical products should be designed so that they break down into harmless substances after use.

Example: the development of new **degradable plastic** products that will not take up space in land-fill sites and will not persist in the environment for centuries to come (see Chapter 12 and Figure 18.21). Recycling of metals, plastics, paper and glass also helps to conserve energy and resources, as well as reducing waste.

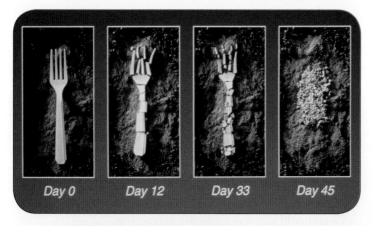

Day 0   Day 12   Day 33   Day 45

Figure 18.21 Chemists are devising new plastics that can be broken down by bacteria in the soil or by light.

SAQ

11 Research task: find out how changing catalysts improved the overall atom economy in the production of ibuprofen.

[Hint]

[Answer]

Australian brewery turns on waste

A new scheme to generate electricity from the waste water produced during the brewing of beer is being introduced in Brisbane, Australia. On 5th May 2007, *The Times* newspaper told how waste (a solution containing sugar, starch and alcohol) from Foster's brewery, had been used by bacteria to produce energy (see Figure 18.22). The bacteria effectively 'clean' the water and the energy is used in a kind of microbial fuel cell to generate electricity. Researchers at the University of Queensland believe the process can eventually be used in breweries, wineries and food-processing plants around the world.

The fuel cell has been successfully trialled in the laboratory and is now being scaled up on Foster's site to generate enough electricity for a large house. Once any snags have been ironed out they plan to treat all the waste like this, producing enough energy for 2000 homes.

The system is 'green' as it will be a renewable energy source produced from waste and will reduce the electricity previously used by Foster's in treating the waste water before its disposal. Not only that, in drought-torn Australia, the added benefit is the production of water that is good enough to drink.

Figure 18.22 The waste water from the Foster's brewing process can be treated with bacteria to generate electricity.

International initiatives

We have already seen how countries are trying to tackle global warming by signing up to the Kyoto Protocol (see page 227 and Figure 18.23).

A number of other international initiatives have also taken place aimed at enforcing, supporting and encouraging more sustainable chemicals management. Such initiatives will influence chemical producers worldwide. Some of the first initiatives to control or eliminate harmful chemicals were:

- the Montreal Protocol on Ozone Depleting Substances
- the Stockholm Convention on Persistent Organic Pollutants (POPs)
- the Rio 'Earth Summit' of 1991 which addressed the issue of chemical safety as a key priority for sustainable development. This in turn led to other international collaborations such as 'The United Nations Environment Programme (UNEP)', agreed in February 2006.

Figure 18.23 Ministers at the Kyoto conference on climate change in 1997.

Summary

Glossary

- The concentration of carbon dioxide in the lower atmosphere (troposphere) depends on photosynthesis, plant and animal respiration, the dissolving of carbon dioxide in surface waters and the quantity of carbon dioxide emitted during the combustion of fossil fuels.

- The greenhouse effect is a natural phenomenon which keeps the Earth's surface warm.

- Increased emissions of carbon dioxide, methane, CFCs and other gases are enhancing the greenhouse effect and causing the Earth's surface to warm up. This global warming may result in climate change.

- The greenhouse effect of a gas depends on its concentration in the atmosphere and its ability to absorb infrared radiation.

- Ozone in the stratosphere absorbs harmful ultraviolet radiation from the Sun. However, reactions of chlorine radicals released from CFCs are destroying ozone in the stratosphere.

- HFCs and HCFCs are temporary replacements for CFCs.

- Nitrogen oxides from fossil fuel combustion are major pollutants.

- Nitrogen monoxide influences the balance between ozone and oxygen in the stratosphere.

- The pollutants in car exhaust gases include carbon monoxide, nitrogen(II) oxide and unburnt hydrocarbons. If these gases are not removed by catalytic converters, they lead to the formation of low-level ozone and photochemical smog. Both low-level ozone and photochemical smog can be harmful to humans, animals and plants.

continued

- Catalytic converters are an effective way of reducing harmful emissions from motor vehicles.

- Green Chemistry enables us to maintain and improve living standards through sustainable development that will safeguard the Earth for future generations. Its main principles are:
  - using industrial processes that reduce or eliminate hazardous chemicals and use fewer chemicals
  - designing processes with a high atom economy to minimise waste materials
  - using renewable resources, such as plant-based substances
  - seeking alternative energy sources such as solar energy, rather than consuming finite resources, such as fossil fuels, that will eventually run out
  - ensuring that any waste products are non-toxic, and can be recycled or broken down into harmless substances in the environment (biodegraded).

Questions

1 In the year 2000, the mass of CO_2 emitted in the UK was equivalent to 1 kg per person in every hour.

    a Calculate the volume of 1 kg of carbon dioxide (in dm^3). Assume that 1 mole of CO_2 occupies 24 dm^3. [2]

    b The UK has set a target to cut CO_2 emissions by 60% of the 2000 value by 2050. Calculate the reduction needed in the volume of CO_2 emissions (in dm^3) each hour per person if the target is to be met. [1]

OCR Chemistry AS (2811) Jan 2005 [Total 3]

Hint

Answer

2 *Arcton* 133 is a CFC with the molecular formula $C_2H_2ClF_3$.

    a When *Arcton* 133 is released into the atmosphere, its molecules can absorb energy. The C–Cl bond breaks, forming radicals.

$$C_2H_2ClF_3 \longrightarrow C_2H_2F_3\bullet + Cl\bullet$$

      i What source of energy is required for this reaction to take place? [1]

      ii Chlorine free radicals catalyse the breakdown of ozone, O_3. Write two equations to show how this happens. [2]

      iii Write an equation for the overall reaction in **a ii**. [1]

    b In some applications, CFCs are being replaced by hydrocarbons such as alkanes.

      i What is the M_r of *Arcton* 133, $C_2H_2ClF_3$? [1]

      ii The formulae of some alkanes are shown below.

        C_5H_{12} C_6H_{14} C_7H_{16} C_8H_{18} C_9H_{20}

        Choose the molecular formula of the alkane whose M_r is most similar to that of *Arcton* 133. [1]

      iii Suggest why hydrocarbons are replacing CFCs. [1]

      iv Apart from cost, suggest one possible disadvantage of using a hydrocarbon instead of a CFC. [1]

OCR Chemistry AS (2813) Jan 2001 [Total 8]

Answer

continued

3 Many chemical reactions occur in the atmosphere.
   a Car engines produce carbon monoxide and nitrogen monoxide near to the Earth's surface.
      Explain how carbon monoxide and nitrogen monoxide are formed in the car engine. [2]
   b In the upper atmosphere, reactions occur involving chlorine radicals, Cl.
      Equations for two such processes are given below.

$$Cl + O_3 \longrightarrow ClO + O_2 \qquad \text{equation } 1$$
$$ClO + \underline{\hspace{1cm}} \longrightarrow Cl + O_2 \qquad \text{equation } 2$$

      i   Complete equation 2. [1]
      ii Write the overall equation for the two processes shown in equations 1 and 2. [1]
      iii Describe how the chlorine radicals, Cl, are formed in the upper atmosphere. [2]
      iv State one undesirable result of ozone depletion in the upper atmosphere for life on Earth. [1]

OCR Chemistry AS (2813) June 2006 [Total 7]

Answer

4 This question looks at some aspects of the use of petrol as a fuel for cars.
   Petrol contains octane, C_8H_{18}. Two of the stages that occur when petrol, containing octane, is used in a car engine are shown below.

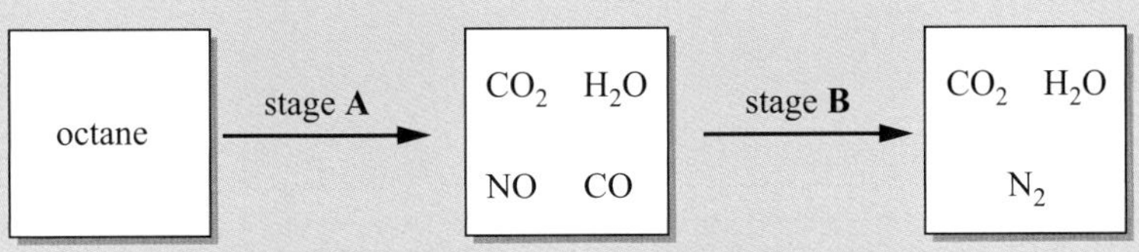

Hint

   a Stage A includes the complete combustion of octane. Write the equation for
      this reaction. [2]
   b Stage B requires a catalyst.
      i   Name two metals generally present in the catalyst. [1]
      ii Describe how the catalyst works. [3]
      iii Using the substances shown above, write the equation for the reaction that occurs in stage B. [2]
   c If stage B does not happen, further reactions occur and pollution levels rise.
      Suggest one pollutant whose level in the atmosphere would rise. [1]

OCR Chemistry AS (2813) Jan 2005 [Total 9]

Answer

5 Both methane and octane undergo incomplete combustion in a car engine. As a result of this, unburnt
   hydrocarbons and carbon monoxide, CO, occur in the exhaust gases. Nitrogen monoxide, NO, is also formed
   inside the engine. All three pollutants can be removed by fitting a catalytic converter to the exhaust system.
   a State one environmental consequence of each of the following emissions.
      ● unburnt hydrocarbons
      ● CO
      ● NO [3]
   b The catalytic converter is positioned as close to the engine as possible, so that it heats up quickly.
      Why does the converter work best when it is hot? [1]

OCR Chemistry AS (2813) May 2002 [Total 4]

Answer

continued

6 Oxides of nitrogen such as NO and NO_2 are gases that pollute the atmosphere. They are produced during the combustion of petrol in car engines.

a The table below lists the enthalpy changes of formation of NO and NO_2.

Hint

compound	$\Delta H_f^\ominus / kJ\,mol^{-1}$
NO	+90
NO_2	+33

   i Define the term *standard enthalpy change of formation*. [2]

  ii Write an equation, including state symbols, to represent the standard enthalpy change of formation of $NO_2(g)$. [2]

 iii Use the data in the table to calculate the enthalpy change for the following reaction.

$$2NO(g) + O_2(g) \longrightarrow 2NO_2(g)$$ [3]

b Oxides of nitrogen only form under high energy conditions such as during lightning strikes or in internal combustion engines. Suggest why this is so. [1]

c Modern cars have 'catalytic converters' in their exhausts to convert nitrogen oxides into less harmful substances. One reaction that occurs is as follows.

$$\_\_\_\_CO(g) + \_\_\_\_NO_2(g) \longrightarrow \_\_\_\_CO_2(g) + \_\_\_\_N_2(g)$$

   i Balance this equation by writing the appropriate numbers in the spaces. [1]

  ii Explain how this type of catalyst provides a different pathway for the reaction with a lower activation energy. [2]

 iii Suggest why a catalytic converter is designed to have a large surface area. [2]

OCR Chemistry AS (2813) Jan 2001   [Total 13]

Answer

Appendix: Periodic Table

Key

relative atomic mass
atomic symbol
name
atomic (proton) number

1	2												3	4	5	6	7	0
						1.0 **H** hydrogen 1												4.0 **He** helium 2
6.9 **Li** lithium 3	9.0 **Be** beryllium 4												10.8 **B** boron 5	12.0 **C** carbon 6	14.0 **N** nitrogen 7	16.0 **O** oxygen 8	19.0 **F** fluorine 9	20.2 **Ne** neon 10
23.0 **Na** sodium 11	24.3 **Mg** magnesium 12												27.0 **Al** aluminium 13	28.1 **Si** silicon 14	31.0 **P** phosphorus 15	32.1 **S** sulfur 16	35.5 **Cl** chlorine 17	39.9 **Ar** argon 18
39.1 **K** potassium 19	40.1 **Ca** calcium 20	45.0 **Sc** scandium 21	47.9 **Ti** titanium 22	50.9 **V** vanadium 23	52.0 **Cr** chromium 24	54.9 **Mn** manganese 25	55.8 **Fe** iron 26	58.9 **Co** cobalt 27	58.7 **Ni** nickel 28	63.5 **Cu** copper 29	65.4 **Zn** zinc 30		69.7 **Ga** gallium 31	72.6 **Ge** germanium 32	74.9 **As** arsenic 33	79.0 **Se** selenium 34	79.9 **Br** bromine 35	83.8 **Kr** krypton 36
85.5 **Rb** rubidium 37	87.6 **Sr** strontium 38	88.9 **Y** yttrium 39	91.2 **Zr** zirconium 40	92.9 **Nb** niobium 41	95.9 **Mo** molybdenum 42	[98] **Tc** technetium 43	101.1 **Ru** ruthenium 44	102.9 **Rh** rhodium 45	106.4 **Pd** palladium 46	107.9 **Ag** silver 47	112.4 **Cd** cadmium 48		114.8 **In** indium 49	118.7 **Sn** tin 50	121.8 **Sb** antimony 51	127.6 **Te** tellurium 52	126.9 **I** iodine 53	131.3 **Xe** xenon 54
132.9 **Cs** caesium 55	137.3 **Ba** barium 56	138.9 **La** lanthanum 57	178.5 **Hf** hafnium 72	180.9 **Ta** tantalum 73	183.8 **W** tungsten 74	186.2 **Re** rhenium 75	190.2 **Os** osmium 76	192.2 **Ir** iridium 77	195.1 **Pt** platinum 78	197.0 **Au** gold 79	200.6 **Hg** mercury 80		204.4 **Tl** thallium 81	207.2 **Pb** lead 82	209.0 **Bi** bismuth 83	[209] **Po** polonium 84	[210] **At** astatine 85	[222] **Rn** radon 86
[223] **Fr** francium 87	[226] **Ra** radium 88	[227] **Ac** actinium 89	[261] **Rf** rutherfordium 104	[262] **Db** dubnium 105	[266] **Sg** seaborgium 106	[264] **Bh** bohrium 107	[277] **Hs** hassium 108	[268] **Mt** meitnerium 109	[271] **Ds** darmstadtium 110	[272] **Rg** roentgenium 111								

Elements with atomic numbers 112–116 have been reported but not fully authenticated

140.1 **Ce** cerium 58	140.9 **Pr** praseodymium 59	144.2 **Nd** neodymium 60	144.9 **Pm** promethium 61	150.4 **Sm** samarium 62	152.0 **Eu** europium 63	157.2 **Gd** gadolinium 64	158.9 **Tb** terbium 65	162.5 **Dy** dysprosium 66	164.9 **Ho** holmium 67	167.3 **Er** erbium 68	168.9 **Tm** thulium 69	173.0 **Yb** ytterbium 70	175.0 **Lu** lutetium 71
232.0 **Th** thorium 90	[231] **Pa** protactinium 91	238.0 **U** uranium 92	[237] **Np** neptunium 93	[242] **Pu** plutonium 94	[243] **Am** americium 95	[247] **Cm** curium 96	[245] **Bk** berkelium 97	[251] **Cf** californium 98	[254] **Es** einsteinium 99	[253] **Fm** fermium 100	[256] **Md** mendelevium 101	[254] **No** nobelium 102	[257] **Lr** lawrencium 103

Answers to SAQs

Chapter 1

1 **a** U-235 has 92 protons, 92 electrons and 143 neutrons.
U-238 has 92 protons, 92 electrons and 146 neutrons.

   b K^+-40 has 19 protons, 18 electrons and 21 neutrons.
Cl^--37 has 17 protons, 18 electrons and 20 neutrons.

2 Relative atomic mass of neon
$$= \frac{(90.9 \times 20) + (0.3 \times 21) + (8.8 \times 222)}{100} = 20.2$$

3 **a** $24.3 + (2 \times 35.5) = 95.3$
   b $63.5 + 32.1 + (4 \times 16.0) = 159.6$
   c $(2 \times 23.0) + 12.0 + (3 \times 16.0) + 10[(2 \times 1.0) + 16.0]$
$$= 286.0$$

Chapter 2

1 **a** $\dfrac{35.5}{35.5} = 1$ mol Cl atoms

   b $\dfrac{71}{2 \times 35.5} = 1$ mol Cl_2 molecules

2 **a** 6×10^{23} Cl atoms
   b 1 mol Cl_2 molecules = 2 mol Cl atoms
$$= 2 \times 6 \times 10^{23} = 1.2 \times 10^{24} \text{ atoms}$$

3 **a** $CO_2 = 12.0 + (2 \times 16.0) = 44.0$ g
$$\therefore \text{ mass } 0.1 \text{ mol } CO_2 = 0.1 \times 44.0$$
$$= 4.40 \text{ g}$$
   b $CaCO_3 = 40.1 + 12.0 + 3 \times 16.0 = 100.1$ g
$$\therefore \text{ mass } 10 \text{ mol } CaCO_3 = 10 \times 100.1$$
$$= 1001 \text{ g}$$

4 **a** From equation, mole ratio $H_2 : Cl_2 = 1 : 1$
$$\therefore \text{ mass ratio} = 2.0 : 71.0 \text{ or } 1 : 35.5$$
   b HCl = $1.0 + 35.5 = 36.5 = 1$ mol HCl
$$\therefore \text{ as } 1 \text{ mol } H_2 \text{ produces } 2 \text{ mol HCl,}$$
0.5 mol H_2 produces 1 mol HCl.
$$\therefore 2.0 \times 0.5 = 1.0 \text{ g } H_2 \text{ produces } 36.5 \text{ g HCl.}$$

5 1000 tonne Fe_2O_3 produce
$$111.6 \times \frac{1000}{159.6} \text{ tonne} = 699.2 \text{ tonne Fe}$$
$$\therefore 1 \text{ tonne Fe requires } \frac{1000}{699.2} \text{ tonne } Fe_2O_3$$
$$= 1.43 \text{ tonne}$$
$$\therefore \text{ mass ore} = 1.43 \times \frac{100}{12} = 11.9 \text{ tonne}$$

6 **a** C_3H_7 **b** HO

7

	Cu	**O**
Amount /mol	$\dfrac{0.635}{63.5} = 0.0100$	$\dfrac{0.080}{16.0} = 0.00500$
Ratio /mol	2	1

$\therefore$ empirical formula is Cu_2O.

8 **a** FeO
   b Fe_3O_4

9 **a** $MgBr_2$
   b HI
   c CaS
   d Na_2SO_4
   e KNO_3
   f NO_2

10 **a** Potassium carbonate
   b Aluminium sulfide
   c Lithium nitrate
   d Calcium phosphate
   e Silicon dioxide

11 **a** $2Al + Fe_2O_3 \longrightarrow Al_2O_3 + 2Fe$
   b $2C_8H_{18} + 25O_2 \longrightarrow 16CO_2 + 18H_2O$

or $C_8H_{18} + {}^{25}/_2 O_2 \longrightarrow 8CO_2 + 9H_2O$

   c $2Pb(NO_3)_2 \longrightarrow 2PbO + 4NO_2 + O_2$

12 **a** $Cl_2(aq) + 2Br^-(aq) \longrightarrow 2Cl^-(aq) + Br_2(aq)$
   b $Fe^{3+}(aq) + 3OH^-(aq) \longrightarrow Fe(OH)_3(s)$

13 **a** Amount nitric acid
$$= \frac{25}{1000} \times 0.1 = 2.5 \times 10^{-3} \text{ mol}$$
   b Volume $= \dfrac{50}{1000} = 5 \times 10^{-2} \text{ dm}^3$
$$\therefore \text{ concentration} = \frac{0.125}{5 \times 10^{-2}}$$
$$= 2.5 \text{ mol dm}^{-3}$$

14 **a** CH_3CO_2H
$$= 12.0 + (3 \times 1.0) + 12.0 + (2 \times 16.0) + 1.0 = 60.0$$
$$\therefore \text{ concentration} = 0.50 \times 60 = 30.0 \text{ g dm}^{-3}$$
   b NaOH = $23.0 + 16.0 + 1.0 = 40.0$
$$\therefore \text{ concentration} = \frac{4.00}{40.0} = 0.100 \text{ mol dm}^{-3}$$

(N.B. Three significant figures in these answers.)

15 **a** Amount KOH $= \dfrac{20}{1000} \times 0.100 = 2 \times 10^{-3}$ mol

$$KOH + HCl \longrightarrow KCl + H_2O$$
$$\therefore \text{ amount KOH} = \text{amount HCl}$$
$$= 2 \times 10^{-3} \text{ mol}$$

$$\text{Volume HCl} = \frac{25.0}{1000} = 2.5 \times 10^{-2}\,\text{dm}^3$$

$$\therefore \text{ concentration HCl} = \frac{2 \times 10^{-3}}{2.5 \times 10^{-2}}$$

$$= 0.08\,\text{mol}\,\text{dm}^{-3}$$

b $36.5 \times 0.08 = 2.92\,\text{g}\,\text{dm}^{-3}$

16 Amount $HNO_3 = \dfrac{24}{1000} \times 0.050$

$$= 1.20 \times 10^{-3}\,\text{mol}$$

$\therefore$ stoichiometric mole ratio
nitric acid : iron hydroxide is
$1.20 \times 10^{-3} : 4.00 \times 10^{-4}$ i.e. 3 : 1
Iron hydroxide contains three hydroxide ions to exactly neutralise three HNO_3 molecules. So equation is

$$3HNO_3(aq) + Fe(OH)_3(s) \longrightarrow Fe(NO_3)_3(aq) + 3H_2O(l)$$

17 a Number of moles He $= \dfrac{2.4}{24} = 0.10\,\text{mol}$

b 0.5 mol propane $= 0.5 \times 24 = 12\,\text{dm}^3$
1.5 mol butane $= 1.5 \times 24 = 36\,\text{dm}^3$
$\therefore$ total volume $= 48\,\text{dm}^3$

Chapter 3

1 $HNO_3(l) \xrightarrow{\text{water}} H^+(aq) + NO_3^-(aq)$

2 $H_2SO_4(l) \xrightarrow{\text{water}} 2H^+(aq) + SO_4^{2-}(aq)$

3 $KOH(s) \xrightarrow{\text{water}} K^+(aq) + OH^-(aq)$

4 $Ca^{2+}(OH)^-{}_2(aq) + 2H^+NO_3^-(aq) \longrightarrow$
$$Ca^{2+}(NO_3)^-{}_2(aq) + 2H_2O(l)$$
The acid is neutralised as the OH^- ions accept the H^+ ions, forming water. The salt is calcium nitrate.

5 a Excess copper(II) oxide is added to a beaker of sulfuric acid.

b The mixture is stirred so that the copper(II) oxide reacts with the sulfuric acid.

c When the reaction is complete the unreacted copper(II) oxide is removed by filtration.

d The copper(II) sulfate solution is heated to evaporate off most of the water.

e The remaining water is allowed to evaporate slowly, leaving crystals of copper(II) sulfate.

6 a $Cu^{2+}CO_3^{2-}(s) + 2H^+Cl^-(aq) \longrightarrow$
$$Cu^{2+}Cl^-{}_2(aq) + CO_2(g) + H_2O(l)$$
Salt formed is copper(II) chloride.

b $K^+OH^-(aq) + H^+{}_2SO_4^{2-}(aq) \longrightarrow$
$$K^+{}_2SO_4^{2-}(aq) + 2H_2O(l)$$
Salt formed is potassium sulfate.

c $Mg^{2+}O^{2-}(s) + 2H^+NO_3^-(aq) \longrightarrow$
$$Mg^{2+}(NO_3)^-{}_2 + H_2O(l)$$
Salt formed is magnesium nitrate.

d $2NH_3(aq) + H^+{}_2SO_4^{2-}(aq) \longrightarrow (NH_4)^+{}_2SO_4^{2-}(aq)$
Salt formed is ammonium sulfate.

7 Measure $25\,\text{cm}^3$ of potassium hydroxide using a graduated pipette and put into a conical flask.
Add indicator, e.g. phenolphthalein.
Fill a burette with the hydrochloric acid.
Add the acid to the potassium hydroxide until the indicator just changes colour.
Repeat to obtain concordant results.
Repeat without adding indicator, using the volumes of acid and alkali from the titration.
Evaporate the salt solution until only a little liquid remains.
Leave the solution to crystallise.
Titration is used. Since potassium hydroxide is soluble in water it is not possible to add excess potassium hydroxide to the hydrochloric acid, followed by filtration to remove the excess.

8 $CuCl_2.2H_2O$ (or $CuCl_2H_4O_2$)

9 a 0.800 g

b 0.0444 mol

c 0.0111 mol

d $MnSO_4.4H_2O$ (or $MnSO_4H_8O_4$)

10 a Add excess calcium carbonate to nitric acid solution.
When effervescence has stopped, filter to remove excess calcium carbonate.
Evaporate the salt solution until only a little liquid remains.
Leave the solution to crystallise.

b Heat so as to drive off the water of crystallisation without decomposing the salt, until no further decrease in mass is noticed.

c $Ca(NO_3)_2$ (or CaN_2O_6)

d $Ca(NO_3)_2.4H_2O$ (or $CaN_2O_{10}H_8$)

Chapter 4

1 a +4   **b** +3

2 a This is not a redox reaction since the oxidation state of every element is the same before and after the reaction.
Before: Ca is +2, C is +4, O is −2, H is +1, Cl is −1.
After: Ca is +2, C is +4, O is −2, H is +1, Cl is −1.

b This is a redox reaction. The oxidation number of manganese decreases from +7 to +2, so manganese is reduced. The oxidation number of some of the chloride ions increases from −1 to 0, so some of the chloride ions are oxidised.
Before: K is +1, Mn is +7, O is −2, H is +1, Cl is −1.
After: K is +1, Mn is +2, O is −2, H is +1, Cl is −1 and 0.

3 $Mg \longrightarrow Mg^{2+} + 2e^-$ (or $Mg - 2e^- \longrightarrow Mg^{2+}$)

4 a copper(I) chloride
b copper(II) chloride

5 a sodium sulfate(IV)
b sodium sulfate(VI)

6 Sodium chlorate(III) is $Na^+ClO_2^-$, sodium chlorate(V) is $Na^+ClO_3^-$, and sodium chlorate(VII) is $Na^+ClO_4^-$.

7 a $2Ca(s) + O_2(g) \longrightarrow 2Ca^{2+}O^{2-}(s)$
b Calcium is oxidised. Each calcium atom loses two electrons. The oxidation number of calcium increases from 0 to +2.
c Oxygen is reduced. Each oxygen atom gains two electrons. The oxidation number of oxygen decreases from 0 to –2.

8 a $Zn(s) + H^+_2SO_4^{2-}(aq) \longrightarrow Zn^{2+}SO_4^{2-}(aq) + H_2(g)$
(Note: the charges on the ions are not normally shown in full equations, this is just to help you answer the question)
b Zinc is oxidised. Each zinc atom loses two electrons. The oxidation number of zinc increases from 0 to +2.
c Hydrogen is reduced. Each hydrogen ion gains one electron. The oxidation number of hydrogen decreases from +1 to 0.

Chapter 5

1 All the isotopes have the same number and arrangement of electrons and this controls their chemical properties.

2 a Sodium has 11 electrons in all. There is one electron in its outer shell ($n = 3$) and this is the easiest to remove. The second ionisation energy shows the energy required to remove an electron from the next inner (filled) shell ($n = 2$).
The ninth electron to be removed is in shell $n = 2$ and the tenth is in shell $n = 1$, which is closest to the nucleus.
b The first electron is in the outer shell $n = 3$.
The relatively low increases from the second to the ninth ionisation energies show that eight electrons are in the same shell $n = 2$. The tenth and eleventh electrons are in the shell $n = 1$.

3 Group 2. The first and second ionisation energies are fairly close in value. There is a large increase between the second and third ionisation energies, which shows that the second and third electrons are in different shells. This indicates that there are two electrons in the outer shell.

4 a 2, 6, 10, 14
b 2, 8, 18, 32
c on the left, on the right, in the middle, at the bottom.

5

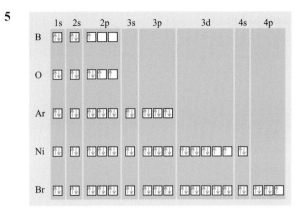

Chapter 6

1 At negative electrode: $Cu^{2+} + 2e^- \longrightarrow Cu$
At positive electrode: $2Br^- \longrightarrow Br_2 + 2e^-$

2 a

b

c

d

3 a $Mg^{2+}(OH)^-_2$ i.e. $Mg(OH)_2$
b $Na^+_2SO_4^{2-}$ i.e. Na_2SO_4
c $(NH_4)^+_2CO_3^{2-}$ i.e. $(NH_4)_2CO_3$
d $Ca^{2+}(NO_3)^-_2$ i.e. $Ca(NO_3)_2$

4 Hydrogen

two hydrogen atoms (1)   hydrogen molecule each hydrogen is now (2)

Chlorine

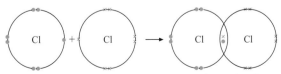

chlorine atoms (2,8,7)   chlorine molecule: each chlorine is now 2,8,8

Sulfur hexafluoride

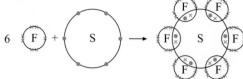

six
fluorine
atoms (2,7)

sulfur
atom
(2,8,6)

sulfur hexafluoride
molecule

Methane

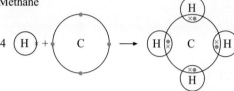

four
hydrogen
atoms (1)

carbon
atom
(2,4)

methane molecule: each
hydrogen now shares two
electrons with carbon

Water

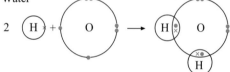

two
hydrogen
atoms (1)

oxygen
atom
(2,6)

water molecule: hydrogen and
oxygen both fill their outer shells
by sharing electrons

Ammonia

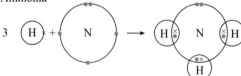

three
hydrogen
atoms (1)

nitrogen
atom
(2,5)

ammonia molecule: hydrogen
and nitrogen both fill their
outer shells by sharing electrons

Hydrogen chloride

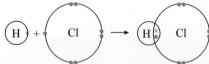

hydrogen
atom
(1)

chlorine
atom
(2,8,7)

hydrogen chloride molecule
hydrogen and chlorine both
fill their outer shells by
sharing electrons

Boron trifluoride

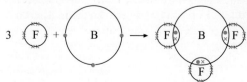

three
fluorine
atoms (2,7)

boron
atom (2,3)

boron trifluoride
molecule

5 a carbon dioxide

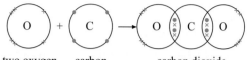

two oxygen
atoms (2,6)

carbon
atom (2,4)

carbon dioxide
molecule

nitrogen

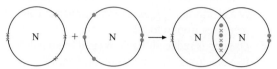

two nitrogen
atoms (2,5)

nitrogen molecule

oxygen

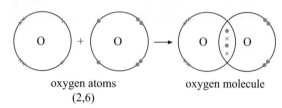

oxygen atoms
(2,6)

oxygen molecule

b oxygen
c carbon dioxide
d nitrogen

6 a

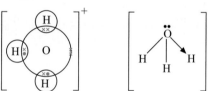

b

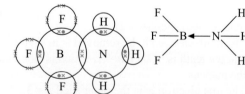

7 a O=O

H—F
δ+ δ−

c

Br
δ−
C—H
H   H δ+

d

δ+
S—Cl
Cl δ−

Thus, **a** is non-polar, and **b**, **c** and **d** are polar.

8 a i

iii

ii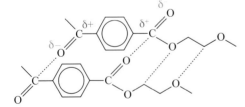

iv

b i triangular pyramid
ii tetrahedral
iii non-linear
iv non-linear

9 a Copper provides better heat transfer as it has a thermal conductivity that is five times higher than that of iron (stainless steel has a lower thermal conductivity than iron).

b Copper has more than three times the density of aluminium. The electrical conductivity of copper is 1.5 times that of aluminium. Aluminium cables will be lighter than copper whilst still being good conductors of electricity. The lighter cables enable less massive (and less unsightly) pylons to be used. As the tensile strength of aluminium is low, aluminium cables are reinforced with a steel core to increase their strength.

c Copper has the highest electrical and thermal conductivities. Its high thermal conductivity helps to keep equipment such as transformers cool.

10 As water molecules are free to rotate, the positive charge on the rod repels the positive end of a water molecule whilst attracting the negative end. The overall effect is thus an attraction. The effect will be the same if the charge on the rod is negative rather than positive.

11

Dotted lines show the dipole–dipole forces. (Note: Extrusion through spinnerets causes more molecules to line up closely, increasing the intermolecular forces (and hence the strength of the fibre) by the closer contact.)

12 a Underlying increase is due to increasing van der Waals' forces as the number of electrons present in the molecules rises.

b The value for water based on this underlying trend would be about $18 \, \text{kJ} \, \text{mol}^{-1}$.

c The much higher value observed for water is due to the presence of hydrogen bonds which are much stronger intermolecular forces than van der Waals' forces.

13 The length of hydrogen bonds causes water molecules to be further apart in ice than they are in water. Ice therefore has a lower density than water.

14 Washing-up liquid lowers the surface tension of water. This reduces the hydrogen bonding at the surface to the point where it is no longer sufficient to keep the needle afloat.

15 a Underlying increase is due to increasing van der Waals' forces as the number of electrons present in the molecules rises.

b The much higher value observed for ammonia is due to the presence of hydrogen bonds, N–H⋯N.

16

Chapter 7

1 The properties predicted for eka-silicon are close to those now known for germanium.

2 a Elements may have several isotopes – atoms with the same number of protons (i.e. atomic number) but different numbers of neutrons and hence different masses. The mass of an isotope of an element is the same as its nucleon number. This equals the number of protons plus the number of neutrons. The relative atomic mass of the element is the 'weighted average' of the nucleon numbers of its isotopes taking into account their relative abundances in a naturally occurring sample of the element.

b Tellurium and iodine have several isotopes each. The weighted average A_r of tellurium is higher than the average for iodine. Thus, in a table based on relative atomic masses only, tellurium would have been placed after iodine.

3 a C $1s^2 \, 2s^2 \, 2p^2$
Si $1s^2 \, 2s^2 \, 2p^6 \, 3s^2 \, 3p^2$

b Both outer-shell configurations are $s^2 \, p^2$.

c Ge is also in Group 4 so outer shell is $s^2 \, p^2$ (actually $4s^2 \, 4p^2$).

4 The noble gases exist only as individual atoms, not in molecules. They don't form covalent bonds.

5 P_4, S_8 and Cl_2 all have simple molecular structures with weak van der Waals' forces between the molecules and these are fairly easily separated at relatively low temperatures. The number of electrons in P_4 and S_8 molecules, however, are much higher than in Cl_2 molecules. More energy is needed to move P_4 or S_8 molecules into the vapour phase than Cl_2 molecules. S_8 and P_4 do not boil until they reach a higher temperature than the boiling point of Cl_2.

6 In general, if the attractive forces between the particles are high, more energy is needed to overcome these forces and the melting point is high.

a These elements all have a metallic structure. The metallic bonding is stronger moving from sodium to aluminium as there are more outer-shell electrons available to be mobile and take part in the bonding.

b Silicon has a giant covalent lattice structure like diamond. The melting point is high as the bonding is very strong.

c These elements exist as non-polar small molecules (sulfur and chlorine) or as separate atoms (argon). Only weak van der Waals' forces are present, so the melting points are low.

7 a Electronic configurations are:
Na $1s^2 2s^2 2p^6 3s^1$
Mg $1s^2 2s^2 2p^6 3s^2$
Al $1s^2 2s^2 2p^6 3s^2 3p^1$
Mg has a higher nuclear charge than Na. This makes it more difficult to remove a 3s electron and thus Mg has a higher first ionisation energy than Na. Al has a lower first ionisaton energy than Mg as the electron being removed is in a 3p orbital, a little further from the nuclear charge (and of higher energy) than the 3s orbital.

b Si $1s^2 2s^2 2p^6 3s^2 3p^2$
P $1s^2 2s^2 2p^6 3s^2 3p^3$
S $1s^2 2s^2 2p^6 3s^2 3p^4$
P has a higher nuclear charge than Si. This makes it more difficult to remove a 3p electron and thus P has a higher first ionisation energy than Si. The first ionisation energy of S is slightly lower than that of P because two of the electrons in the 3p subshell of S share the same orbital (they spin-pair). The repulsion between the electrons in the pair makes it easier to remove one of them.

8 Francium is the most likely: its outer electron is in a 7s orbital, distant from the nucleus and well screened by several filled inner shells. Compared with other elements whose outer electron is in the 7th shell (e.g. radium) it has a lower nuclear charge.

9 a Between the fifth and sixth successive ionisation energies.

b Group 7 (the element is fluorine).

Chapter 8

1 a i The metallic radii increase from element to element down the group.

ii The first ionisation energies decrease from element to element down the group.

b i The radii increase as additional shells of electrons are added going down the group from magnesium to barium.

ii The electron removed is further from the nucleus and shielded by more inner filled shells of electrons. The distance and shielding effects together are able to reduce the effect of the increasing nuclear charge from element to element down the group. Hence the ionisation energy decreases from element to element down the group.

c As less energy is needed to remove an electron going down the group from magnesium to barium, the electronegativity will also decrease from element to element down the group. As the atoms get larger down the group, the outer shell gets further from the attractive force of the nucleus and is shielded from this force by more complete inner shells of electrons. These counteract the increasing nuclear charge going down the group.

2 MgO is giant ionic. The ionic bonding is strong, giving a very high melting point.

3 Magnesium oxide.
0.33 g

4 a $2Sr(s) + O_2(g) \longrightarrow 2SrO(s)$

b The metal burns with a red flame and a white solid is formed.

c Each strontium atom loses two electrons and changes oxidation state from 0 to +2. Strontium is oxidised. Each oxygen atom gains two electrons and changes oxidation state from 0 to –2. Oxygen is reduced.

d Two electrons are lost from each metal atom in the reaction. Down the group, the first two ionisation energies decrease from magnesium to barium. Consequently, the reactivity of the metals increases down the group as less energy is required to remove the two electrons.

5 Calcium hydroxide is more soluble in water than magnesium hydroxide. Calcium hydroxide therefore produces a solution with a much higher pH – around pH 10 or even higher. People strongly dislike the taste of solutions of high pH.

6 a Calcium carbonate dissolves rapidly in dilute hydrochloric acid with the evolution of carbon dioxide.
$$CaCO_3(s) + 2HCl(g) \longrightarrow$$
$$CaCl_2(aq) + H_2O(l) + CO_2(g)$$

b The calcium hydroxide contains hydroxide ions which will neutralise the acidity in the soil.

c $Mg(OH)_2(s) + 2HCl(g) \longrightarrow MgCl_2(aq) + 2H_2O(l)$

7 a thermal decomposition

b $SrCO_3 \longrightarrow SrO + CO_2$

c Strontium carbonate is harder to decompose than calcium carbonate but easier to decompose than barium carbonate.

8 **a** $BaO + H_2O \longrightarrow Ba(OH)_2$
   b barium hydroxide
   c The solution contains a high concentration of hydroxide ions.

Chapter 9

1 **a** Solid. At_2 molecules are larger than I_2 molecules (At_2 molecules have more electrons than I_2 molecules) therefore van-der-Waals' forces will be stronger. Iodine is a solid at room temperature, therefore so is astatine.
   b **i** $Cl_2 + 2Na \longrightarrow 2NaCl$ and
         $F_2 + 2Na \longrightarrow 2NaF$
      ii The reaction involving fluorine. Smaller fluorine atoms have a more negative electron affinity than larger chlorine atoms. The extra electron attracted into fluorine's outer shell is nearer to the attractive force of the nucleus and experiences less shielding effect from inner electrons than chlorine atoms when forming chloride ions. Fluorine atoms therefore become fluoride ions with greater ease than chlorine atoms become chloride ions.
      iii Both are giant ionic.
   c 0, −1, +7, +4

2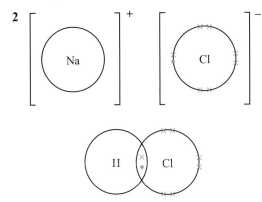

3 **a** Answers should refer to organochlorine compounds which can be carcinogens and mutagens. Levels of these are low, and therefore risk is low. However, exposure over very many years is likely.
   b Disease outbreaks, e.g. cholera.
   c Answers should consider the ethics of taking decisions away from individuals, and the likely consequences of the decision.

4 The halogens have covalent bonds and they are non-polar molecules. Polar molecules dissolve best in water, which is itself polar. Non-polar molecules dissolve best in non-polar solvents.

5 **a** The orange cyclohexane layer would turn purple:
       Br_2(in cyclohexane) $+ 2I^-$(aq) $\longrightarrow$
              $2Br^-$(aq) $+ I_2$(in cyclohexane)

   b No change.
   c Given its position in Group 7, we would expect astatine to be darker in colour than iodine. The orange cyclohexane would turn this dark colour of astatine:
       Br_2(in cyclohexane) $+ 2At^-$(aq) $\longrightarrow$
              $2Br^-$(aq) $+ At_2$(in cyclohexane)

6 **a** $Br_2 + 2NaOH \longrightarrow NaBr + NaOBr + H_2O$
   b 0 before, −1 and +1 after
   c disproportionation

Chapter 10

1 alanine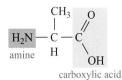

   glucose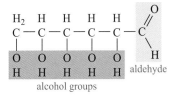

   fructose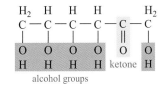

   oleic acid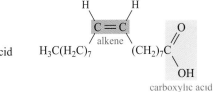

2 **a**
```
            H
            |
        H — C — H
        H   |
        |   |
    H — C — C — Cl
        |   |
        H   |
        H — C — H
            |
            H
```
   b $(CH_3)_3CCl$ **c**

   d C_4H_9Cl **e**

3 a A heptane
B 3-methylhexane
C pentan-3-ol
D pentan-2-one
E 2-methylbutanoic acid
F 2,2-dimethylpropanal

b i

$$CH_3CH_2\overset{\displaystyle O}{\underset{\displaystyle H}{C}}$$

ii

$$H_3C\overset{\displaystyle H}{\underset{\displaystyle OH}{C}}CH_3$$

iii

$$H_3C-\overset{}{\underset{H_2}{C}}-\overset{\displaystyle O}{\underset{\displaystyle \underset{\displaystyle CH_3}{C-CH_3}}{C}}\overset{}{\underset{}{H}}$$

iv

$$H_3C-\underset{H_2}{C}-\underset{H_2}{C}-NH_2$$

4 a

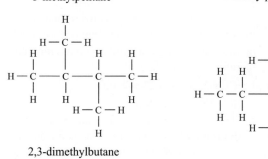

b

c

5

boiling point –0.4 °C   boiling point –11.6 °C

6

$$H-\overset{H}{\underset{H}{C}}-\overset{H}{\underset{H}{C}}-\overset{H}{\underset{H}{C}}-\overset{H}{\underset{H}{C}}-\overset{H}{\underset{H}{C}}-\overset{H}{\underset{H}{C}}-H$$

hexane

3-methylpentane   2-methylpentane

2,3-dimethylbutane

2,2-dimethylbutane

7 a

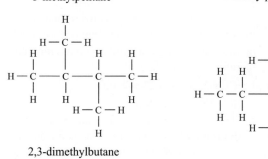

cis   *trans*

b

Br   Br
C=C
H   H
Z (or *cis*)

Br   H
C=C
H   H
no
Z / *E*
isomers

Br   H
C=C
H   Br
E (or *trans*)

Br   CH₃
C=C
H   H
Z (or *cis*)

Br   CH₃
C=C
H   CH₃
no
Z / *E*
isomers

Br   H
C=C
H   CH₃
E (or *trans*)

8

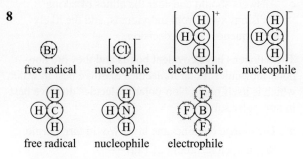

Br
free radical

[Cl]⁻
nucleophile

[HCH]⁺ (electrophile)

[HCH]⁻ (nucleophile)

HCH
free radical

HNH
nucleophile

FBF
electrophile

Free radicals have seven outer-shell electrons, electrophiles have six and nucleophiles have eight.

9 1 mol of butan-1-ol will produce 1 mol of 1-bromobutane. The quantity of butan-1-ol will determine the yield as the other reagents are in excess. 1 mol of butan-1-ol, C_4H_9OH, has a relative molecular mass of

$(4 \times 12.0) + (9 \times 1.0) + (1 \times 16.0) + 1.0 = 74.0\,g$

1 mol of 1-bromobutane, C_4H_9Br, has a relative molecular mass of

$(4 \times 12.0) + (9 \times 1.0) + 79.9 = 136.9\,g$

Hence

maximum yield of 1-bromobutane

$= \dfrac{10.0 \times 136.9}{74.0} = 18.5\,g$

percentage yield $= \dfrac{12.0}{18.5} \times 100 = 64.9\%$

10 a Atom economy

$= \dfrac{(2 \times 12.0) + (4 \times 19.0)}{(2 \times 12.0) + (4 \times 19.0) + [2 \times (1.0 + 19.0)]} \times 100\%$

$= \dfrac{100}{140} \times 100\%$

$= 71.4\%$

b In a substitution reaction there will always be the desired product plus another product (waste) formed whereas addition reactions can have one product only (the desired product) as a result of the reaction.

Chapter 11

1 Both graphs show a (gradually diminishing) increase in the melting and boiling points of the alkanes with increasing chain length.

2 a The lack of polarity of alkane molecules means that only weak instantaneous dipole–induced dipole (van der Waals') forces are present between molecules.

b As the number of electrons increases in the molecule (with increasing numbers of atoms), the strength of these van der Waals' forces also increases. More energy is needed to separate the atoms when melting the solid or boiling the liquid, so the melting and boiling points rise with increasing number of carbon atoms.

3 Possible reactions include:

$CH_3CH_2CH_2CH_2CH_3$

$\longrightarrow CH_3CH_2CH_3 + H_2C{=}CH_2$

$\longrightarrow CH_3CH_2CH_2CH{=}CH_2 + H_2$

$\longrightarrow CH_3CH_2CH{=}CHCH_3 + H_2$

$\longrightarrow CH_3CH_3 + CH_3CH{=}CH_2$

$\longrightarrow CH_2{=}CHCH{=}CHCH_3 + 2H_2$

4 a i $C_8H_{18}(l) + 8\tfrac{1}{2}O_2(g) \longrightarrow 8CO(g) + 9H_2O(l)$

ii $C_8H_{18}(l) + 12\tfrac{1}{2}O_2(g) \longrightarrow 8CO_2(g) + 9H_2O(l)$

b i $12.5 - 8.5 = 4$ moles $O_2(g)$

ii Volume of oxygen $= 4 \times 24.0 = 96.0\,dm^3$.
Volume of air $= 96.0 \times 100/20 = 480\,dm^3$

5 $C(s) + O_2(g) \longrightarrow CO_2(g)$

$CH_4(g) + 2O_2(g) \longrightarrow CO_2(g) + 2H_2O(l)$

$C_8H_{18}(l) + 12\tfrac{1}{2}O_2(g) \longrightarrow 8CO_2(g) + 9H_2O(l)$

$CH_3OH(l) + 1\tfrac{1}{2}O_2(g) \longrightarrow CO_2(g) + 2H_2O(l)$

$2H_2(g) + O_2(g) \longrightarrow 2H_2O(l)$

6 A kilogram of hydrogen contains 500 moles of $H_2(g)$, whereas a kilogram of methane contains 62.5 moles of $CH_4(g)$.

7 a The volume of a liquid is much smaller than the volume of a gas of the same mass.

b A spherical shape gives the lowest surface area for the container of any given volume of liquid or gas. This saves material for making the container, and helps to keep the surface area of the contents, which can be affected by heating from the Sun, as small as possible.

8 $2NO_2(g) + H_2O(l) \longrightarrow HNO_2(aq) + HNO_3(aq)$

nitric(III) acid   nitric(V) acid

$SO_2(g) + H_2O(l) \longrightarrow H_2SO_3$

sulfuric(IV) acid

$SO_3(g) + H_2O(l) \longrightarrow H_2SO_4$

sulfuric(VI) acid

9 a $Br_2(l)$ and $Cl_2(g)$ only – the others are in aqueous solution and are already ionised.

b $C_4H_{10}(g) + Br_2(l) \longrightarrow C_4H_9Br(l) + HBr(g)$

$C_4H_{10}(g) + Cl_2(g) \longrightarrow C_4H_9Cl(l) + HCl(g)$

Chapter 12

1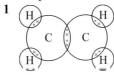

molecule is planar

2 a **D** can exist as a *cis / trans* isomer.

b

cis-pent-2-ene   *trans*-pent-2-ene

3 a The positive carbon atom has six electrons in its outer shell. It gains two more by accepting a lone pair from the bromide ion.

b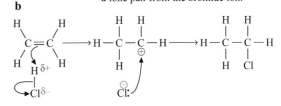

c CH_2ClCH_2Br

4 a

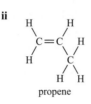

chloroethene → poly(chloroethene)

phenylethene → poly(phenylethene)

b i poly(propene)

ii propene

Chapter 13

1 a Ethanol molecules and water molecules can form hydrogen bonds with each other.

b Pentan-1-ol molecules and water molecules can form hydrogen bonds with each other but the C_5H_{11} sections in the pentan-1-ol molecules disrupt hydrogen bonding between other water molecules.

c Ethanol molecules can form hydrogen bonds with each other, propane molecules can't.

d Propan-1-ol molecules have a larger hydrocarbon portion than ethanol molecules, so they form stronger van der Waals' forces in addition to the hydrogen bonding.

2 Energy is absorbed when a bond is broken. In order from strongest to weakest:
$E(O–H) > E(C–H) > E(C–O) > E(C–C)$

3 a C–O

b The oxygen atom is very electronegative compared to hydrogen or carbon.

c An electrophile is an electron pair receiver. A nucleophile is an electron pair donor.

4

		Alcohol	Molecular model	Structural formula	Classification
	a	pentan-1-ol		$CH_3CH_2CH_2CH_2CH_2OH$	primary
	b	pentan-2-ol		$\overset{\displaystyle OH}{\underset{\displaystyle |}{CH_3CH_2CH_2CHCH_3}}$	secondary
	c	2-methylbutan-2-ol		$CH_3\overset{\displaystyle CH_3}{\underset{\displaystyle OH}{CH_3CH_2CCH_3}}$	tertiary
	d	3-methylbutan-1-ol		$\overset{\displaystyle CH_3}{\underset{\displaystyle |}{CH_3CHCH_2CH_2OH}}$	primary
	e	3-methylbutan-2-ol		$\overset{\displaystyle CH_3}{\underset{\displaystyle |}{CH_3CH—CHCH_3}}$ with OH	secondary
	f	2-methylbutan-1-ol		$\overset{\displaystyle CH_3}{\underset{\displaystyle |}{CH_3CH_2CHCH_2OH}}$	primary

5 a Propan-1-ol should be heated gently with acidified potassium dichromate(VI), the propanal should be distilled out immediately.

b $CH_3CH_2CH_2OH + [O] \longrightarrow CH_3CH_2CHO + H_2O$

c Propan-1-ol should be refluxed with acidified potassium dichromate(VI), the propanoic acid should be distilled out after at least 15 minutes refluxing.

d $CH_3CH_2CH_2OH + 2[O] \longrightarrow CH_3CH_2COOH + H_2O$
Alternatively:
$CH_3CH_2CH_2OH + [O] \longrightarrow CH_3CH_2CHO + H_2O$
followed by
$CH_3CH_2CHO + [O] \longrightarrow CH_3CH_2COOH$

6 a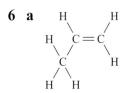

b $CH_3CH(OH)CH_3 \longrightarrow CH_3CH=CH_2 + H_2O$

Chapter 14

1 a $CH_3CH_2CH_2I$: 1-iodopropane
b $CH_3CHBrCH_3$: 2-bromopropane
c $CBrF_2CBrF_2$:
1,2-dibromo-1,1,2,2-tetrafluoroethane

2 Structural isomerism.

$$H-\overset{\overset{\displaystyle H}{|}}{\underset{\underset{\displaystyle H}{|}}{C}}-\overset{\overset{\displaystyle H}{|}}{\underset{\underset{\displaystyle \underset{\displaystyle H}{|}}{\overset{|}{C}-H}}{C}}-\overset{\overset{\displaystyle H}{|}}{\underset{\underset{\displaystyle H}{|}}{C}}-Cl$$

1-chloro-2-methylpropane (primary)

3 a 1-chloropropane is polar and has dipole–dipole intermolecular forces that are stronger than the instantaneous dipole–induced dipole forces (van der Waals' forces) in non-polar butane. More energy is needed to overcome the intermolecular forces in 1-chlorobutane, so its boiling point is higher.

b 1-chloropropane attracts water molecules by dipole–dipole forces that are weaker than the hydrogen bonds in water. An input of energy would be required for 1-chloropropane to mix with water and break some of these hydrogen bonds.

4

$$H_3C-\overset{\overset{\displaystyle CH_3}{|}}{\underset{\underset{\displaystyle CH_3}{|}}{C}}-Br + OH^- \longrightarrow H_3C-\overset{\overset{\displaystyle CH_3}{|}}{\underset{\underset{\displaystyle CH_3}{|}}{C}}-OH + Br^-$$

2-methylpropan-2-ol

Chapter 15

1 a

$$CH_3-\overset{\displaystyle \overset{\textstyle O}{\|}}{C}\diagdown_{\textstyle O-H}$$

b C=O at $1710\,cm^{-1}$, strong and sharp.

2 In Figure 15.5, **a** is butanone because a strong, sharp absorption is present at $1710\,cm^{-1}$, characteristic of the C=O bond in butanone; **b** shows a strong, broad absorption at $3200–3500\,cm^{-1}$, characteristic of the O–H bond in an alcohol.

3 a ^{90}Zr, ^{91}Zr, ^{92}Zr, ^{94}Zr, ^{96}Zr
b $A_r(Zr)$
$$= \frac{(51.5\times90)+(11.2\times91)+(17.1\times92)+(17.4\times94)+(2.8\times96)}{100}$$
$$= 91.3$$

4 The methoxymethane will have a peak at 31 for the CH_3O^+ fragment, ethanol will not. Ethanol will have a peak at 29 corresponding to the $C_2H_5^+$ fragment and a peak at 17 corresponding to the OH^+ fragment, methoxymethane will not. Both methoxymethane and ethanol will have a peak at 15 corresponding to the CH_3^+ fragment.

Chapter 16

1 Exothermic: crystallisation; magnesium oxide formation. Endothermic: evaporation; copper oxide from copper carbonate.

2 a i $2C(s) + 3H_2(g) \longrightarrow C_2H_6(g)$;
$\Delta H_f^{\ominus} = -84.7\,kJ\,mol^{-1}$
ii $2Al(s) + \tfrac{3}{2}O_2(g) \longrightarrow Al_2O_3(s)$;
$\Delta H_f^{\ominus} = -1669\,kJ\,mol^{-1}$

b

Enthalpy H

$2C(s) + 3H_2(g)$

$\Delta H_f^{\ominus} = -84.7\,kJ\,mol^{-1}$

$C_2H_6(g)$

Reaction pathway

3 a Figure 16.6a: either $\Delta H_r^{\ominus}$ or $\Delta H_c^{\ominus}$
Figure 16.6b: $\Delta H_r^{\ominus}$

b i $C_8H_{18}(l) + 12\tfrac{1}{2}O_2(g) \longrightarrow$
$8CO_2(g) + 9H_2O(l)$; $\Delta H_c^{\ominus} = -5512\,kJ\,mol^{-1}$
ii $C_2H_5OH(l) + 3O_2(g) \longrightarrow$
$2CO_2(g) + 3H_2O(l)$; $\Delta H_c^{\ominus} = -1371\,kJ\,mol^{-1}$

c One mole of water is formed by burning one mole of hydrogen.

255

4 a The *standard enthalpy change of formation* is the enthalpy change when one mole of a compound is formed from its elements under standard conditions; the compound and the elements must be in their standard states.

b The *standard enthalpy change of combustion* is the enthalpy change when one mole of an element or compound reacts completely with oxygen under standard conditions.

5 $\Delta H_c^{\ominus}(H_2)$ is calculated directly from experimental measurements. $\Delta H_f^{\ominus}(H_2O)$ is found from bond energies, which are average values calculated from measurements in a number of different experiments.

6 The value in the data book was calculated from much more accurate experimental data. Some of the energy transferred from the burning propanol would not change the temperature of the water but would be 'lost' in heating the apparatus and surroundings.

7 The reaction that produces the enthalpy change is the same in each case, no matter which acid and alkali are used. Only $H^+(aq)$ and $OH^-(aq)$ are involved:

$$H^+(aq) + OH^-(aq) \longrightarrow H_2O(l)$$

8 a and b

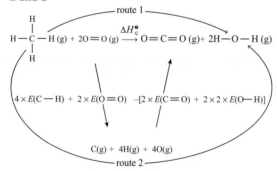

c By Hess's law the enthalpy change for route 1 = enthalpy change for route 2. Hence $\Delta H_c^{\ominus}$

$= 4 \times E(C–H) + 2 \times E(O=O) + [2 \times -E(C=O)$
$\qquad\qquad\qquad\qquad\qquad + 2 \times 2 \times -E(O–H)]$
$= (4 \times 410 + (2 \times 500) + [(2 \times -805) + (2 \times 2 \times -465)]$
$= -830 \, kJ \, mol^{-1}$

d The experimental value is more accurate. The bond energies used are based on average values.

9 a

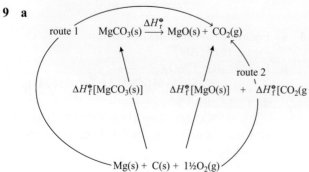

b The enthalpy change for route 1
$= \Delta H_f^{\ominus}[MgCO_3(s)] + \Delta H_r^{\ominus}$
The enthalpy change for route 2
$= \Delta H_f^{\ominus}[MgO(s)] + \Delta H_f^{\ominus}[CO_2(g)]$
Applying Hess's law:
$\Delta H_f^{\ominus}[MgCO_3(s)] + \Delta H_r^{\ominus}$
$= \Delta H_f^{\ominus}[MgO(s)] + \Delta H_f^{\ominus}[CO_2(g)]$
or
$(-1096) + \Delta H_r^{\ominus} = (-602) + (-394);$
$\Delta H_r^{\ominus} = +100 \, kJ \, mol^{-1}$

10 a and b

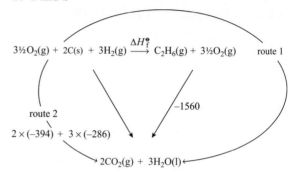

c Applying Hess's law:
$\Delta H_f^{\ominus} + (-1560) = 2 \times (-394) + 3 \times (-286)$
$\Delta H_f^{\ominus} = 2 \times (-394) + 3 \times (-286) - (-1560)$
$= -86 \, kJ \, mol^{-1}$

Chapter 17

1 Increase concentration of hydrogen peroxide solution, increase temperature of hydrogen peroxide solution, add a catalyst (manganese(IV) oxide) to hydrogen peroxide solution, increase surface area of catalyst by using it in powder form.

2 $2C_2H_2(g) + 5O_2(g) \longrightarrow 4CO_2(g) + 2H_2O(g) \text{ or (l)}$

3 In the first situation the acid has a higher concentration. Therefore the H^+ ions are closer together, they will collide with the magnesium ribbon more often, causing a faster rate of reaction. In the first situation the acid has a higher temperature. Therefore the H^+ ions have more kinetic energy, which increases the number of collisions in a given time and the proportion of successful collisions, and so increases reaction rate.

4 A spark or flame might provide the activation energy.

5 To provide a very large surface area for reaction.

6 The system is not closed, water will evaporate.

7 a The position of equilibrium would move towards the reactants. A new position of equilibrium will be established with a lower concentration of the ester (and water).

b Use a large excess of ethanol or ethanoic acid. (Preferably whichever is cheaper!) Remove the ethyl ethanoate as it is formed (the ester may be distilled from the mixture).

8 a No change occurs as there are equal numbers of gaseous molecules on either side of the equation.
b There are fewer gaseous molecules on the right-hand side of the equation. There is a reduction in volume moving from reactants to products so more $NH_3(g)$ is formed.

9 a The yield of ammonia would be too low.
b The rate of the reaction would be too slow.
c The costs would be too high (buying better compressors, building a plant to resist high pressure, increased maintenance costs, increased labour costs).
d The yield of ammonia would be too low.

Chapter 18

1 In winter there is less photosynthesis due to the lower temperature, less sunlight and the loss of leaves from deciduous plants. As a result, less carbon dioxide is removed from the atmosphere by photosynthesis. At the same time, more fossil fuels are burned to keep us warm, putting more carbon dioxide into the atmosphere. Thus carbon dioxide levels are increased in winter.
At night, plant respiration takes place, releasing carbon dioxide, but there is virtually no photosynthesis to remove carbon dioxide. So the carbon dioxide level increases.

2 Photosynthesis is rapid in rainforests. This removes excess carbon dioxide from the atmosphere and helps maintain a balance. When rainforest trees are cut down they may be burned, releasing carbon dioxide. Removal of rainforests will lead to an increase in atmospheric carbon dioxide and hence global warming.

3 13 billion tonnes more carbon equivalent burned.
This will produce $13 \times \frac{44}{12} = 47.7$ billion tonnes of carbon dioxide, assuming complete combustion.

4 A considerable amount of the carbon dioxide released into the atmosphere dissolves in the oceans and is removed by phytoplankton or reacts with the water. Cold water in northern oceans sinks to great depths, taking the dissolved carbon dioxide with it. Thus the oceans effectively act as a sink for the removal of some of the carbon dioxide in the atmosphere.

5 CCS will require extra energy to retrieve, liquefy and pump the CO_2 to its storage reservoir (as well as the costs of compressors, pipelines, etc.).

6 a Production:
$$O_2 \xrightarrow{UV} O+O$$
$$O+O_2 \longrightarrow O_3$$
b Removal:
$$NO\bullet(g)+O_3(g) \longrightarrow NO_2\bullet(g)+O_2(g)$$
$$NO_2\bullet(g)+O(g) \longrightarrow NO\bullet(g)+O_2(g)$$

7 Lack of reactivity.

8 Hydrofluorocarbons (HFCs) and hydrochlorofluorocarbons (HCFCs) are now being used as substitutes for CFCs. They contain C–H bonds which are broken down in the lower atmosphere (troposphere). This initiates breakdown of the entire molecule and the chlorine is thus unable to reach the upper atmosphere (stratosphere). Their disadvantage is that they are potent greenhouse gases.

9 A secondary pollutant is one which is formed by the chemical reactions of emitted pollutants.

10 For petrol vehicles:
CO emissions in 1992 = 2.72 g per kilometre
CO emissions in 2005 = 1.00 g per kilometre
reduction = 1.72 g per kilometre
For diesel vehicles:
CO emissions in 1992 = 2.72 g per kilometre
CO emissions in 2005 = 0.50 g per kilometre
reduction = 2.22 g per kilometre

11 The original method for making ibuprofen used six steps. The catalyst was aluminium chloride ($AlCl_3$). Although it acts as a catalyst, the $AlCl_3$ cannot be recovered. It is acidic in contact with water and must be disposed of. However, the newer three-step method uses two true catalysts, hydrogen fluoride and Raney Nickel (an alloy of nickel and aluminium), which can be recovered and re-used many times.

Glossary

acid a chemical species which can donate a proton, H^+. Strong acids dissociate fully into ions; weak acids only partially dissociate into ions.

activation energy the energy barrier which must be surmounted before reaction can occur.

addition the joining of two molecules to form a single product molecule.

addition polymerisation forming a polymer by a repeated addition reaction.

addition reactions a reaction in which two molecules join to form a single product molecule.

alkali a soluble base that releases OH^- ions in aqueous solution.

alkaline earth metals the elements found in Group 2 of the Periodic Table.

anhydrous without water of crystallisation.

atom economy is the

$$\frac{\text{molecular mass of the desired products}}{\text{sum of molecular masses of all products}} \times 100\%$$

atomic number the number of protons in the nucleus of each atom of an element.

atomic radius half the distance between the nuclei of two covalently bonded atoms.

base a substance that readily accepts a proton (H^+) from an acid.

biodiesel a fuel made by processing vegetable oils extracted from crops such as oilseed rape.

biofuels renewable fuels obtained from organic material.

bond enthalpy the amount of energy needed to break one mole of a particular bond in one mole of gaseous molecules.

carbocation an organic molecule which has lost an atom or a group of atoms from a carbon atom, creating a single positively charged ion.

carbon capture methods being developed to trap and store carbon dioxide that would otherwise be released to the atmosphere.

catalyst a catalyst increases the rate of a reaction but is not itself used up during the reaction.

catalytic converter a device fitted to the exhaust system of petrol and diesel engines which reduces the emission of pollutants by the use of heterogeneous catalysts.

closed system a closed system can only transfer energy to or from its surroundings. Substances cannot be exchanged.

compound a substance made up of two or more elements chemically joined together.

cracking the thermal decomposition of an alkane into a smaller alkane and an alkene.

dative covalent bond (coordinate bond) a covalent bond where both electrons come from one atom.

d-block elements a block of elements found between Groups 2 and 3 in the Periodic Table.

degradable plastics plastics that will rot away once discarded.

dehydration a reaction involving the removal of a water molecule.

disproportionation a type of redox reaction in which the same species is both reduced and oxidised. It can be thought as a 'self reduction–oxidation' reaction.

double covalent bond two shared pairs of electrons that bond two atoms together.

dynamic equilibrium an equilibrium is dynamic at the molecular level; both forward and reverse processes occur at the same rate; a closed system is required and macroscopic properties remain constant.

electrolysis the decomposition of a substance caused by the passing of a d.c. electric current.

electronegativity describes the ability of an atom to attract the bonding electrons in a covalent bond.

electrons tiny negatively charged sub-atomic particles, found in orbitals around the nucleus of an atom.

electrophile an atom (or group of atoms) which is attracted to an electron-rich centre or atom, where it accepts a pair of electrons to form a new covalent bond.

elements substances made up of only one type of atom. The elements are listed in the Periodic Table.

elimination when a small molecule is removed from a larger molecule.

empirical formula the simplest whole-number ratio of the elements present in a compound.

endothermic term used to describe a reaction in which heat energy is absorbed from the surroundings (enthalpy change is positive).

enthalpy ΔH is the term used by chemists for heat energy transferred during reactions.

enthalpy cycle a diagram displaying alternative routes between reactants and products which allows the determination of one enthalpy change from other known enthalpy changes using Hess's law.

exothermic term used to describe a reaction in which heat energy is transferred to the surroundings (enthalpy change is negative).

free radical an atom or group of atoms with an unpaired electron.

functional group an atom or group of atoms which gives rise to a homologous series. Compounds in the same homologous series show similar chemical properties.

general formula a formula which may be written for each homologous series (C_nH_{2n+2} for alkanes).

global warming the increase in average temperature of the Earth's surface caused by an enhanced greenhouse effect due to increased concentration of greenhouse gases (e.g. carbon dioxide) in the atmosphere.

greenhouse effect natural phenomenon by which some gases present in the atmosphere absorb infrared radiation emitted from the Earth's surface and then re-emit some of this infrared radiation back to the Earth's surface.

Hess's law the total enthalpy change for a chemical reaction is independent of the route by which the reaction takes place, provided initial and final conditions are the same.

heterolytic fission when a bond breaks to form a positive ion and a negative ion.

homologous series a series of organic molecules with the same functional group.

homolytic fission when a bond breaks to form two free radicals.

hydrated with water of crystallisation.

hydrocarbon a compound containing only carbon and hydrogen.

hydrogen bond a weak intermolecular bond formed between molecules containing a hydrogen atom bonded to one of the most electronegative elements (N, O or F).

hydrolysis the breakdown of a compound by water.

initiation the first step in a free-radical substitution in which the free radicals are generated by heat or ultraviolet light.

ion a positively or negatively charged atom or (covalently bonded) group of atoms.

ionic bonding the electrostatic attraction between oppositely charged ions.

ionisation energy the first ionisation energy is the energy needed to remove one electron from each atom in one mole of gaseous atoms or ions of an element.

isomerisation the reaction which takes place in an oil refinery to convert a straight-chain alkane to a branched-chain isomer.

isomers compounds with the same molecular formula but with different arrangements of atoms.

isotopes atoms of an element with the same number of protons but different numbers of neutrons.

Le Chatelier's principle when any of the conditions affecting the position of a dynamic equilibrium are changed, then the position of that equilibrium will shift to minimise that change.

mass number the total number of protons and neutrons in the nucleus of an atom.

mass spectrometer an analytical instrument in which atoms and/or molecules are ionised, deflected and detected. It can be used to find relative isotopic abundances of elements and to identify unknown organic compounds.

mole the unit of amount of substance (abbreviation: mol). One mole of a substance is the mass that has the same number of particles (atoms, molecules, ions or electrons) as there are atoms in exactly 12 g of carbon-12.

molecular formula shows the total number of atoms present in a molecule of the compound.

monomer the small molecule used to build a polymer molecule.

neutron a sub-atomic particle found in the nucleus of an atom. It carries no charge and has the same mass as a proton.

nucleophile a chemical that can donate a pair of electrons with the subsequent formation of a covalent bond.

nucleus the small, dense core at the centre of an atom, containing protons and neutrons (hence a nucleus is always positively charged).

oxidation the loss of electrons.

oxidation number (oxidation state) a number (with a positive or negative sign) assigned to the atoms of each element in an ion or compound. Oxidation states are determined using a set of rules devised by chemists.

photochemical reactions reactions which are brought about by the action of electromagnetic radiation on matter.

photochemical smog a whitish-yellow haze containing nitrogen oxides, hydrocarbons, peroxy compounds, ozone and aldehydes, produced by the action of sunlight on nitrogen oxides and hydrocarbons in the troposphere.

polar covalent bond a covalent bond in which the two bonding electrons are not shared equally by the two bonded atoms. The atom that gets the bigger share of the two electrons becomes $\delta-$. The atom that gets the smaller share of the two electrons becomes $\delta+$.

precipitate an insoluble solid formed when two solids react, e.g. white silver chloride formed when silver nitrate is added to sodium chloride solution.

propagation the stage in a free-radical substitution which constitutes the two reaction steps of the chain reaction.

protons positively charged sub-atomic particles, found in the nucleus of an atom.

radicals reactive atoms or molecules with an unpaired electron; also called *free radicals*.

rate of reaction the amount in moles of a reactant which is used up or product which is formed in a given time.

redox reactions which involve reduction and oxidation processes.

reduction the gain of electrons.

re-forming the conversion of alkanes to cycloalkanes or arenes.

relative atomic mass A_r, the weighted average mass of the atoms of an element taking into account the relative abundance of its naturally occurring isotopes, measured on a scale on which carbon-12 is given a mass of exactly 12.

relative formula mass the mass of one formula unit of a compound relative to an atom of carbon-12.

relative isotopic mass the mass of an isotope of an atom of an element relative to an atom of carbon-12.

relative molecular mass the mass of a molecule of a compound relative to an atom of carbon-12.

salt a compound produced when the H^+ ion of an acid is replaced by a metal ion or an ammonium ion.

saturated contains only C–C single bonds.

single covalent bond a shared pair of electrons that bonds two atoms together.

stoichiometry the stoichiometry (or stoichiometric ratio) for a reaction shows the mole ratio of reactants and products in the balanced equation for the reaction.

stratosphere the region of the atmosphere that extends from 10–16 km to 60 km above the Earth's surface.

structural isomerism where two or more structural isomers have the same molecular formula but different structural formulae.

sublimes when a substance turns directly from a solid into a gas without passing through the liquid phase.

substitution reaction when an atom (or group of atoms) is replaced by a different atom (or group of atoms).

successive ionisation energies the sequence of first, second, third, fourth, etc. ionisation energies needed to remove the first, second, third, fourth, etc. electrons from each atom in one mole of gaseous atoms of an element.

termination the step at the end of a free-radical substitution reaction which occurs when reactants are significantly depleted.

thermal decomposition the breakdown of a compound by heat.

titration measurement of the exact amount of one solution needed to react with a fixed amount of another solution.

transition elements elements in the d-block that can form at least one ion with a partially filled d subshell.

triple covalent bond three shared pairs of electrons that bond two atoms together.

troposphere the region of the atmosphere that extends from the Earth's surface to a height of 10–16 km.

unsaturated contains one or more C=C double bonds.

van der Waals' forces the weak forces of attraction between molecules based on instantaneous and induced dipoles.

volatility a measure of the ease with which a solid or liquid evaporates to a gas. Volatility increases as boiling point decreases.

water of crystallisation water molecules incorporated into the crystal structure of a salt.

***Z / E* isomerism** occurs because a C=C bond cannot freely rotate. In some alkenes (with additional groups either side of the double bond) two isomers (*Z* and *E*) are possible.

Index